Advanced Radio Frequency Antennas for Modern Communication and Medical Systems

Edited by Albert Sabban

Published in London, United Kingdom

IntechOpen

Supporting open minds since 2005

Advanced Radio Frequency Antennas for Modern Communication and Medical Systems
http://dx.doi.org/10.5772/intechopen.87691
Edited by Albert Sabban

Contributors
Volkan Akan, Erdem Yazgan, Adnane Latif, Gregory Mitchell, David Turowski, Oussama Gassab, Arab Azrar, Sara Bouguerra, Haipeng Li, Peng Xie, Guang-Ming Wang, Yawei Wang, Xiangjun Gao, Steve W. Y. Mung, Man Ho Tsoi, Quyen Tran Cao, Amor Nafkha, Bastien Trotobas, Yves Louet, Diego Caratelli, Abraham Loutridis, Simay Kazıcı, Oleg V. Stukach, Arman B. Mirmanov, Thanh-Huong Nguyen, Albert Sabban

First published in London, United Kingdom, 2020 by IntechOpen
IntechOpen is the global imprint of INTECHOPEN LIMITED, registered in England and Wales, registration number: 11086078, 5 Princes Gate Court, London, SW7 2QJ, United Kingdom
Printed in Croatia

British Library Cataloguing-in-Publication Data
A catalogue record for this book is available from the British Library

Additional hard and PDF copies can be obtained from orders@intechopen.com

Advanced Radio Frequency Antennas for Modern Communication and Medical Systems
Edited by Albert Sabban
p. cm.
Print ISBN 978-1-83968-345-9
Online ISBN 978-1-83968-346-6
eBook (PDF) ISBN 978-1-83968-347-3

For EU product safety concerns:
IN TECH d.o.o., Prolaz Marije Krucifikse Kozulić 3, 51000 Rijeka, Croatia, info@intechopen.com or visit our website at intechopen.com.

Meet the editor

Dr. Albert Sabban holds a Ph.D. in Electrical Engineering from the Faculty of Electrical and Computer Engineering, University of Colorado at Boulder, USA (1991), and an MBA from the Faculty of Management, Haifa University, Israel (2005). He is currently a Senior Lecturer and researcher at the Department of Electrical and Electronic Engineering at Kinneret and Ort Braude Engineering Colleges. From 2007 to 2010, Albert Sabban was an RF and Antenna specialist at biomedical hi-tech companies where he designed wearable compact systems and antennas for medical systems. In 1976 he joined RAFAEL in Israel where he worked as a senior researcher, group leader, and project leader in the electromagnetic department until 2007. From 2008 to 2010 he worked as a RF Specialist and project leader at hi-tech biomedical companies. He has published over 100 research papers and holds a patent in the antenna area. He wrote four books on wearable compact systems and antennas for communication and medical systems. He wrote a book on electromagnetics and microwave theory for graduate students, and a book on wide band microwave technologies for communication and medical applications. He also wrote two chapters in books on microstrips and wearable printed antennas for medical applications.

Contents

Preface

The main objective of this book is to present novel Radio Frequency (RF) antennas for 5G, IoT, and medical applications. The book is divided into four sections that present the main topics of novel Radio Frequency antennas.

Section 1: Introduction

Section 2: Novel Antennas for 5G, IoT, and Medical Applications

Section 3: Novel RF Antennas Technologies

Section 4: Advanced Antenna Arrays

The rapid growth in development of cellular wireless communication systems over the last twenty years has resulted in most of world population owning smartphones, smart watches, I-pads, and other RF communication devices. Efficient compact wideband antennas are crucial in RF communication devices.

Section two describes innovations in the development of antennas for 5G, IoT, and medical applications. Several types of antennas are presented in this section, such as planar antennas, phased array antennas, UWB coplanar fed antennas, and beamforming phased array.

Section three presents novel RF antennas for space applications, Vivaldi elliptical antennas and additive manufacturing for antennas applications.

Section four presents advanced arrays antennas. MIMO antennas, reconfigurable Pabry-Perot cavity antenna and time modulated linear array are included in Section three.

Each chapter in this book covers significant information to enable RF engineers, biomedical engineers, system engineers, students, and scientist from all areas to follow and understand the topics presented in the book.

I want to thank all the chapter authors for their excellent job in writing.

Dr. Albert Sabban
Kinneret Academic College,
Israel

Section 1

Introduction

Chapter 1

Introductory Chapter: Novel Radio Frequency Antennas

Albert Sabban

1. Introduction

Antennas are part of radio and television broadcasting, point-to-point radio communication systems, wireless LAN, cell phones, radar, medical systems, and spacecraft communication. Low-profile compact antennas are crucial in the development of wireless communication and wearable biomedical systems. Compact, low-profile, and light-weight printed antennas are the best solution for communication, IoT, and medical systems. Printed antennas' low production costs are crucial in development of low-cost communication systems. Moreover, the advantage of an integrated compact low-cost feed network is attained by integrating the antenna feed network with the antennas on the same printed board. Wireless communication and medical industry are in continuous growth in the last few years. Printed antennas are used in communication systems that employ MIC, MEMS, LTCC, and MMIC technologies.

2. Introduction to RF antennas

Antennas are major components in communication systems [1–19]. Mobile antenna systems are presented in [11]. Transmitting antennas efficiently radiate electromagnetic fields and match RF systems to space. Antennas may transmit or receive electromagnetic fields. Transmitting antennas convert electric current to electromagnetic fields. Receiving antennas convert electromagnetic fields to electric current. In transmitting antennas, an alternating current is created in the elements by applying a voltage at the antenna feed network, causing the antenna to radiate an electromagnetic energy. In receiving mode, an electromagnetic field from an outer source induces an alternating current in the receiving antenna and generates a voltage at the antenna's feed network. Antennas are used in outdoor and indoor communication systems. Antennas can be used in under water communication systems. Antennas may be implanted inside human body and operate as wearable sensors. Receiving antennas such as parabolic and horn antennas incorporate shaped reflective surfaces to receive the electromagnetic fields and focus the fields to the conductive receiving elements.

The book consists of four sections presenting several types of novel antennas.

Section 1: Introduction

Section 2: Novel Antennas for 5G, IoT, and Medical Applications

Section 3: Novel RF Antennas Technologies

Section 4: Advanced Antenna Arrays

The design and electrical performance of several novel antennas are presented in this book.

IntechOpen

3. Type of antennas

Small antennas for communication systems.

Monopole—quarter wavelength wire antenna.

Dipole—Dipole antenna consists of two quarter wavelength wires. Dipole is a half wavelength wire antenna. The monopole and dipole antennas couple to the electric field of the electromagnetic wave in the region near the antenna.

Slot antenna—half wavelength slot: A *slot antenna* consists of a metal surface with a *slot* cut out.

Biconical antennas—biconical half wavelength wire antenna: Biconical antenna is a wideband antenna made of two conical conductive objects.

Loop antennas—Loop antenna is known as a magnetic loop. The loop antenna behaves as an inductor. The loop antenna couples the electromagnetic magnetic field in the region near the loop antenna. Monopole and dipole antennas couple to the electric field.

Helical antennas—helical wire antennas.

Printed antennas—antennas printed on a dielectric substrate.

Aperture antennas for base station communication systems

- Horn and open waveguide
- Reflector antennas
- Antenna arrays
- Microstrip and printed antenna arrays
- Slot antenna arrays

A comparison of directivity and gain values for several antennas is given in **Table 1**.

Phased arrays: Phased array antennas are electrically steerable. The physical antenna can be stationary. Phased arrays, smart antennas, incorporate active components for beam steering.

Antenna type	Directivity (dBi)	Gain (dBi)
Isotropic radiator	0	0
Dipole λ/2	2	2
Dipole above ground plane	6–4	6–4
Microstrip antenna	7–8	6–7
Yagi antenna	6–18	5–16
Helix antenna	7–20	6–18
Horn antenna	10–30	9–29
Reflector antenna	15–60	14–58

Table 1.
Antenna directivity versus antenna gain.

3.1 Steerable antennas

- Arrays with switch-able elements. The array may be partially electronically and mechanically steerable arrays.

- Hybrid antenna systems—for fully electronically steerable arrays. The design can be based on digital beam forming (DBF). These systems may consist of phase and amplitude shifters for each radiating element.

- In digital beam forming, the steering is performed directly in a digital level. DBF allows the most powerful and flexible control of the antenna beam steering.

4. Monopole antenna for communication systems

Monopole antenna is usually a one-quarter wavelength-long conductor mounted above a ground plane or the earth as shown in **Figure 1**. Based on the image theory behind the ground plane, the monopole image is located. The monopole antenna and the monopole image form a dipole antenna. Monopole antenna is half a dipole that radiates electromagnetic fields above the ground plane. The impedance of 0.5 λ monopole antenna is around 37 Ω. The beam width of a monopole, 0.25 λ long, is around 40°. The directivity of a monopole, 0.25 λ long, is around 3 dBi to 5dBi. Usually in wireless communication systems, very short monopole antenna is employed. The impedance of 0.05 λ monopole antenna is around 1 Ω with capacitive reactance. The beam width of a monopole, 0.05 λ long, is around 45°. The directivity of a monopole, 0.05 λ long, is around 3 dBi.

Inverted monopole antenna is shown in **Figure 2a**. Loaded monopole antenna is shown in **Figure 2b**. Monopole antennas may be printed on a dielectric substrate as part of wearable communication devices.

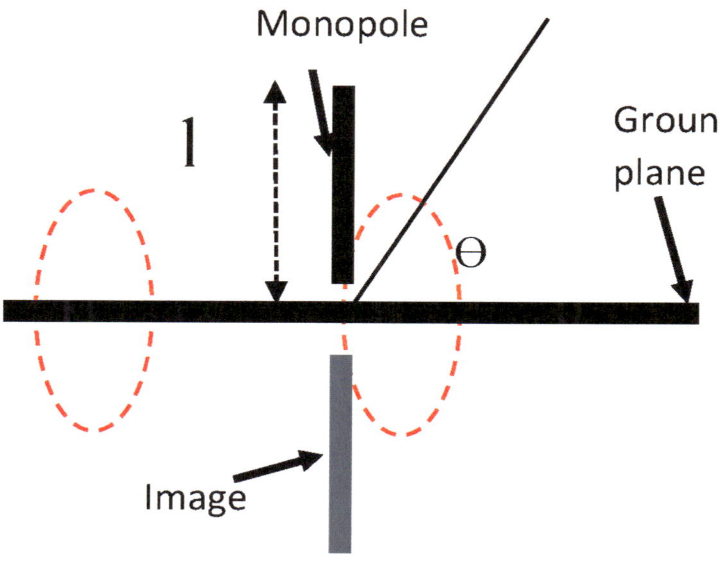

Figure 1.
Monopole antenna.

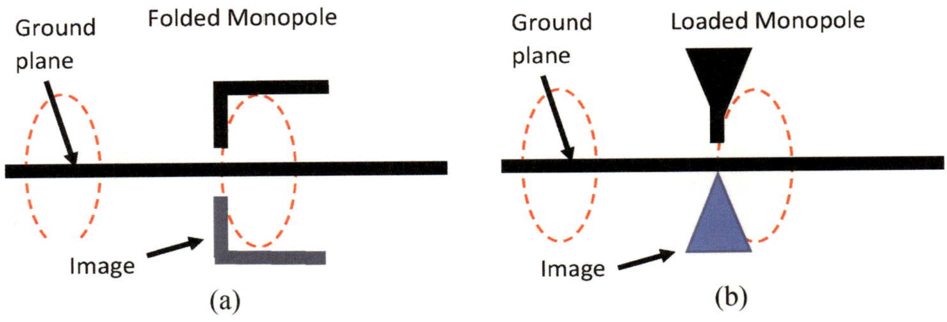

Figure 2.
(a) Inverted monopole antenna. (b) Loaded monopole antenna.

5. Evaluation of basic antenna parameters

Monopole and dipole antennas are the most basic antennas. Evaluation of the electrical parameters of dipole antenna is used to define basic antenna parameters such as radiation pattern, directivity, and antenna impedance. Dipole antenna is a small wire antenna. It consists of two conductors excited by a voltage fed via a transmission line as presented in **Figure 3**. The center conductor of the transmission line is connected to one of the conductors and the outer conductor is connected to the second conductor. The half-wave dipole consists of two conductors, in which each conductor is approximately quarter wavelength long. We can compute the electromagnetic fields radiated from the dipole by defining a potential function, Eq. (1).

5.1 Dipole antenna

The electric potential function is φ_l. The magnetic potential function is A. The potential function is given in Eq. (1).

$$\varphi_l = \frac{1}{4\pi\varepsilon_0}\int_c \frac{\rho_l e^{j(\omega t-\beta R)}}{R}\,\mathrm{dl}$$
$$A_l = \frac{\mu_0}{4\pi}\int_c \frac{\mathrm{ie}^{j(\omega t-\beta R)}}{R}\,\mathrm{dl} \tag{1}$$

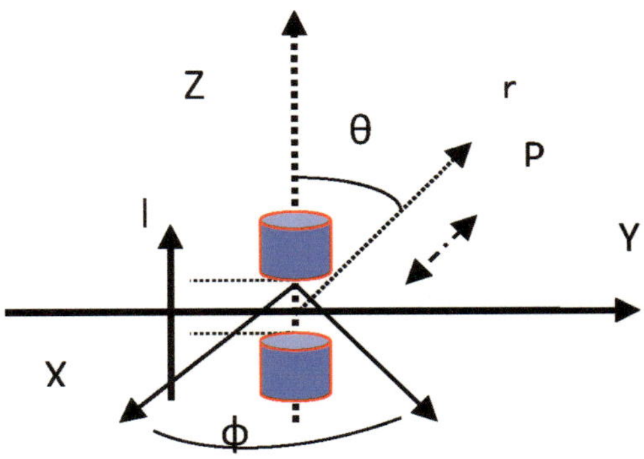

Figure 3.
Dipole antenna.

5.2 Radiation from small dipole

The length of a small dipole is small compared to its wavelength and is called elementary dipole. The current along a small dipole is uniform. We can compute the electromagnetic fields radiated from the dipole in spherical coordinates by using the potential function given in Eq. (1). The electromagnetic field at a point P(r, θ, φ) is listed in Eq. (3). The electromagnetic fields in Eq. (3) vary as $\frac{1}{r}$, $\frac{1}{r^2}$, $\frac{1}{r^3}$. For r << 1, the dominant component of the field varies as $\frac{1}{(r)^3}$ and is written in Eq. (4). These fields are the dipole near fields. In the near field, the waves are standing waves and the energy oscillates in the antenna near zone and is not radiated to the free space. The real part of the pointing vector equals to zero. At r >> 1, the dominant component of the field varies as 1/r as written in Eq. (5). These fields are the dipole far fields. In the far fields, the electromagnetic fields vary as $\frac{1}{r}$ and sinθ. Wave impedance in free space is given in Eq. (6).

6. Dipole radiation pattern

The antenna radiation pattern represents the radiated fields in space at a point P (r, θ, φ) as a function of θ and φ. The antenna radiation pattern is three dimensional. When φ is constant and θ varies, we get the E plane radiation pattern. When φ varies and θ is constant, usually θ = π/2, we get the H plane radiation pattern.

6.1 Dipole E plane radiation pattern

The dipole E plane radiation pattern is given in Eq. (2) and presented in **Figure 4**.

$$|E_\theta| = \eta_0 \frac{l\beta I_0 |\sin\theta|}{4\pi r} \quad (2)$$

At a given point P(r, θ, φ), the dipole E plane radiation pattern is given in Eq. (7).

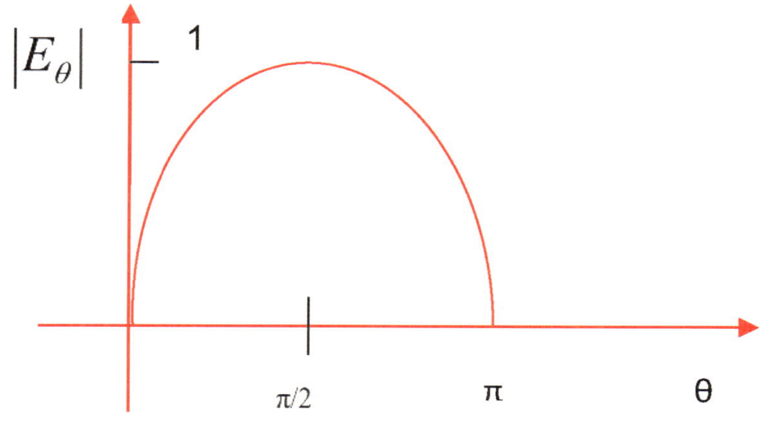

Figure 4.
Dipole E plane radiation pattern.

$$
\begin{aligned}
E_r &= \eta_0 \frac{\text{lI}_0 \cos\theta}{2\pi r^2}\left(1 - \frac{j}{\beta r}\right) e^{\,j(\omega t - \beta r)} \\
E_\theta &= j\eta_0 \frac{\beta \text{lI}_0 \sin\theta}{4\pi r}\left(1 - \frac{j}{\beta r} - \frac{1}{(\beta r)^2}\right) e^{\,j(\omega t - \beta r)} \\
H_\varphi &= j\frac{\beta \text{lI}_0 \sin\theta}{4\pi r}\left(1 - \frac{j}{\beta r}\right) e^{\,j(\omega t - \beta r)} \\
H_r &= 0 \;\; H_\theta = 0 \quad E_\phi = 0
\end{aligned} \tag{3}
$$

$$
\begin{aligned}
I &= I_0 \cos\omega t \\
E_r &= -j\eta_0 \frac{\text{lI}_0 \cos\theta}{2\pi\beta r^3} e^{\,j(\omega t - \beta r)} \\
E_\theta &= -j\eta_0 \frac{\text{lI}_0 \sin\theta}{4\pi\beta r^3} e^{\,j(\omega t - \beta r)} \\
H_\varphi &= \frac{\text{lI}_0 \sin\theta}{4\pi r^2} e^{\,j(\omega t - \beta r)}
\end{aligned} \tag{4}
$$

$$
\begin{aligned}
E_r &= 0 \\
E_\theta &= j\eta_0 \frac{l\beta I_0 \sin\theta}{4\pi r} e^{\,j(\omega t - \beta r)} \\
H_\varphi &= j\frac{l\beta I_0 \sin\theta}{4\pi r} e^{\,j(\omega t - \beta r)}
\end{aligned} \tag{5}
$$

$$
\frac{E_\theta}{H_\varphi} = \eta_0 = \sqrt{\frac{\mu_0}{\varepsilon_0}} \tag{6}
$$

$$
\begin{aligned}
|E_\theta| &= \eta_0 \frac{l\beta |I_0| \sin\theta|}{4\pi r} = A|\sin\theta| \\
&\text{Choose} \quad A = 1 \\
|E_\theta| &= |\sin\theta|
\end{aligned} \tag{7}
$$

Dipole E plane radiation pattern in spherical coordinate system is shown in **Figure 5**.

6.2 Dipole H plane radiation pattern

For $\theta = \pi/2$, the dipole H plane radiation pattern is given in Eq. (8) and presented in **Figure 6**.

$$
|E_\theta| = \eta_0 \frac{l\beta I_0}{4\pi r} \tag{8}
$$

The dipole H plane radiation pattern in xy plane is a circle with r = 1. At a given point P(r, θ, φ), the dipole H plane radiation pattern is given in Eq. (9). The radiation pattern of a vertical dipole is omnidirectional. It radiates equal power in all azimuthal directions perpendicular to the axis of the antenna.

6.3 Antenna radiation pattern

An antenna radiation pattern is shown in **Figure 7**. The antenna main beam is defined between the points that the maximum relative field level E decays to

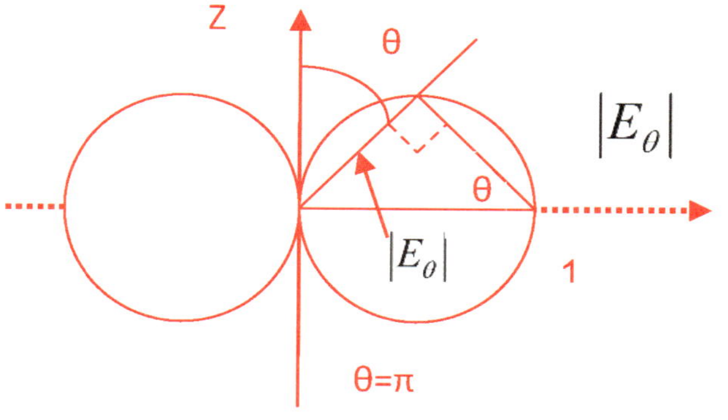

Figure 5.
Dipole E plane radiation pattern in spherical coordinate system.

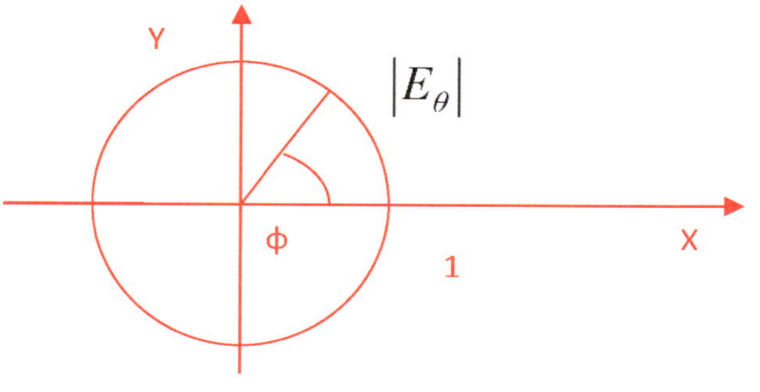

Figure 6.
Dipole H plane radiation pattern for θ = π/2.

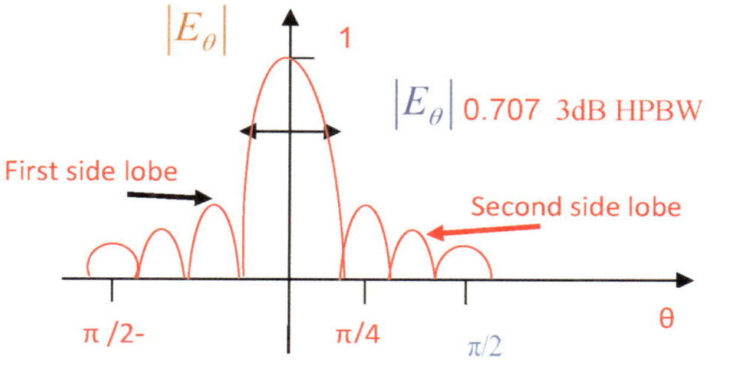

Figure 7.
Antenna typical radiation pattern.

0.707E. Half of the radiated power, −3 dB points, is concentrated in the antenna main beam. The antenna main beam is defined as the 3 dB beam width. Radiation to undesired direction is concentrated. The antenna side lobes present radiation to undesired direction.

$$
\begin{gathered}
|E_\theta| = \eta_0 \frac{l\beta I_0 |\sin\theta|}{4\pi r} = A \\
\text{Choose} \quad A = 1 \\
|E_\theta| = 1
\end{gathered}
\tag{9}
$$

For a dipole, the power intensity varies as $\left(\sin^2\theta\right)$. At θ = 45° and θ = 135° the radiated power equals to half the power radiated toward θ = 90°. The dipole beam width is θ = (135–45) =90°.

7. Dipole directivity

Directivity is defined as the ratio between the amounts of energy propagating in a certain direction compared to the average energy radiated to all directions over a sphere as written in Eqs. (10) and (11).

$$D = \frac{P(,)maximal}{P(,)average} = 4\frac{P(,)maximal}{Prad} \tag{10}$$

$$P(,)average = \frac{1}{4}P(,)sindd = \frac{Prad}{4} \tag{11}$$

The radiated power from a dipole is calculated by computing the pointing vector P as given in Eq. (12).

$$\begin{gathered} P = 0.5(E \times H) = \frac{15\pi I_0^2 l^2 \sin^2\theta}{r^2\lambda^2} \\ W_T = \int_s P \cdot \mathrm{ds} = \frac{15\pi I_0^2 l^2}{\lambda^2}\int_0^\pi \sin^3\theta d\theta \int_0^{2\pi} d\varphi = \frac{40\pi^2 I_0^2 l^2}{\lambda^2} \end{gathered} \tag{12}$$

The overall radiated energy is WT. WT is computed as written in Eq. (12), by integration of P over an imaginary sphere surrounding the dipole. The power flow of an isotropic element equal to the overall radiated energy divided by the area of the sphere, $4\pi r^2$, see Eq. (13). The dipole directivity at θ = 90° is 1.5 as given in Eq. (14). For small antennas or for antennas without losses, D = G, losses are negligible. For a given θ and φ for small antennas, the approximate directivity is given by Eq. (15).

$$\oint_s \mathrm{ds} = r^2 \int_0^\pi \sin\theta d\theta \int_0^{2\pi} d\phi = 4\pi r^2$$

$$P_{\mathrm{iso}} = \frac{W_T}{4\pi r^2} = \frac{10\pi I_0^2 l^2}{r^2\lambda^2} \tag{13}$$

$$D = \frac{P}{P_{\mathrm{iso}}} = 1.5\sin^2\theta$$

$$G_{\mathrm{dB}} = 10\log_{10}G = 10\log_{10}1.5 = 1.76\mathrm{dB} \tag{14}$$

$$\begin{gathered} D = \frac{41253}{\theta_{3\mathrm{dB}}\phi_{3\mathrm{dB}}} \\ G = \xi D \quad \xi = \mathrm{Efficency} \end{gathered} \tag{15}$$

Antenna losses degrade the antenna efficiency. Antenna losses consist of conductor loss, dielectric loss, radiation loss, and mismatch losses. For resonant small antennas, ξ =1. For reflector and horn antennas, the efficiency varies between $\xi = 0.5$ and.

$\xi = 0.7$. The beam width of a small dipole, 0.1 λ long, is around 90°. The 0.1 λ dipole impedance is around 2 Ω. The beam width of a dipole, 0.5 λ long, is around 80°. The impedance of 0.5 λ dipole is around 73 Ω.

8. Antenna impedance

Antenna impedance determines the efficiency of transmitting and receiving energy in antennas. The dipole impedance is given in Eq. (16).

$$R_{rad} = \frac{2W_T}{I_0^2}$$
$$\text{For a Dipole: } R_{rad} = \frac{80\pi^2 l^2}{\lambda^2} \tag{16}$$

8.1 Loop antennas for wireless communication systems

Loop antennas are compact, low-profile, and low-cost antennas. Loop antennas are employed in wearable wireless communication systems. The loop antenna is referred to as the dual of the dipole antenna, see **Figure 8**. A small dipole has magnetic current flowing (as opposed to electric current as in a regular dipole), the loop antenna fields are similar to that of a small loop. The short dipole has a

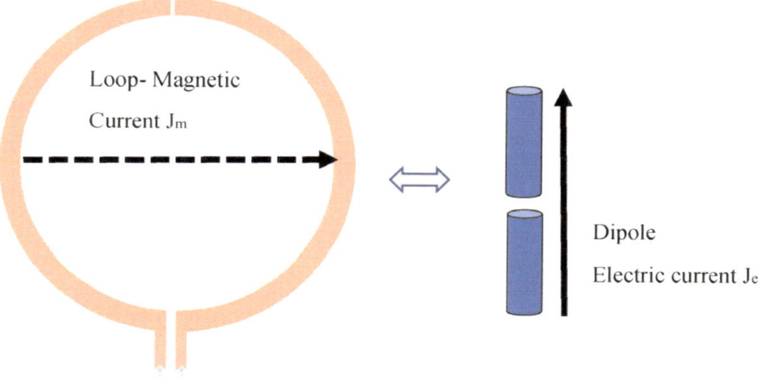

Figure 8.
Duality relationship between dipole and loop antennas.

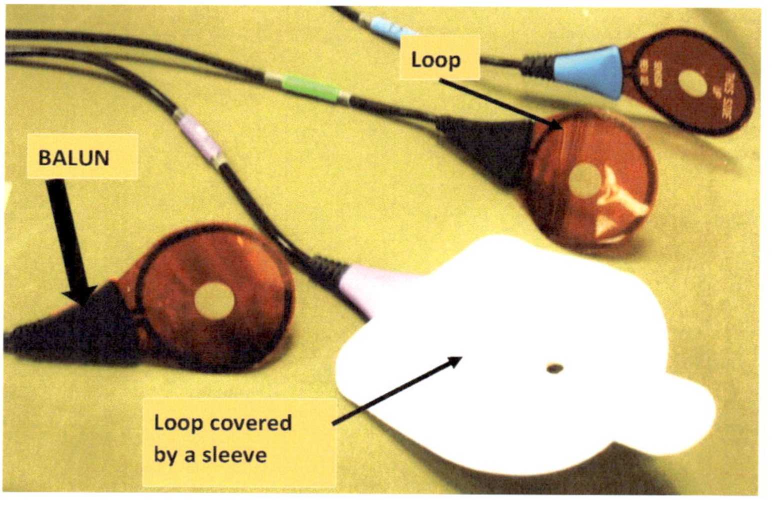

Figure 9.
Photo of loop antennas.

capacitive impedance and the impedance of a small loop is inductive. A photo of loop antennas is shown in **Figure 9**.

8.2 Printed antennas

Printed antennas have been analyzed and presented in several papers and books in the last 30 years [1–18]. Microstrip antennas are the most popular type of compact printed antennas. Printed Slot, PIFA, and dipole antennas are employed in several wireless communication systems. Printed antennas are used in wireless communication systems, wearable devices, IoT applications, novel 5G communication links, seekers, and medical systems.

Printed antennas possess attractive features such as light weight, compactness, flexibility, and are cheap compared to other similar antennas.

8.3 Applications of wearable antennas

- Medical
- Wireless communication
- WLAN
- GPS
- Military applications

9. Microstrip antennas

Microstrip antennas are printed on a on a dielectric substrate with low dielectric losses. Cross section of the microstriop antenna is shown in **Figure 10**.

Microstrip antennas are thin patches etched on a dielectric substrate εr. The substrate thickness, H, is less than 0.1λ. Microstrip antennas are widely presented in [1–18].

Advantages of microstrip antennas:

- Flexible
- Low-profile compact uniform arrays
- Compact
- Low-cost antennas

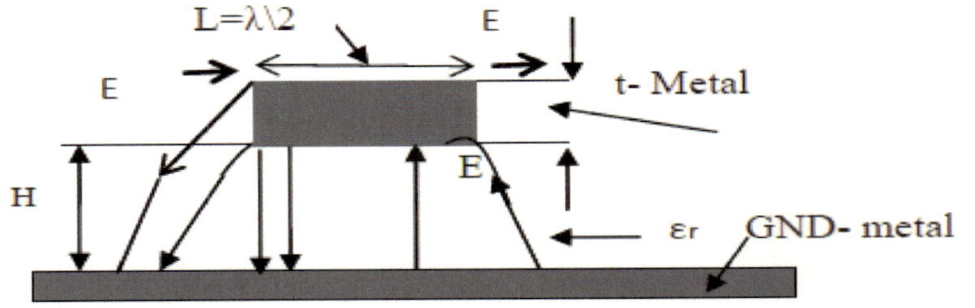

Figure 10.
Microstrip antenna cross section.

Disadvantages of microstrip antennas:

- Narrow bandwidth (usually 0.5–5%). However, wider bandwidth may be achieved by using multilayer structure and novel technologies.

- Limited power handling up to 50 W.

- High feed network losses at high frequencies.

The electric field along the radiating edges is presented in **Figure 11**. The magnetic field is perpendicular to the electric field according to Maxwell's equations. At the edge of the strip (X/L = 0 and X/L = 1), the magnetic field drops to zero, because there is no conductor to carry the electromagnetic current, it is maximum in the patch center. The electric field is at maximum level (with opposite polarity) at the patch edges (X/L = 0 and X/L = 1) and zero at the patch center. The ratio of electric to the magnetic field is proportional to the impedance that we measure when we feed the patch. Microstrip antennas may be fed by a microstrip line or by a coaxial line or probe feed. By varying the location of the antenna feed point between the patch center and the patch edge, we can get a 50 Ω impedance or any other desired impedance. Microstrip antenna may have any arbitrary shape such as square, triangle, circle, ring, rectangular, and fractal shape as presented in **Figure 12**.

The antenna dimension W is given by Eq. (17). The antenna bandwidth is given in Eq. (18).

$$W = \frac{c}{2f\sqrt{\epsilon_{eff}}} \tag{17}$$

$$BW = \frac{H}{\sqrt{\epsilon_{eff}}} \tag{18}$$

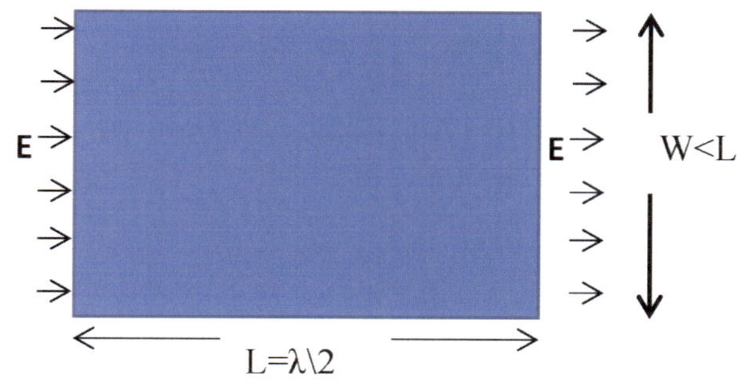

Figure 11.
Rectangular microstrip antenna.

Figure 12.
Microstrip antenna shapes.

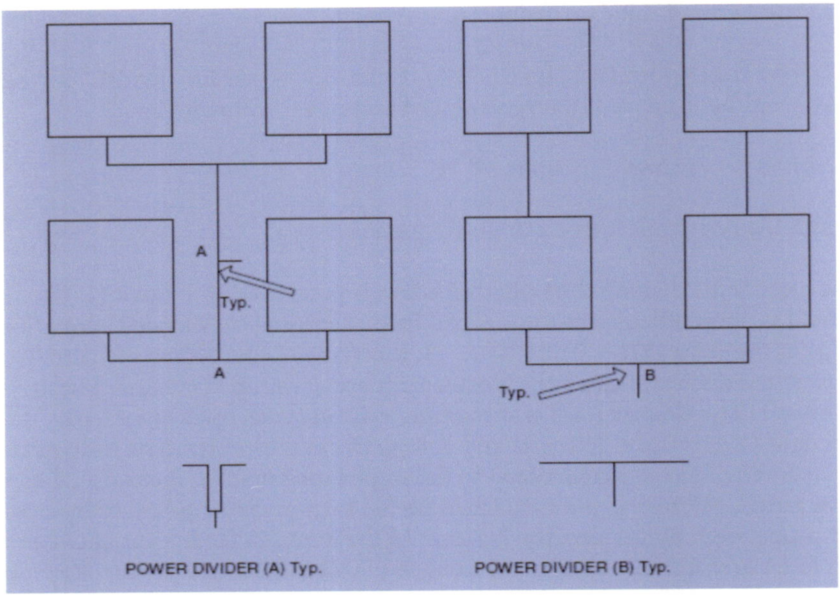

Figure 13.
MM wave microstrip antenna array. (a) Parallel array feed network. (b) Parallel-series array feed network.

The gain of microstrip patch is between 0 and 5 dBi. The microstrip patch gain is a function of the antenna configuration, structure, and dimensions. Microstrip antenna arrays are used to get gain up to 30 dB. In microstrip patch arrays, a low-profile and low-cost feed and matching network is attained by integrating the antenna feed network with the radiating patches on the same substrate. Microstrip antenna feed networks are shown in **Figure 13**. A parallel feed network is shown in **Figure 13a**. A parallel–series feed network is shown in **Figure 13b**.

10. Conclusions

This chapter presents electrical parameters of several basic antennas. Antennas are part of radio and television broadcasting, point-to-point radio communication systems, wireless LAN, cell phones, radar, medical systems, and spacecraft communication. Compact wideband efficient antennas are crucial in communication systems.

DOI: http://dx.doi.org/10.5772/intechopen.93142

Author details

Albert Sabban
Kinneret and Ort Braude Colleges, Israel

*Address all correspondence to: sabban@mx.kinneret.ac.il

References

[1] Sabban A. Wearable Communication Systems and Antennas for Commercial, Sport, and Medical Applications. Bristol UK: IET Publication; December 2018

[2] Sabban A. Wideband RF Technologies and Antenna in Microwave Frequencies. NJ, USA: Wiley Sons; July 2016

[3] Sabban A. Novel Wearable Antennas for Communication and Medical Systems. FL, USA: Taylor & Francis Group; October 2017

[4] Sabban A, Gupta KC. Characterization of radiation loss from microstrip discontinuities using a multiport network modeling approach. IEEE Transactions on Microwave Theory and Techniques. 1991;**39**(4): 705-712

[5] Sabban A. A new wideband stacked microstrip antenna. In: IEEE Antenna and Propagation Symposium; Houston, Texas, USA; 1983

[6] U.S Patent Inventors: Albert Sabban. Dual polarized dipole wearable antenna. U.S Patent Number: 8203497; 2012

[7] Sabban A. Wideband microstrip antenna arrays. In: IEEE Antenna and Propagation Symposium MELCOM; Tel-Aviv; 1981

[8] Sabban A. Microstrip antenna arrays. In: Nasimuddin N, editor. Microstrip Antennas. Rijeka, Croatia: InTech; 2011. pp. 361-384. ISBN: 978-953-307-247-0. Available from: http://www.intechopen.com/articles/show/title/microstrip-antenna-arrays

[9] Chirwa LC, Hammond PA, Roy S, Cumming DRS. electromagnetic radiation from ingested sources in the human intestine between 150 MHz and 1.2 GHz. IEEE Transactions on Biomedical Engineering. 2003;**50**(4): 484-492

[10] Werber D, Schwentner A, Biebl EM. Investigation of RF transmission properties of human tissues. Advances in Radio Science. 2006;**4**:357-360

[11] Fujimoto K, James JR, editors. Mobile Antenna Systems Handbook. Boston, USA: Artech House; 1994

[12] Gupta B, Sankaralingam S, Dhar S. Development of wearable and implantable antennas in the last decade. In: Microwave Mediterranean Symposium (MMS); Guzelyurt, Turkey; 2010. pp. 251-267

[13] Thalmann T, Popovic Z, Notaros BM, Mosig JR. Investigation and design of a multi-band wearable antenna. In: 3rd European Conference on Antennas and Propagation (EuCAP 2009); Berlin, Germany; 2009. pp. 462-465

[14] Salonen P, Rahmat-Samii Y, Kivikoski M. Wearable antennas in the vicinity of human body. In: IEEE Antennas and Propagation Society Symposium, Vol. 1; Monterey, USA; 2004. pp. 467-470

[15] Kellomaki T, Heikkinen J, Kivikoski M. Wearable antennas for FM reception. In: First European Conference on Antennas and Propagation (EuCAP 2006); Hague, Netherlands; 2006. pp. 1-6

[16] Sabban A. Wideband printed antennas for medical applications. In: APMC 2009 Conference; Singapore; 2009

[17] Lee Y. Microchip. Antenna Circuit Design for RFID Applications. USA: Microchip Technology Inc. Microchip AN 710c; 2003

[18] Sabban A. Low Visibility Antennas for Communication Systems. New York, USA: Taylor & Francis Group; 2015

[19] Sabban A. Multiport network model for evaluating radiation loss and coupling among discontinuities in microstrip circuits [PhD thesis]. Boulder, CO, USA: University of Colorado at Boulder; 1991

Section 2

Novel Antennas for 5G, IoT, and Medical Applications

Chapter 2

Wideband Wearable Antennas for 5G, IoT, and Medical Applications

Albert Sabban

Abstract

Wearable compact antennas are a major part of every wearable 5G communication system, IoT, and biomedical systems. Several types of printed antennas are employed as wearable antennas. Printed dipole, microstrip antennas, printed loops, slot antennas, and PIFA antennas are employed as wearable antennas. Compact efficient antennas significantly affect the electrical performance of wearable communication systems. In several communication and medical systems, the polarization of the received signal is not known. The polarization of the received signal may be vertical, horizontal, or circular polarized. In these systems, it is crucial to use dual-polarized receiving antennas. The antennas presented in this chapter may be linearly or dually polarized. Design trade-offs, simulation results, and measured results on human body of small wideband printed antennas with high efficiency are presented in this chapter. For example, the low-volume dually polarized antenna dimensions are 50 × 50 × 0.5 mm. The antenna beamwidth is around 100°. The antennas gain is around 0–4 dBi. Metamaterial technology is used to improve the electrical performance of wearable antennas. The proposed antennas may be used in wearable wireless communication and medical RF systems. The antennas' electrical performance on human body is presented in this chapter.

Keywords: wearable antennas, 5G communication system, IoT, biomedical systems, metamaterial technology, metamaterial antennas, microstrip antennas

1. Introduction

Microstrip antennas are widely used in communication system and seekers. Microstrip antennas possess attractive features that are crucial to 5G communication, IoT, and medical systems. These antennas are compact, flexible, lightweight, and relatively cheap. In addition, we can integrate the RF modules with the antennas on the same substrate. Printed antennas have been widely presented in the literature in the last 20 years [1–9]. Electromagnetic fields' transmission losses of human tissues have been investigated in the papers [10, 11]. However, the effect of human body on the impedance and efficiency of wearable antennas was not always presented [12, 13]. Printed wearable antennas have been presented in the last 10 years [1–20]. A review of wearable antennas designed and developed for several applications at different frequencies over the last 10 years is listed [15]. Wearable meander line antennas are presented in [12]. These antennas function in the frequency range between 750 and 2600 MHz. A textile antenna performance near human body is presented at 2.4 GHz, see [13]. The effect of human body on

IntechOpen

portable RF antennas is studied in [16]. In this chapter, the authors determine that the antennas' length in free space is larger by 10–20% from the length of wearable antennas. Measurement of the antenna gain in this paper shows that a wide dipole (1.16 × 0.1 m) has −13 dBi gain. Wearable antennas for cellular applications are presented in [12–16]. Electrical specifications of medical devices are different from the electrical specifications for cellular devices. Medical wearable sensors are presented in [21–48]. Wearable devices support the development of personal medical sensors and systems with real-time response to help improve patient's health. Wearable medical sensors and devices can measure the sweat rate, body temperature, heartbeat, and blood pressure, perform gait analysis, and measure other body health parameters of the patient wearing these sensors, see Refs. [21–49]. In this chapter, novel wideband compact wearable antennas for 5G communication and medical systems are presented. Numerical results in free space and near the human body are presented.

2. Printed wearable antennas for 5G and medical applications

Wearable antennas should be compact, have lightweight, are low cost, and are flexible. Microstrip antennas, printed loops, printed dipoles, slot antennas, and PIFA antennas are compact, low cost, conformable, and have lightweight. These antennas are a good choice to be employed as wearable antennas for IoT and medical applications.

Applications of wearable antennas:

- 5G Communication Systems
- Medical
- Wireless Communication
- IoT
- WLAN
- HIPER LAN
- GPS
- Military Applications

2.1 Double-layer printed wearable dipole antennas

Single-layer printed dipole antennas have a narrow bandwidth less than 1% for VSWR better than 2:1. The length of the dipole may be between quarter wavelength to half wavelength. The antenna directivity is around 0 dBi and the beam width is around 90°–100°. The antenna bandwidth may be improved by printing the antenna feed network on a dielectric substrate and by printing the radiating dipole on a second layer. The electromagnetic fields are coupled to the radiating dipole. The bandwidth of the double-layer printed dipole may be between 1 and 5% for VSWR better than 2:1 as a function of the dipole configuration and the layers thickness. The printed dipole antenna consists of two layers. The first layer consists of a 0.8-mm substrate with 3.5 as dielectric constant. The second layer consists of a 0.8-mm substrate with 2.2 as dielectric constant. The substrate thickness determines the antenna bandwidth. However, thinner antennas are flexible. The antenna dimensions are designed to operate on the patient's body by using electromagnetic software [50]. The double-layer antenna is shown in **Figure 1**. The directivity of the antenna at 420 MHz is around 4 dBi as shown in **Figure 2**.

A double-layer 460 MHz dipole antenna is shown in **Figure 3**. The antenna dimensions are 20 × 4 cm. The directivity of the antenna at 460 MHz is around 5 dBi as presented in **Figure 4**. The antenna beamwidth is around 120°.

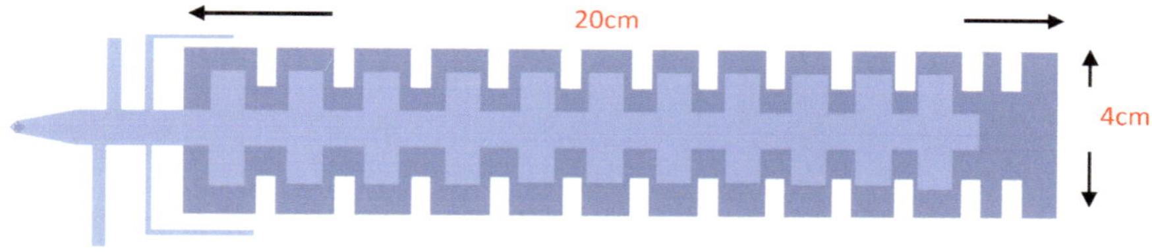

Figure 1.
Wearable double-layer 420 MHz printed dipole antenna.

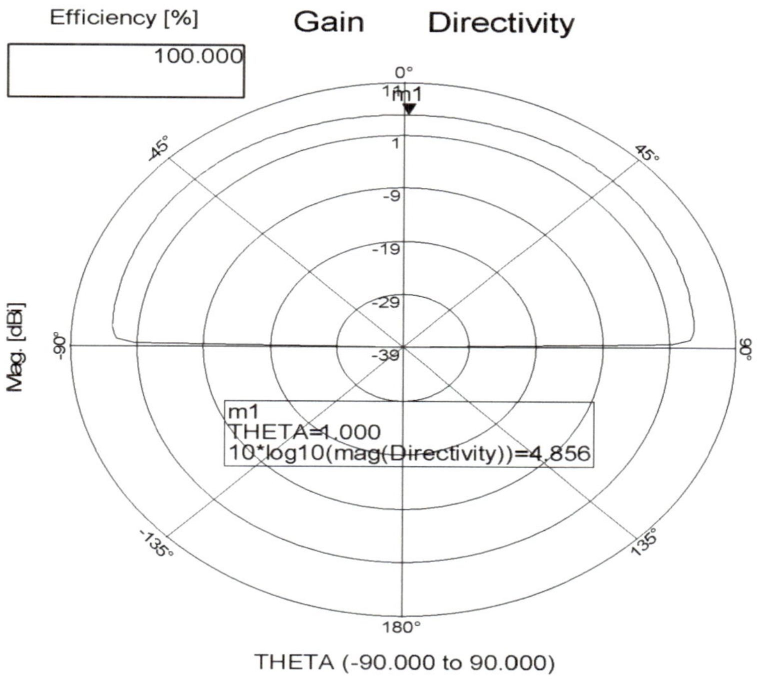

Figure 2.
Radiation pattern of a wearable double-layer printed dipole antenna.

Figure 3.
Wearable double-layer 460 MHz printed dipole antenna.

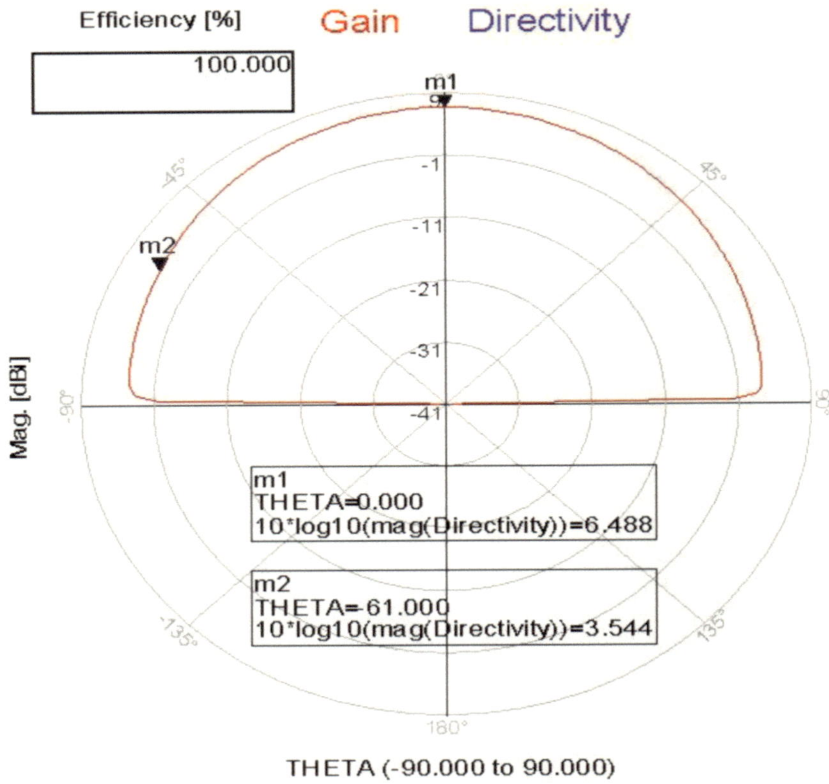

Figure 4.
Radiation pattern of a wearable double-layer printed dipole at 460 MHz.

3. Printed wearable dual-polarized dipole antennas

In several communication and medical systems, the polarization of the received signal is not known. The polarization of the received signal may be vertical, horizontal, or circular polarized. In these systems, it is crucial to use dual-polarized receiving antennas. Two wearable antennas are presented in this section; the first is a dual-polarized printed dipole. The second antenna is a dual-polarized, folded, printed microstrip dipole. The compact, printed, loaded dipole antenna is horizontally polarized. The antenna dimensions have been designed to operate on the patient's body by employing electromagnetic software [50]. The antenna consists of two layers. The first layer consists of a 0.08-cm dielectric substrate with 3.5 as relative dielectric constant. On this layer, the antenna feed network is printed. The radiating elements are printed on the second layer which consists of a 0.08-cm dielectric substrate with 2.2 as relative dielectric constant. Thicker antennas have a wider bandwidth. However, thinner antennas are more flexible with a narrower bandwidth. The printed slot antenna is vertically polarized. The printed dipole and the slot antenna provide dual orthogonal polarizations. The wearable antenna current distribution and dimensions are shown in **Figure 5**.

The radiating dipole dimensions are $21 \times 4 \times 0.16$ cm. The wearable antenna may be employed in medical and IoT systems. The antenna may be attached to the patient clothes, in the front or in the back zone. The antenna has been analyzed by using Key-sight momentum software [50]. The antenna bandwidth is around 15% for VSWR better than 3:1. The antenna -3 dB beamwidth is 100°. The antenna gain

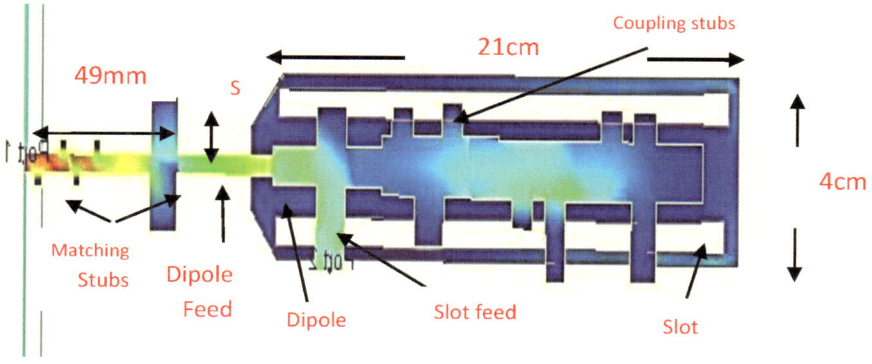

Figure 5.
Current distribution of the dual-polarized wearable antenna.

is around 2 dBi. The simulated S_{11} and S_{22} parameters are shown in **Figure 6**. **Figure 7** presents the antenna's measured S_{11} parameters. The simulated radiation patterns are shown in **Figure 8**. There is a good agreement between the measured and computed results. The co-polar radiation is in the yz plane. The cross-polar radiation is in the xz plane. The antenna cross-polarization value may be adjusted by varying the feed lines and matching stubs' locations. The dimensions and current distribution of the folded antenna are shown in **Figure 9**. The radiating element

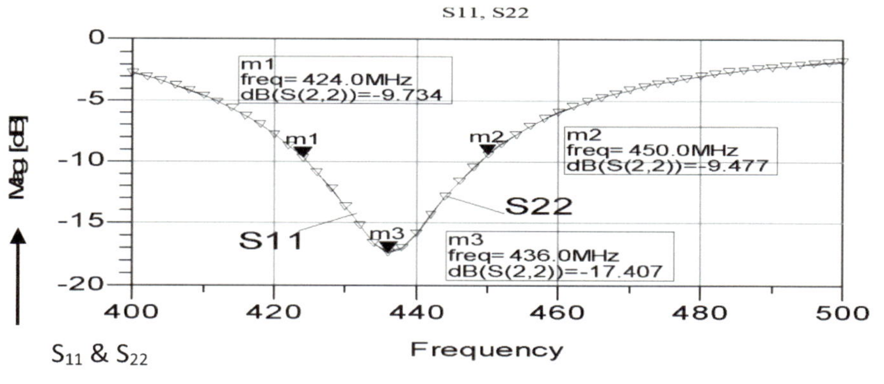

Figure 6.
Computed S_{11} and S_{22} results of the dual-polarized dipole on human body.

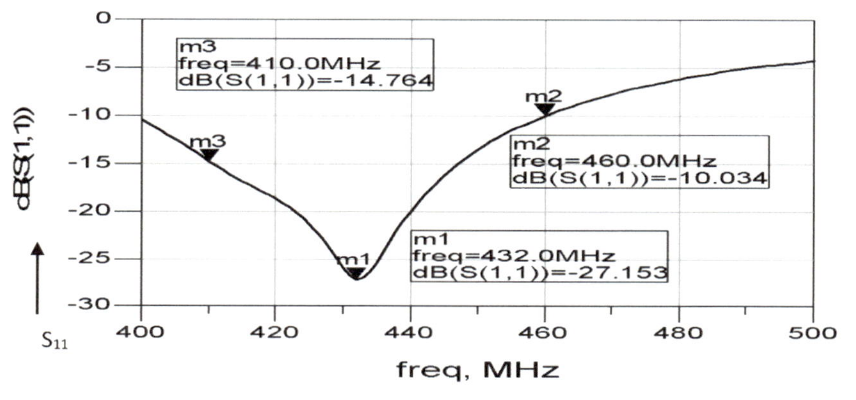

Figure 7.
Measured S_{11} of the wearable dual-polarized dipole antenna on human body.

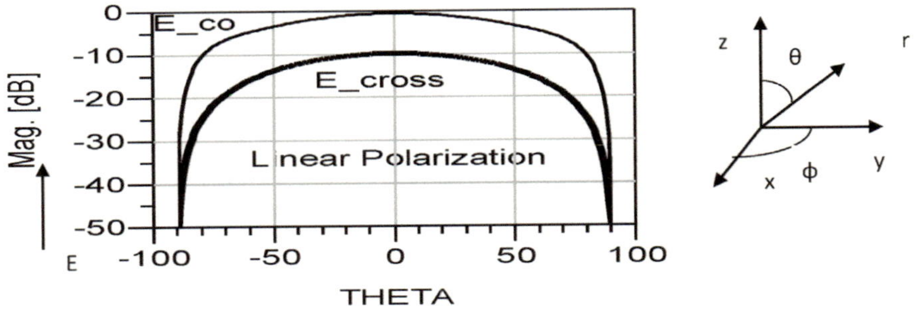

Figure 8.
Radiation pattern of the dual-polarized wearable antenna.

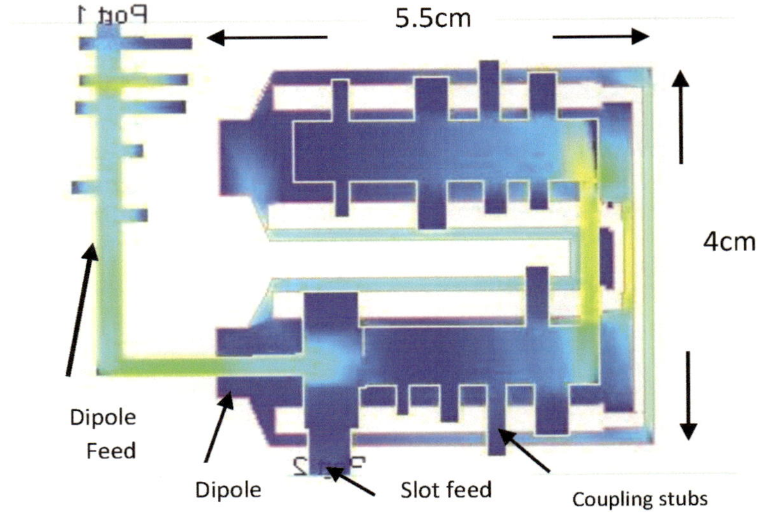

Figure 9.
Current distribution of the folded wearable dipole antenna, 6 × 5 × 0.16 cm.

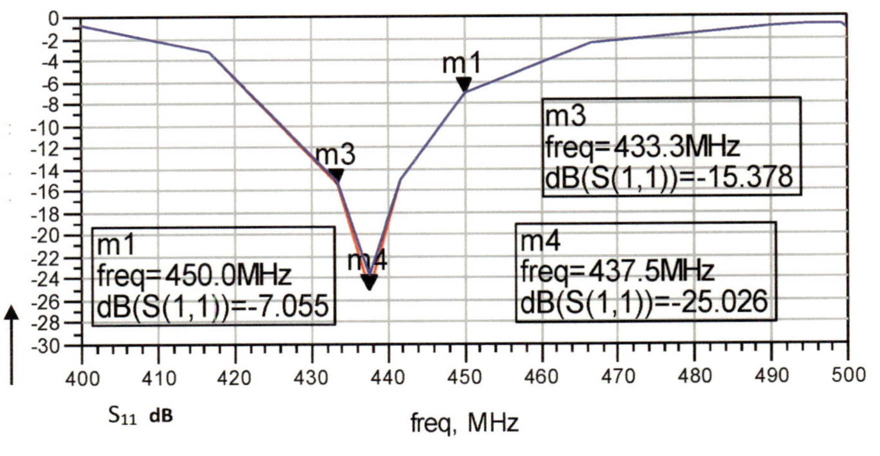

Figure 10.
Folded antenna's computed S_{11} and S_{22} results on human body.

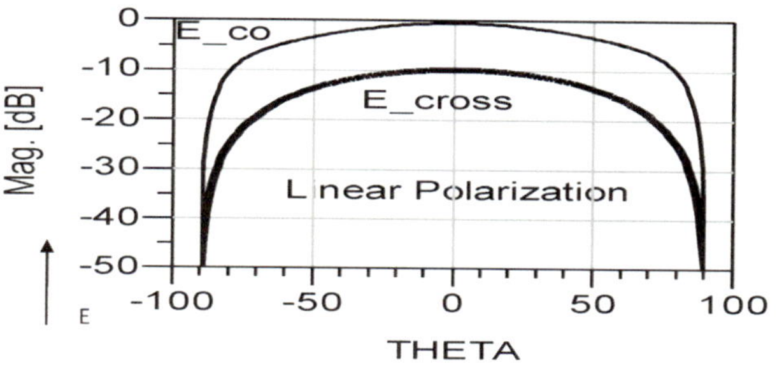

Figure 11.
Radiation pattern of the folded dual-polarized antenna.

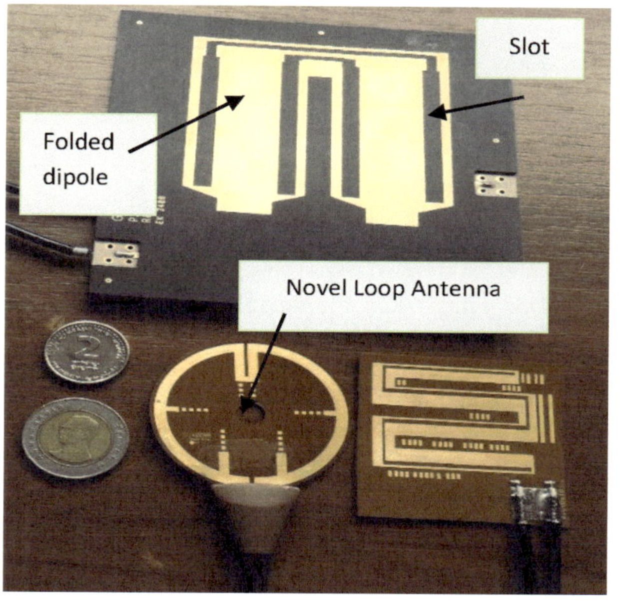

Figure 12.
Photo of wearable antennas.

dimensions are 55 × 40 × 1.6 mm. **Figure 10** presents the antenna's simulated S_{11} and S_{22} parameters. The folded dipole radiation pattern is shown in **Figure 11**. The antennas' radiation characteristics on human body were measured by using a phantom. The phantom liquid presents the body tissue's electrical characteristics. The phantom diameter is 40 cm and has 1.5 m length. The phantom liquid is a mix of 55% water, 44% sugar, and 1% salt. The wearable antenna was placed on the phantom during the measurements of the antenna's electrical characteristics. S_{11} and S_{12} parameters were measured on the patient's body by using a network analyzer. Photo of wearable antennas is shown in **Figure 12**.

4. Wearable microstrip antennas for 5G, medical, and IoT applications

Printed antennas are usually low profile, compact, flexible, light weight, and low-cost relative to wired antennas. Microstrip antennas may be used as wearable antennas. Printed antennas have been widely presented in the literature in the last 20 years, [1–19]. The most popular type of printed antennas is the microstrip antennas. However, loop, PIFA, slot, and dipole-printed are widely used in RF systems. Printed antennas may be employed in communication mobile phones, IoT, seekers, and in medical systems.

4.1 Wearable microstrip antennas for 5G and medical systems

Microstrip antennas are etched on a low loss dielectric substrate. A cross-sectional view of the microstrip antenna electric fields is presented in **Figure 13**. Microstrip antennas are thin conducting patches etched on a substrate with dielectric constant εr and thickness H. Usually, H is less than 0.1 λ. Microstrip antennas are presented in [1–7]. The wearable antenna may be attached to the human body or inserted inside a belt.

Advantages of microstrip antennas:

- Light weight and low volume.
- Flexible, Conformal structures are possible.
- Low cost relative to conventional wired antennas.
- Easy to fabricate a large uniform arrays and phased arrays.

These features are very important for wearable communication systems.

Disadvantages of microstrip antennas:

- Limited bandwidth (usually 1–5%). However, wider bandwidth is possible with increased antenna structure complexity.
- Low power handling less than 50 W depends on substrate thickness.
- Limited gain up to 30 dBi, 16 × 16 arrays.
- High feed network losses at high frequencies, above 12 GHz.

The patch magnetic field is perpendicular to the E-field. There is no conductor to carry the RF current so at the edge of the strip (X/L = 0 and X/L = 1), the H-field drops to zero and is maximum in the center. The E-field is zero at the center and at maximum value (and opposite polarity) at the edges (X/L = 0 and X/L = 1), see **Figure 14**. The ratio of E- to H-field is proportional to the patch impedance. Microstrip antennas may be fed by a coaxial probe feed or by a microstrip feed line.

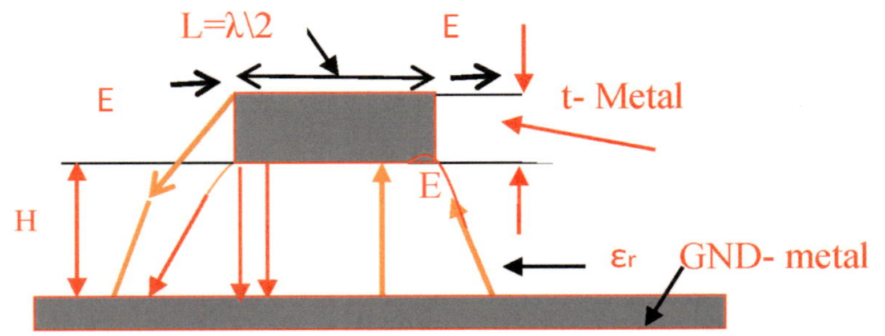

Figure 13.
Microstrip antenna electric fields, a cross-sectional view.

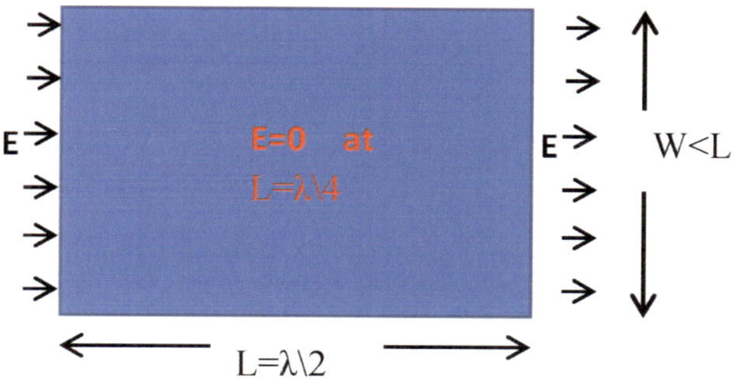

Figure 14.
Fields of rectangular microstrip antenna.

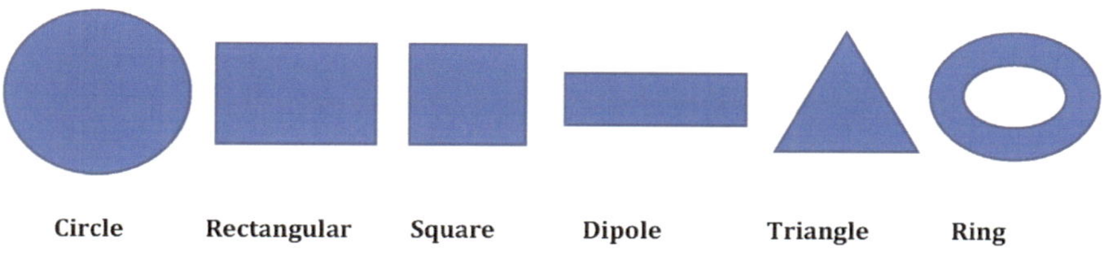

Figure 15.
Microstrip antenna popular configurations.

By adjusting the location of the antenna feed point, we can achieve any impedance and match the antenna to the RF system, usually 50 Ω. The antenna shape may be rectangular, square, triangle, circle, or any arbitrary shape as presented in **Figure 15**.

The antenna dimension, W, is given by Eq. (1) and is a function of the effective dielectric constant and resonant frequency:

$$W = \frac{c}{2f\sqrt{\epsilon_{eff}}} \tag{1}$$

The antenna bandwidth is given in Eq. (2):

$$BW = \frac{H}{\sqrt{\epsilon_{eff}}} \tag{2}$$

The gain of patch antenna is the function of the antenna effective area and can be between 0 and 5 dBi. We may increase printed antenna gain by using antenna array configuration. In low-cost microstrip antenna arrays, the RF feed network may be integrated to the radiating elements on the same substrate. Microstrip arrays feed networks are shown in **Figure 16**. A parallel feed network is illustrated in **Figure 16(a)**. A parallel series feed network is illustrated in **Figure 16(b)**.

4.2 Transmission line model of patch antennas

In the transmission line model (TLM), the patch antenna functions as two narrow slots connected by a microstrip line, as illustrated in **Figure 17**. TLM model provides a good physical understanding of the electrical characteristics of patch antennas. The electric field along and underneath the patch is given in Eq. (3) and is a function of z. In the design of a wearable patch antenna, the body electrical parameter should be considered to achieve an accurate design.

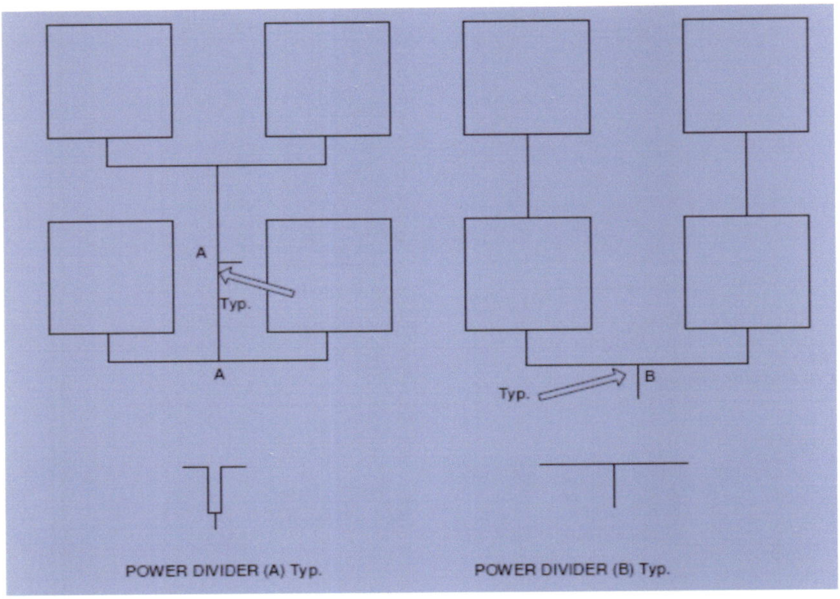

Figure 16.
Configuration of integrated microstrip array and feed network. (a) Parallel feed network. (b) Parallel series feed network.

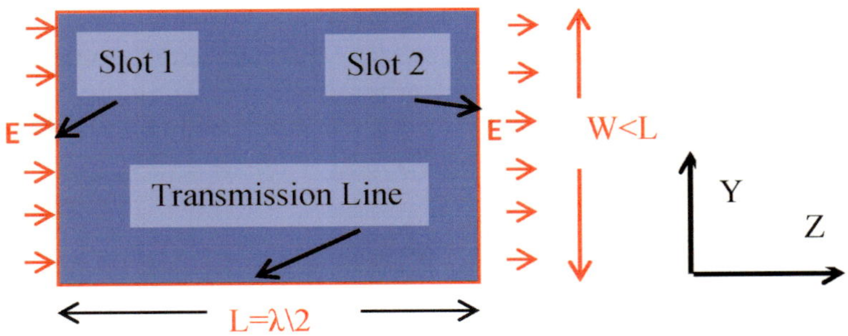

Figure 17.
Patch transmission line model, two slots connected by a microstrip line.

$$E_x \sim \cos\left(\frac{\pi z}{L_{eff}}\right) \tag{3}$$

At the patch edges z = 0 and z=L_{eff}, the electric field is maximum. At the patch center z=$\frac{L_{eff}}{2}$, the electric field is equal to zero. For $\frac{H}{\lambda_0} < 0.1$, the electric field distribution along the x-axis is uniform. The slot admittance is given in Eqs. (4) and (5):

$$G = \frac{W}{120\lambda_0}\left[1 - \frac{1}{24}\left(\frac{2\pi H}{\lambda_0}\right)^2\right] for \ \frac{H}{\lambda_0} < 0.1 \tag{4}$$

$$B = \frac{W}{120\lambda_0}\left[1 - 0.636\, ln\left(\frac{2\pi H}{\lambda_0}\right)^2\right] for \ \frac{H}{\lambda_0} < 0.1 \tag{5}$$

Here, R represents the radiation losses; G = 1/R; and B represents the capacitive nature of the slot. At resonance, for any position of the feed point along the patch, the susceptances of both slots cancel out at the feed point. However, the patch admittance is a function of the feed point position along the z-axis as given in Eq. (6). At the feed point, the slot admittance is transformed by the equivalent length of the transmission line. The width, W, of the microstrip antenna controls the input impedance. For a square patch antenna fed by a microstrip line, the input impedance is around 300 Ohms. By increasing the width, the impedance can be reduced. Larger widths can increase the patch bandwidth.

$$Y(l_1) = Z_0 \frac{1 + j\frac{Z_L}{Z_0}\tan\beta l_1}{\frac{Z_L}{Z_0} + j\tan\beta l_1} = Y_1 \tag{6}$$

$$Y_{in} = Y_1 + Y_2$$

4.3 Excitation of higher order modes in microstrip antennas

To prevent excitation of higher-order modes, the thickness of the substrate should be less than a tenth of the wavelength. We can calculate the cutoff frequency of the higher-order mode by using Eq. (7):

$$f_c = \frac{c}{4H\sqrt{\varepsilon - 1}} \tag{7}$$

4.4 Microstrip effective dielectric constant

As shown in **Figures 13** and **14**, the edges of microstrip line and antenna part of the fields propagate in air and the other part of the fields propagates in the dielectric substrate. The effective dielectric constant is usually higher than $\frac{\varepsilon_r+1}{2}$ and is less than the substrate's dielectric constant. The effective dielectric constant of the microstrip line may be calculated by using Eqs. (8) and (9) as function of W/H:

For $\left(\frac{W}{H}\right) < 1$,

$$\varepsilon_e = \frac{\varepsilon_r + 1}{2} + \frac{\varepsilon_r - 1}{2}\left[\left(1 + 12\left(\frac{H}{W}\right)\right)^{-0.5} + 0.04\left(1 - \left(\frac{W}{H}\right)\right)^2\right] \tag{8}$$

For $\left(\frac{W}{H}\right) \geq 1$,

$$\varepsilon_e = \frac{\varepsilon_r + 1}{2} + \frac{\varepsilon_r - 1}{2}\left[\left(1 + 12\left(\frac{H}{W}\right)\right)^{-0.5}\right] \tag{9}$$

This calculation ignores the strip thickness and frequency dispersion. If the substrate thickness is less than tenth of a wavelength their effects are negligible.

4.5 Losses in microstrip antennas

A major part of losses in microstrip line are due to conductor loss. Radiation loss and dielectric losses are lower. Losses in microstrip lines and antennas are the major disadvantage of microstrip antennas and limit the gain and efficiency of microstrip antennas at high frequencies. Losses in microstrip lines and antennas increase significantly at high frequencies as presented in Eqs. (10) and (11).

4.5.1 Conductor loss

Conductor loss may be calculated by using Eq. (10):

$$\begin{aligned} \alpha_c &= 8.686 \log\left(R_S/(2WZ_0)\right)\ dB/Length \\ R_S &= \sqrt{\pi f \mu \rho} \quad Skin \quad Re\,sis\,\tan ce \end{aligned} \tag{10}$$

Conductor losses can be calculated by defining an equivalent loss tangent δc, given by $\delta c = {}^{\delta s}\!/_{h}$, and $\delta s = \sqrt{{}^{2}\!/_{\omega\mu\sigma}}$. The strip conductivity is σ, μ is the free space permeability, and h is the substrate height.

4.5.2 Dielectric loss

The dielectric loss is given in Eq. (11):

$$\begin{aligned} \alpha_d &= 27.3 \frac{\varepsilon_r}{\sqrt{\varepsilon_{eff}}} \frac{\varepsilon_{eff} - 1}{\varepsilon_r - 1} \frac{tg\delta}{\lambda_0} \quad dB/cm \\ tg\delta &= dielectric \quad loss \quad coefficent \end{aligned} \tag{11}$$

Losses in microstrip lines are presented in **Tables 1** and **2**. For example, total loss of a microstrip line presented in **Table 1** at 40 GHz is 0.5 dB/cm. For example, total loss of a microstrip line presented in **Table 2** at 40 GHz is 1.4 dB/cm. We may conclude that losses in microstrip lines limit the applications of microstrip technology at high frequencies.

Frequency (GHz)	Loss tangent loss (dB/cm)	Metal loss (dB/cm)	Total loss (dB/cm)
10	−0.004	−0.23	−0.23
20	−0.009	−0.333	−0.34
30	−0.013	−0.415	−0.43
40	−0.018	−0.483	−0.5

*W = 0.12 mm, Tanδ = 0.0002, 3 um gold, and conductivity = 3.5E7 mhos/meter.

Table 1.
Microstrip line losses for a substrate of 0.127 mm thickness with εr =9.9.*

Frequency (GHz)	Tangent loss (dB/cm)	Metal loss (dB/cm)	Total loss (dB/cm)
10	−0.010	−0.66	−0.67
20	−0.02	−0.96	−0.98
30	−0.03	−1.19	−1.22
40	−0.04	−1.38	−1.42

*W = 0.034 mm, Tanδ = 0.0004, 3 um gold, and conductivity = 3.5E7 mhos/meter.

Table 2.
Microstrip line losses for a GaAs substrate of 0.05 mm thickness with εr =12.88.*

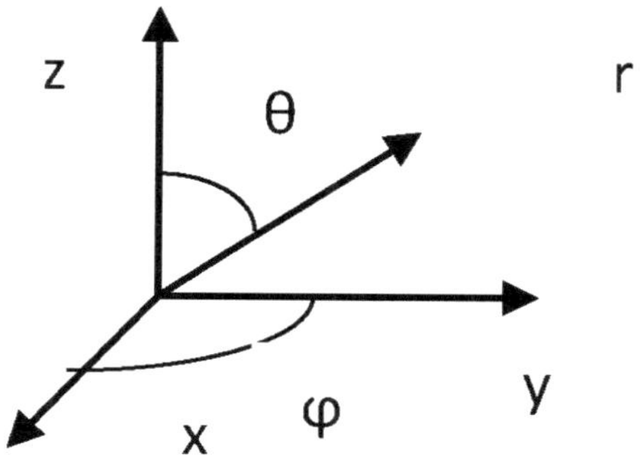

Figure 18.
Cartesian coordinate system.

4.6 Patch radiation pattern

The patch radiation pattern is function of the patch width, W. The coordinate system is presented in **Figure 18**. The normalized radiation pattern may be simulated by using Eqs. (12) and (13):

$$E_\theta = \frac{\sin\left(\frac{_kW}{2}\sin\theta\sin\varphi\right)}{\frac{k_0W}{2}\sin\theta\sin\varphi}\cos\left(\frac{k_0L}{2}\sin\theta\cos\varphi\right)\cos\varphi \quad (411) \tag{12}$$
$$k_0 = {}^{2\pi}/_{\lambda}$$

$$E_\varphi = \frac{\sin\left(\frac{_kW}{2}\sin\theta\sin\varphi\right)}{\frac{k_0W}{2}\sin\theta\sin\varphi}\cos\left(\frac{k_0L}{2}\sin\theta\cos\varphi\right)\cos\theta\sin\varphi \quad (4.12) \tag{13}$$
$$k_0 = {}^{2\pi}/_{\lambda}$$

The magnitude of the fields is given by Eq. (14):

$$f(\theta,\varphi) = \sqrt{E_\theta^2 + E_\varphi^2} \tag{14}$$

5. Wearable stacked microstrip antennas for 5G, IoT, and medical applications

Stacked patch antennas were presented first in [1–7]. Single-layer microstrip antennas have a narrow bandwidth. This disadvantage limits the applications of microstrip antennas. By designing a double-layer microstrip antenna, we may get a wider bandwidth. Two-layer patch antennas may be the best antenna choice for wideband communication systems. On the first layer, the antenna matching network and a resonator are printed. On the second layer, the radiating element is printed. The electromagnetic field is coupled from the resonator to the radiating element. The resonator and the radiating element shapes may be rectangular, square, triangle, circle, or any other shape. The distance between the layers is optimized to get maximum bandwidth with the best antenna efficiency. The spacing between the layers may be foam or a substrate with low dielectric losses. All the antennas' electrical parameters were calculated and optimized by.

using electromagnetic software. A 2.2 GHz square patch with circular polarization stacked antenna was designed. The resonator and the feed network were printed on a substrate with a relative dielectric constant of 2.4 and with a thickness of 0.16 cm. The dimensions of the square resonator are W = L = 4.5 cm. The radiating element was printed on a substrate with a relative dielectric constant of 2.2 and with a thickness of 0.16 cm. The radiator is a square patch with dimensions W = L = 4.8 cm. The antenna is circular polarized. A 3 dB, 90° branch coupler is connected to the antenna feed lines, as shown in **Figure 19**. The antenna bandwidth is 13% for VSWR better than 3:1. The measured antenna beamwidth is 73°. The measured antenna gain is 7.5 dBi at 2.2 GHz. This antenna may be used in wideband communication systems. Comparison of calculated and measured results of stacked patch antennas is listed in **Table 3**. The antennas listed in **Table 3** may be used in wearable communication systems. Results in **Table 3** indicate that the bandwidth of stacked patch antennas may be around 9–15% for VSWR better than 2:1. There is a good agreement between the measured and calculated results. In **Figure 20**, a stacked microstrip antenna is shown. The antenna feed and matching network is printed on FR4 with a dielectric constant of 4.2 and 1.6 mm thickness. The radiator is printed on a dielectric substrate with a dielectric constant of 2.2 and 1.6 mm thickness. The dimensions of the microstrip stacked patch antenna shown in **Figure 20** are 3.3 × 2 × 0.32 cm. The computed S11 parameters are presented in **Figure 21**. Radiation pattern of the microstrip stacked patch is shown in **Figure 22**.

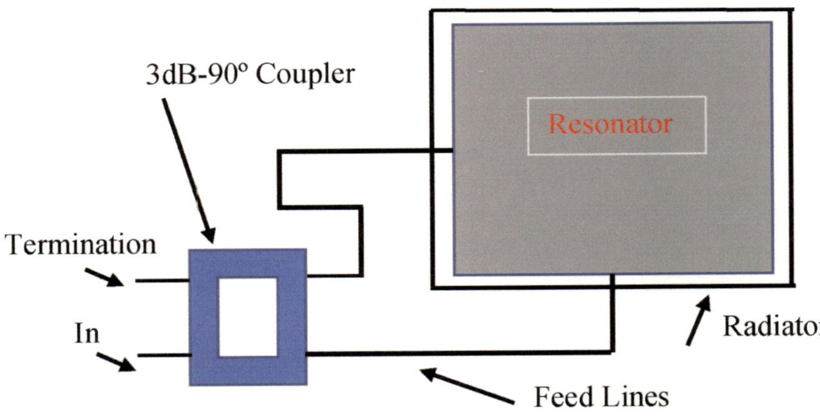

Figure 19.
Feed network of a circular polarized stacked patch antenna.

Antenna	F (GHz)	Bandwidth (%)		Beamwidth		Gain (dBi)		Polarization
		Calc.	Meas.	Calc.	Meas.	Calc.	Meas.	
Square	2.2	11	10	74	72	7.5	7.5	Circular
Circular	2.2	14	15	74	72	7.5	7.9	Linear
Annular disc	2.2	12	11.5	80	78	6.5	6.6	Linear
Rectangular	2.0	10	9	72	72	7.5	7.4	Linear
Circular	2.4	10	9	74	72	7.5	7	Linear
Circular	2.4	10	10	74	72	7.5	7.5	Circular

Table 3.
Comparison of calculated and measured results of stacked microstrip antennas.

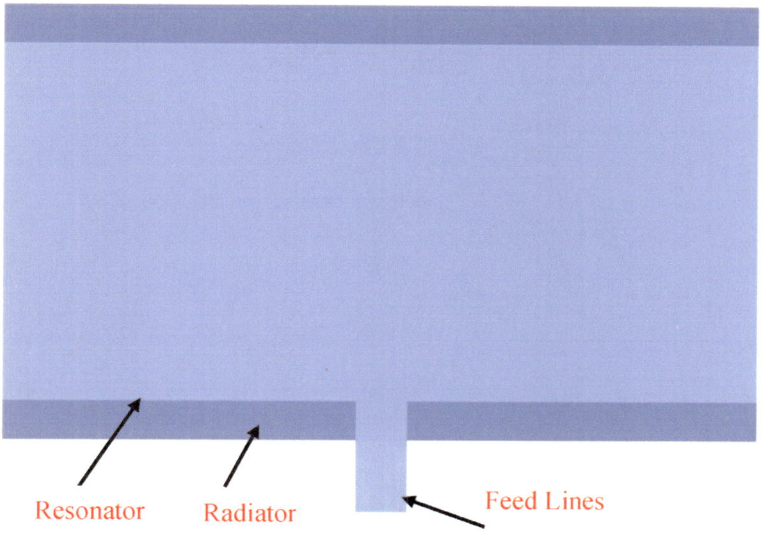

Figure 20.
A microstrip stacked patch antenna for 5G, IoT, and medical applications.

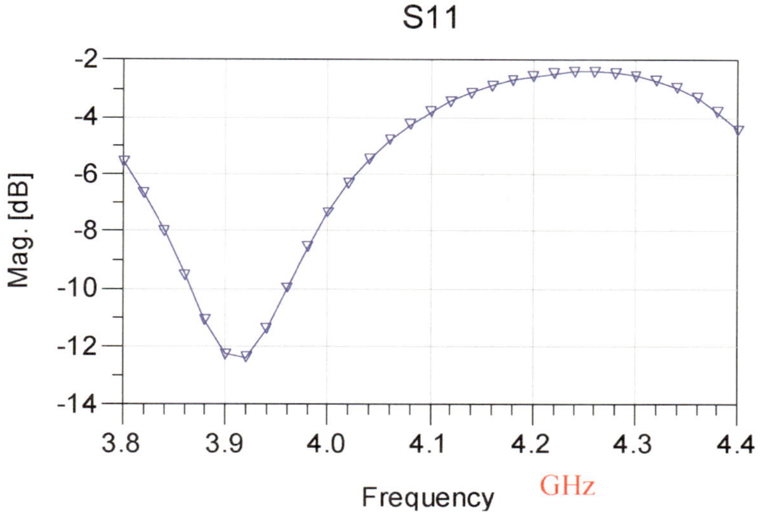

Figure 21.
Computed S11 of the microstrip stacked patch.

The antenna bandwidth is around 7% for VSWR better than 3:1. The antenna bandwidth is improved to 10% for VSWR better than 2.0:1 by adding 8 mm air spacing between the layers. The antenna beamwidth is around 72°.

The antenna gain is around 7 dBi.

5.1 Stacked microstrip 35 GHz antennas arrays

Two Ka-band, stacked patch microstrip antenna arrays, which consist of 256 radiating elements, have been designed on a substrate with εr = 2.2, 0.25 mm thick. The first Type A array with a parallel feed network, is shown in **Figure 16(a)**. The second Type B array is shown in **Figure 16(b)** has more bend discontinuities in the

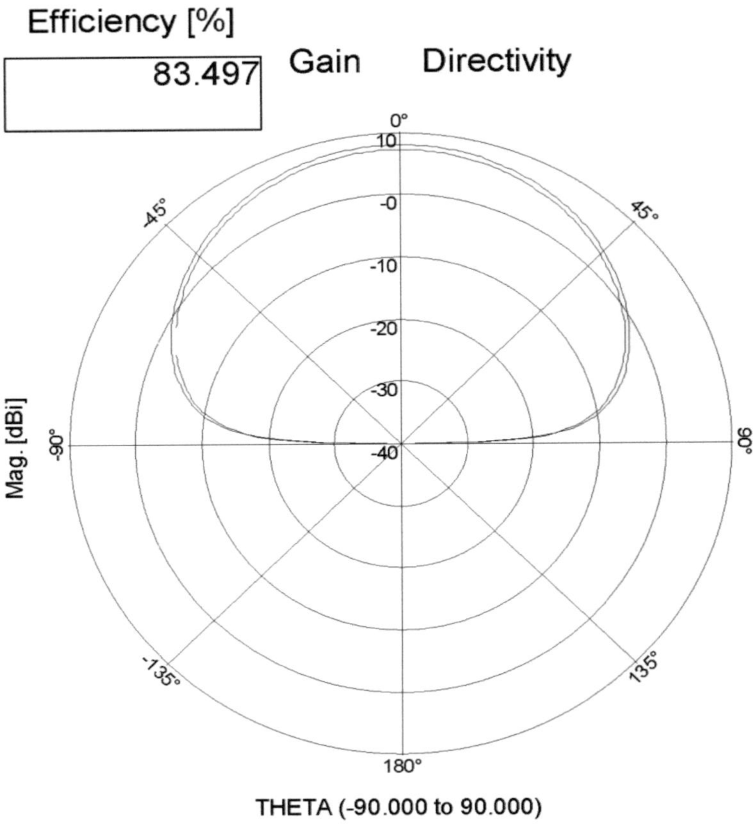

Figure 22.
Radiation pattern of the microstrip stacked patch.

Parameter	Type A	Type B	Type C
Number of radiators	256	256	256
Beamwidth (°)	4.2	4.2	4.2
Computed gain(dBi)	32	32	32
Microstrip line loss(dB)	3.1	3.1	1.5
Radiation loss T-J. (dB)	0.72	0.72	0.72
Radiation loss bends(dB)	0.13	1.17	0.13
Radiation loss steps(dB)	0.12	—	0.12
Mismatch Loss (dB)	0.5	0.5	0.5
Expected Gain(dBi)	27.43	26.5	29.03
Measured Gain(dBi)	27.5	26.5	29.5
Efficiency (%)	34.9	28.2	51

Table 4.
Comparison of electrical performance of 256 stacked patch microstrip antenna arrays.

feeding network than Type A array. In the Type C array, a 10-cm coaxial line was used to replace the same length of microstrip line in the Type A array. Comparison of measured results of the arrays, given in **Table 4**, shows that the gain of the modified array Type C was increased by 1.6 dB. The arrays' measured bandwidth is around 12% for VSWR better than 2:1.

6. Stacked mono-pulse Ku-band patch antenna

A mono-pulse double-layer circular patch antenna was designed at Ku band, 15 GHz. The mono-pulse antenna consists of four circular patch antennas and a feed network as presented in **Figure 23**. The circular resonator and the branch coupler were printed on a substrate with a relative dielectric constant of 2.45 and with a thickness of 0.8 mm. The diameter of the circular microstrip resonator is 0.42 cm. The circular radiator was printed on a substrate with a relative dielectric constant of 2.25 and with a thickness of 0.8 mm. The diameter of the circular patch is 0.45 cm. The comparator consists of three 3 dB, 180° rat-race couplers that are connected to four circular patches via the antenna feed lines, as presented in **Figures 23** and **24**. The strip-line 3 dB, 180° rat-race couplers are printed on a substrate with a relative dielectric constant of 2.2 and thickness of 0.8 mm. The comparator structure and ports are shown in **Figures 23** and **24**. The comparator output ports are: a sum port $\sum$, difference port **Δ**, an azimuth difference port **Δ**Az, and an elevation difference port **Δ**El. The antenna bandwidth is 11% for VSWR better than 2:1. The antenna beam width is around 36°. The computed and measured antenna gain is around 10.5 dBi. The maximum comparator losses are 0.75 dB.

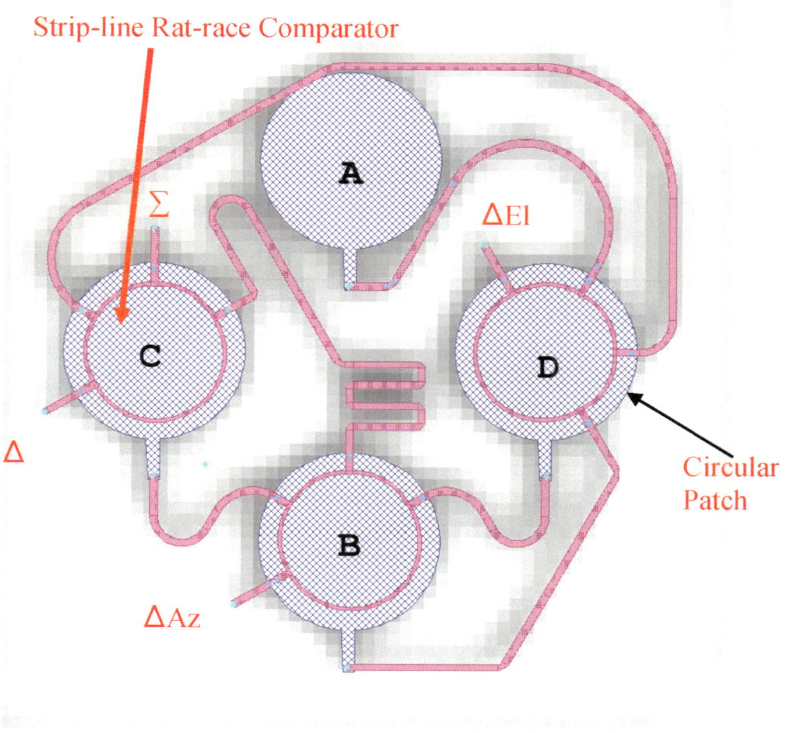

Figure 23.
A microstrip stacked mono-pulse antenna.

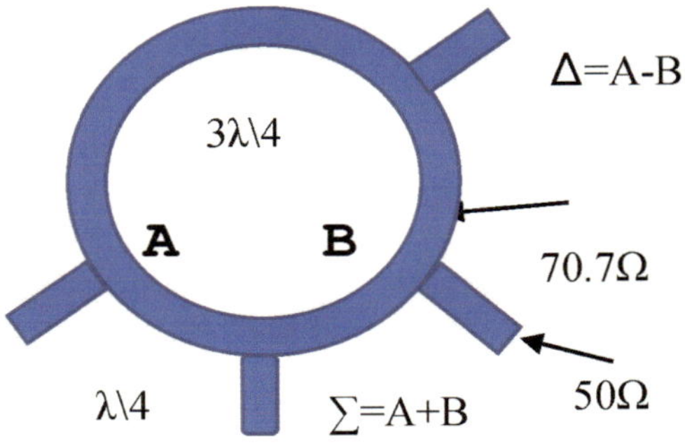

Figure 24.
Rat-race coupler.

6.1 Rat-race coupler

A rat-race coupler is shown in **Figure 24**. The rat-race circumference is 1.5 wavelengths. The distance from A to **Δ** port is 3λ\4. The distance from A to ∑ port is λ\4.

For an equal-split rat-race coupler, the impedance of the entire ring is fixed at 1.41 × Z0, or 70.7 Ω for Z0 = 50 Ω. For an input signal V, the outputs at ports 2 and 4 are equal in magnitude, but 180 degrees out of phase.

7. Wearable Metamaterial antennas for 5G, IoT, and medical applications

Low profile efficient antennas are crucial in the development of commercial compact 5G communication and IoT systems. Communication, IoT, and biomedical industries are in rapid growth in the last years. It is important to develop efficient high gain compact antennas for 5G communication and IoT systems. Metamaterials and fractal structures may be used to improve the efficiency of compact printed antennas. In this chapter metamaterial antennas will be presented.

7.1 Introduction

Small printed antennas suffer from low efficiency. Metamaterial technology is used to design wearable compact antennas with high efficiency. The metamaterial antennas may be used in 5G communication systems, IoT, and medical systems. Design trade-offs, development, and computed and measured results of compact, efficient metamaterial antennas are presented in this chapter. The gain and directivity of the patch antenna with split ring resonators (SRRs) are higher by 2.5 dB than the patch antenna without SRR. The resonant frequency of the antenna with SRR on human body is shifted by 3%. Printed antennas are used in communication systems and are presented in journals and books, as referred in [1–5]. Microstrip and printed antennas have several advantages such as being light weight, compact, flexible, and having low production cost. The main disadvantages of these printed antennas are narrow bandwidth and low efficiency. In Ref. [51], artificial media with negative dielectric permittivity were presented. Materials with dielectric constant and permeability less than 1 are developed by using periodic SRR and metallic posts structures as presented in [51–59]. New wearable printed metamaterial antennas with high efficiency are presented in this chapter.

DOI: http://dx.doi.org/10.5772/intechopen.93492

7.2 Stacked microstrip antenna with SRR

Stacked microstrip patches antennas with and without SRR has been designed, see Refs. [1–5]. The antennas was designed on the same substrate. The antennas are stacked double-layer antennas. The first layer consists of a FR4 substrate with a dielectric constant of 4.2 and 1.6 mm thickness. The second layer consists of a dielectric substrate with a dielectric constant of 2.3 and 1.6 mm thickness. The antenna has been analyzed and optimized by using full wave electromagnetic software. The dimensions of the microstrip stacked patch antenna are 33 × 20 × 3.2 mm as presented in **Figure 25**. The antenna bandwidth is around 6% for VSWR better than 3:1. The antenna beam width is around 74°. The stacked antenna directivity and gain are around 7 dBi. The computed S11 parameters are presented in **Figure 26**. Radiation pattern of the microstrip stacked patch is shown in **Figure 27**. The stacked patch antenna with SRR is presented in **Figure 28**. This antenna has the same structure as the stacked antenna shown in **Figure 25**. The spacing between the SRR rings is 0.25 mm and the ring width is 0.2 mm. Four rows of seven SRRs are placed on the radiating patch. The measured S11 parameters of the antenna with SRR are presented in **Figure 29**. The antenna bandwidth is around 13% for VSWR better than 2.5:1. By adding an air space of 4 mm between the antenna layers, the VSWR was improved to 2:1. The antenna gain is around 9–10 dBi. The antenna's

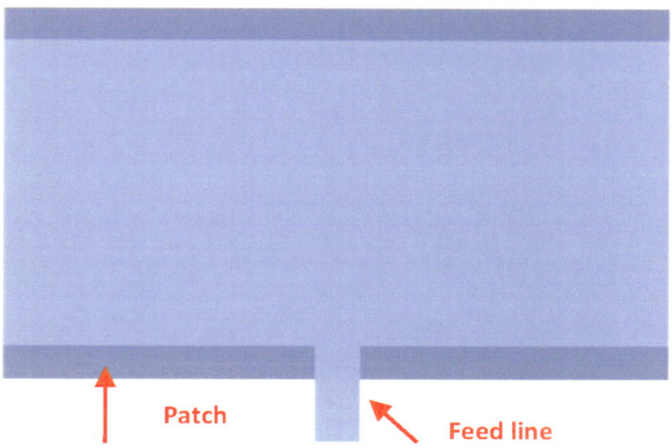

Figure 25.
A microstrip stacked patch antenna.

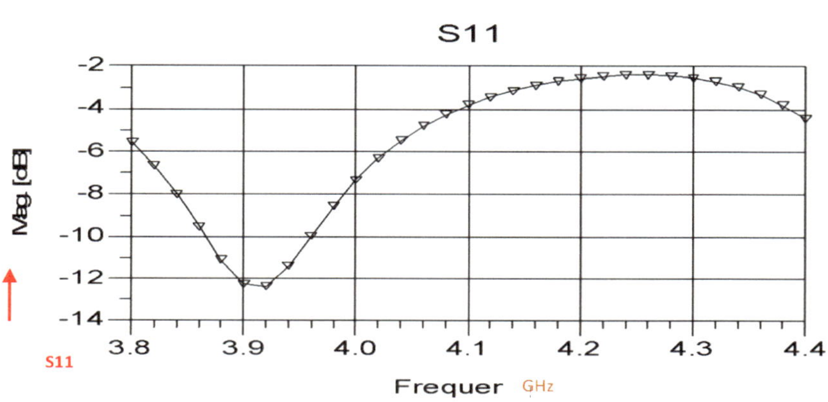

Figure 26.
Computed S11 of the microstrip stacked patch.

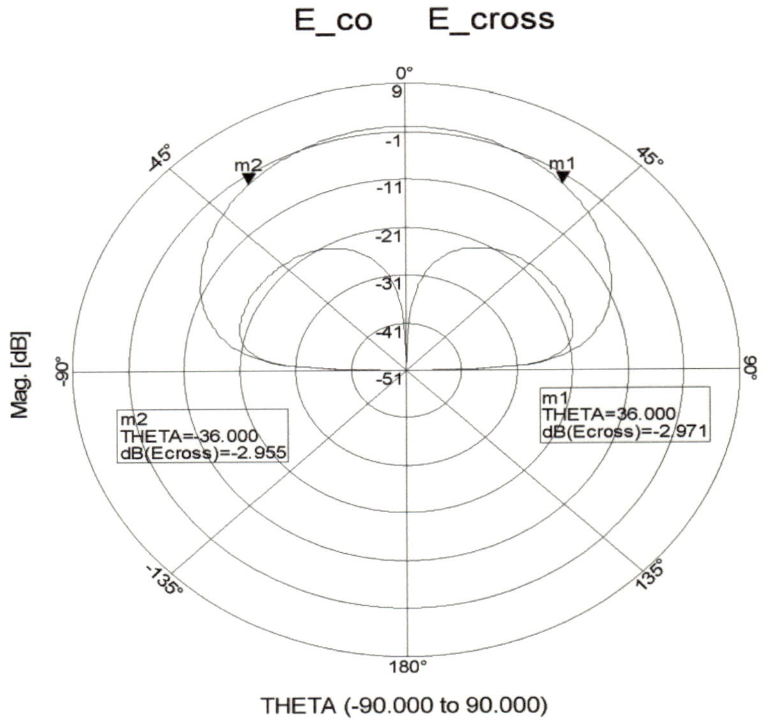

Figure 27.
Radiation pattern of the microstrip stacked patch.

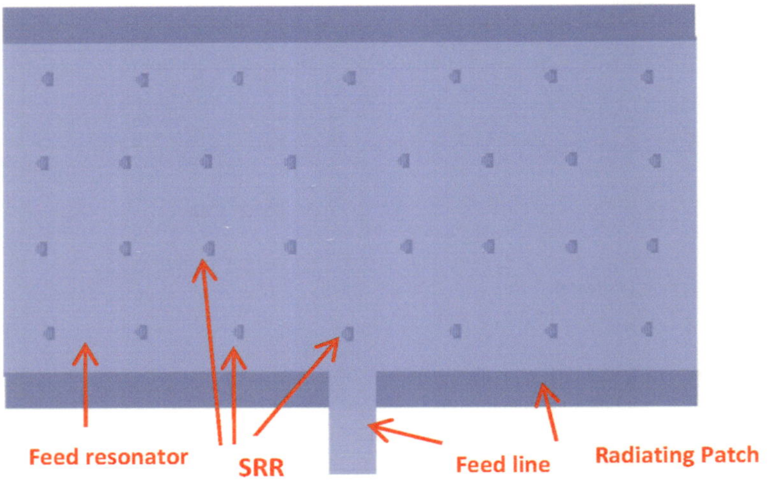

Figure 28.
Printed antenna with split ring resonators.

efficiency is around 95%. The antenna computed radiation pattern is shown in **Figure 30**. There is a good agreement between the measured and computed results. The effective area of a patch antenna without SRR is lower than the effective area of a patch antenna with SRR. The resonant frequency of a patch antenna without SRR is higher by 10% than the resonant frequency of a patch antenna with SRR. The antenna beamwidth is around 70°. The directivity and gain of the stacked antenna with SRR is higher by 2–3 dB than the patch antenna without SRR.

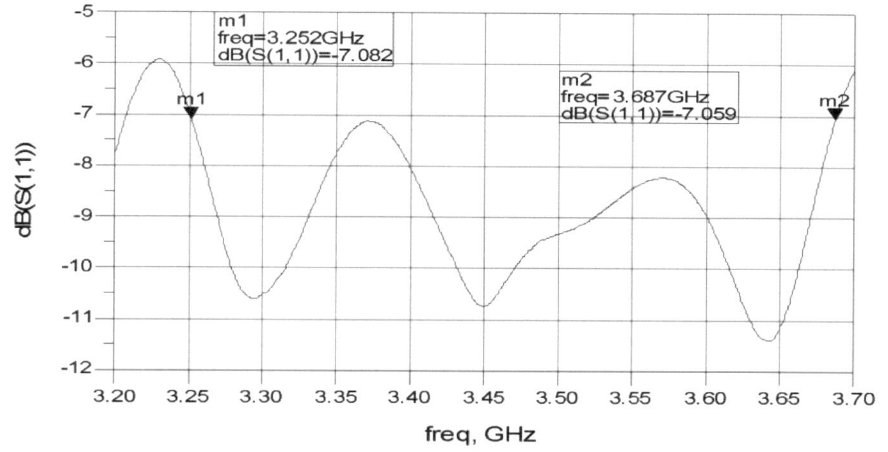

Figure 29.
Patch with split ring resonators for medical and 5G applications, measured S11.

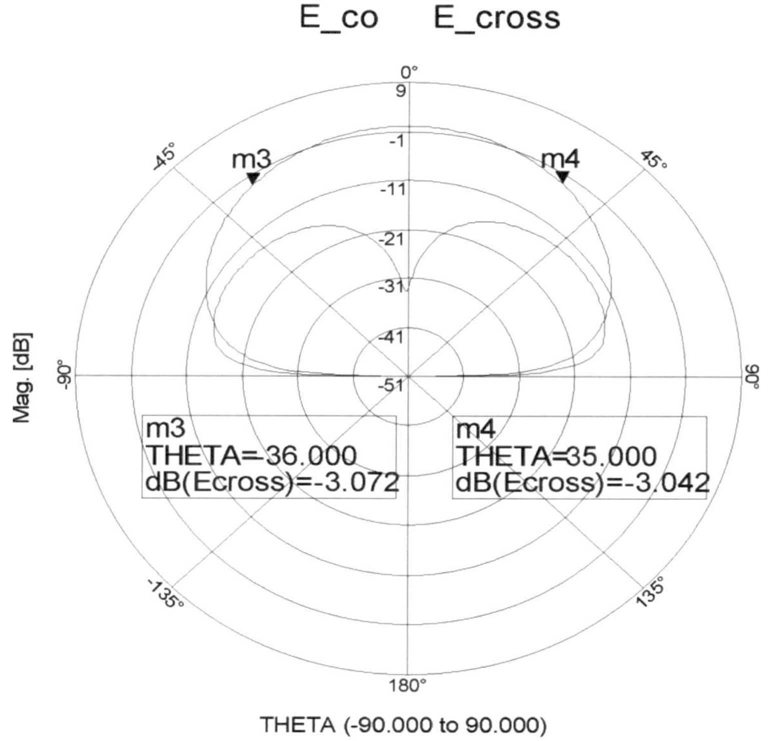

Figure 30.
Radiation pattern for patch with SRR for medical and 5G applications.

7.3 Patch antenna with split ring resonators

A patch antenna with split ring resonators was developed. The antenna is printed on the dielectric substrate with a dielectric constant of 2.2 and with a 1.6 mm thickness. The dimensions of the microstrip patch antenna shown in

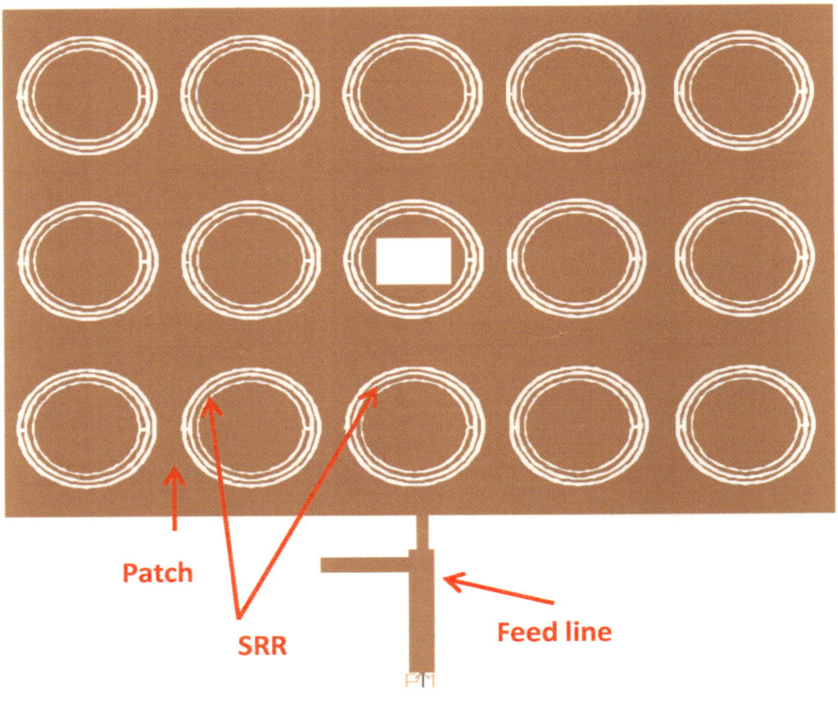

Figure 31.
Patch antenna with 15 Split ring resonators.

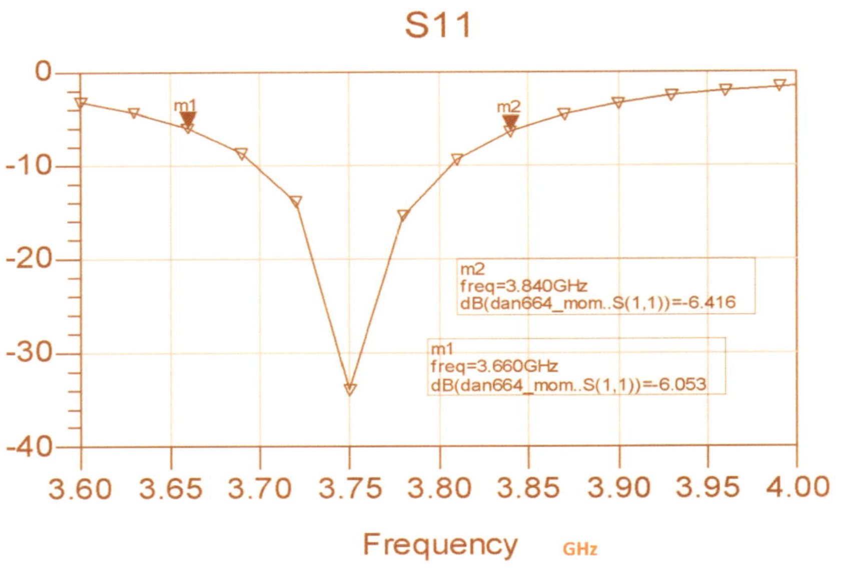

Figure 32.
Patch with 15 split ring resonators, computed S11.

Figure 31 are 36 × 20 × 1.6 mm. The metamaterial antenna bandwidth is around 6% for VSWR better than 2:1. The antenna bandwidth is around 9% for S11 lower than −6 dB. The antenna directivity and gain are around 7.5 dBi. The computed and measured antenna beam width is around 72°. The antenna efficiency is 77.25%. The measured S11 parameters are presented in **Figure 32**. The gain and directivity of the patch antenna with SRR is higher by 2.5 dB than the patch antenna without SRR. A photo of printed metamaterial antennas for medical and IoT applications is shown in **Figure 33**. Metamaterial patch antenna with SRR for 5G, IoT, and medical

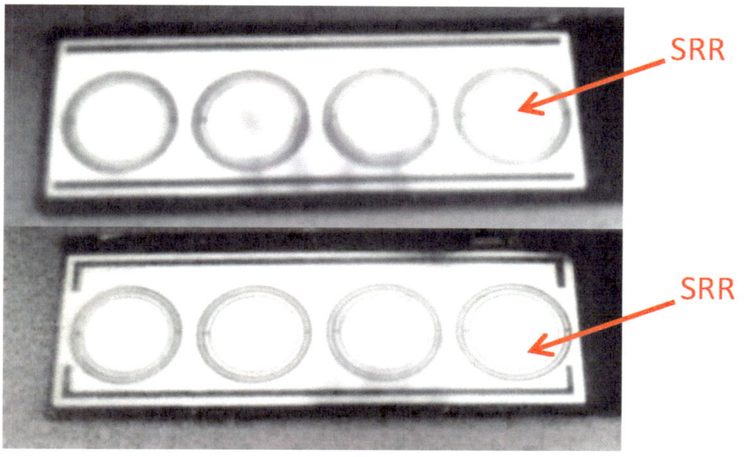

Figure 33.
Photo of printed metamaterial antennas for medical applications.

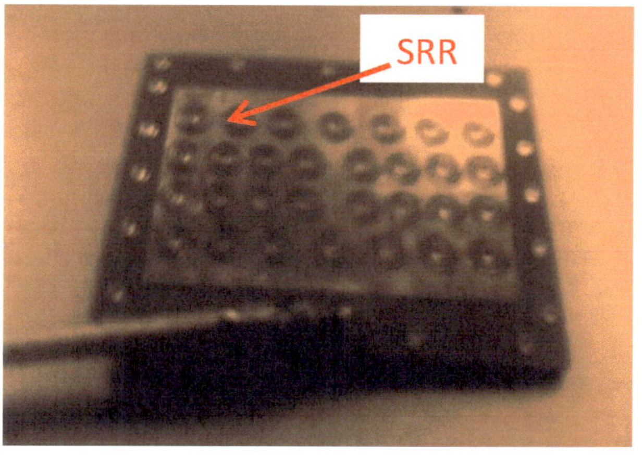

Figure 34.
Photo of metamaterial patch antenna with SRR.

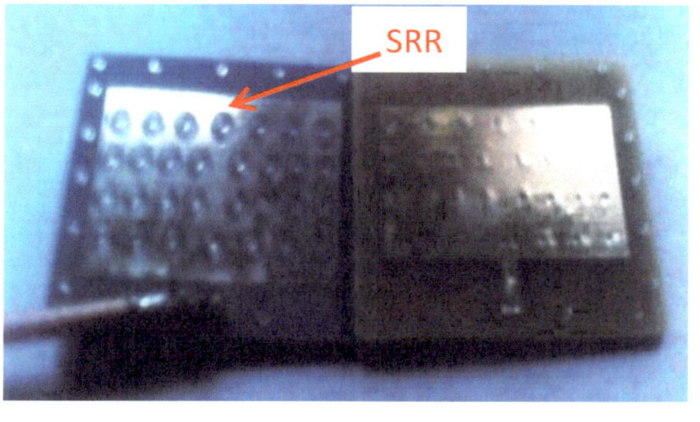

Figure 35.
Meta-material stacked patch antenna with SRR.

applications is shown in **Figure 34**. Metamaterial stacked patch antenna with SRR for 5G, IoT, and medical applications is shown in **Figure 35**.

8. Conclusions

This chapter presents several wideband wearable antennas with high efficiency for medical, IoT and sport applications. Wearable technology provides a useful novel tool to health-care centers and surgical rehabilitation services. Wireless wearable body area network (wireless WBAN) is emerging as a significant option for hospitals, medical centers, and patients. Wearable devices provide a useful network that may improve the long-term context and physiological response of patients and health-care customers. Wearable devices and technology will help to develop personal treatment devices with online and real-time feedback to improve patient's health. Wearable medical devices and sensors can measure heartbeat, blood pressure, body temperature, and sweat rate, perform gait analysis, and measure almost any medical parameters of the patient wearing the medical system.

Design considerations computed and measured results of several wearable printed antennas are described in this chapter. The antenna's electrical characteristics and dimensions were designed to meet the medical system specification. The dimensions of the compact antennas may vary from $260 \times 60 \times 1.6$ mm to $50 \times 50 \times 0.5$ mm to meet the medical system specification. The compact wearable antennas bandwidth is between 9 and 12% for VSWR better than 2:1. The compact wearable antenna beam width varies from 72° to 100° and the wearable antennas gain varies from 0 to 5 dBi as a function of the antenna dimensions.

The length of the antennas without SRR is higher by 5–10% than the length of the antennas with SRR. Moreover, the resonant frequency of the antennas without SRR is higher by 5–10% than the antennas with SRR. The gain and directivity of the patch antenna without SRR is lower by 2–3 dB than the patch antenna with SRR. The resonant frequency of the wearable antennas with SRR on human body may be shifted by 2–5%.

Author details

Albert Sabban
Kinneret College, Emek Hayarden, Israel

*Address all correspondence to: sabban@netvision.net.il

References

[1] Sabban A. Wideband RF Technologies and Antenna in Microwave Frequencies. USA: Wiley Sons; 2016

[2] Sabban A. Novel Wearable Antennas for Communication and Medical Systems. FL, USA: Taylor & Francis Group; 2017

[3] Sabban A. Low Visibility Antennas for Communication Systems. USA: Taylor & Francis Group; 2015

[4] Sabban A. Small wearable meta materials antennas for medical systems. The Applied Computational Electromagnetics Society Journal. April 2016;**31**(4):434-443

[5] Sabban A. Microstrip antenna arrays. In: Nasimuddin N, editor. Microstrip Antennas. Croatia: InTech; 2011. pp. 361-384. ISBN: 978-953-307-247-0. Available from: http://www.intechopen.com/articles/show/title/microstrip-antenna-arrays

[6] Sabban A. New wideband printed antennas for medical applications. IEEE Transactions on Antennas and Propagation. 2013;**61**(1):84-91

[7] Sabban A. Dual polarized dipole wearable antenna. USA: U.S Patent Number: 8203497; 2012

[8] Sabban A. A new wideband stacked microstrip antenna. In: IEEE Antenna and Propagation Symposium. Houston, Texas, USA; June 1983

[9] Sabban A. Wideband microstrip antenna arrays. In: IEEE Antenna and Propagation Symposium MELCOM, Tel-Aviv, June 1981. 1981

[10] Chirwa* Lawrence C, Hammond Paul A, Scott Roy Cumming David RS. "Electromagnetic radiation from ingested sources in the human intestine between 150 MHz and 1.2 GHz", IEEE Transactions on Biomedical Engineering. April 2003;**50**(4):484-492

[11] Werber D, Schwentner A, Biebl EM. Investigation of RF transmission properties of human tissues. Advances in Radio Science. 2006;**4**:357-360

[12] Thalmann T, Popovic Z, Notaros BM, Mosig JR. Investigation and design of a multi-band wearable antenna. In: 3rd European Conference on Antennas and Propagation. Berlin, Germany: EuCAP 2009; 2009. pp. 462-465

[13] Salonen P, Rahmat-Samii Y, Kivikoski M. Wearable antennas in the vicinity of human body. In: IEEE Antennas and Propagation Society Symposium, Vol. 1, Monterey USA, 2004. 2004. pp. 467-470

[14] Fujimoto K, James JR, editors. Mobile Antenna Systems Handbook. Boston USA: Artech House; 1994

[15] Gupta B, Sankaralingam S, Dhar S. Development of wearable and implantable antennas in the last decade. In: Microwave Mediterranean Symposium (MMS), August 2010, Guzelyurt, Turkey. 2010. pp. 251-267

[16] Kellomaki T, Heikkinen J, Kivikoski M. Wearable antennas for FM reception. In: First European Conference on Antennas and Propagation, EuCAP 2006, Hague, the Netherlands. 2006. pp. 1-6

[17] Sabban A. Wideband printed antennas for medical applications. In: APMC 2009 Conference, Singapore, 12/2009. 2009

[18] Lee Y. Antenna Circuit Design for RFID Applications. Az, USA: Microchip Technology Inc., Microchip AN 710c. 2003

[19] Sabban A, Gupta KC. Characterization of radiation loss from

microstrip discontinuities using a multiport network modeling approach. IEEE Transactions on MTT. 1991;**39**(4): 705-712

[20] Sabban A. Multiport network model for evaluating radiation loss and coupling among discontinuities in microstrip circuits [PhD thesis]. Colorado, Boulder, USA: University of Colorado at Boulder; 1991

[21] Mukhopadhyay SC, editor. Wearable Electronics Sensors. Switzerland: Springer; 2015

[22] Bonfiglio A, De Rossi D, editors. Wearable Monitoring Systems, Springer, NY, USA, 2011

[23] Gao T, Greenspan D, Welsh M, Juang RR, Alm A. Vital signs monitoring and patient tracking over a wireless network. In: Proceedings of IEEE-EMBS 27th Annual International Conference of the Engineering in Medicine and Biology, Shanghai, China, 1–5 September 2005. 2005. pp. 102-105

[24] Otto CA, Jovanov E, Milenkovic EA. WBAN-based system for health monitoring at home. In: Proceedings of IEEE/EMBS International Summer School, Medical Devices and Biosensors, Boston, MA, USA, 4–6 September 2006. 2006. pp. 20-23

[25] Zhang GH, Poon CCY, Li Y, Zhang YT. A biometric method to secure telemedicine systems. In: Proceedings of the 31st Annual International Conference of the IEEE Engineering in Medicine and Biology Society, Minneapolis, MN, USA, September 2009. 2009. pp. 701-704

[26] Srinivasan V, Stankovic J, Whitehouse K. Protecting your daily in home activity information from a wireless snooping attack. In: Proceedings of the 10th International Conference on Ubiquitous Computing, Seoul, Korea, 21–24 September 2008. 2008. pp. 202-211

[27] Casas R, Blasco MR, Robinet A, Delgado AR, Yarza AR, Mcginn J, et al. User modelling in ambient intelligence for elderly and disabled people. In: Proceedings of the 11th International Conference on Computers Helping People with Special Needs, Linz, Austria, July 2008. 2008. pp. 114-122

[28] Jasemian Y. Elderly comfort and compliance to modern telemedicine system at home. In: Proceedings of the Second International Conference on Pervasive Computing Technologies for Healthcare, Tampere, Finland, 30 January–1 February 2008. 2008. pp. 60-63

[29] Atallah L, Lo B, Yang GZ, Siegemund F. Wirelessly accessible sensor populations (WASP) for elderly care monitoring. In: Proceedings of the Second International Conference on Pervasive Computing Technologies for Healthcare, Tampere, Finland, 30 January–1 February 2008. 2008. pp. 2-7

[30] Hori T, Nishida Y, Suehiro T, Hirai S. SELF-network: Design and implementation of network for distributed embedded sensors. In: Proceedings of IEEE/RSJ International Conference on Intelligent Robots and Systems, Takamatsu, Japan, 30 October–5 November 2000. 2000. pp. 1373-1378

[31] Mori Y, Yamauchi M, Kaneko K. Design and implementation of the vital sign box for home healthcare. In: Proceedings of IEEE EMBS International Conference on Information Technology Applications in Biomedicine, Arlington, VA, USA, November 2000. 2000. pp. 104-109

[32] Lauterbach C, Strasser M, Jung S, Weber W. Smart clothes self-powered by body heat. In: Proceedings of

Avantex Symposium, Frankfurt, Germany, May 2002. 2002. pp. 5259-5263

[33] Marinkovic S, Popovici E. Network coding for efficient error recovery in wireless sensor networks for medical applications. In: Proceedings of International Conference on Emerging Network Intelligence, Sliema, Malta, 11–16 October 2009. 2009. pp. 15-20

[34] Schoellhammer T, Osterweil E, Greenstein B, Wimbrow M, Estrin D. Lightweight temporal compression of microclimate datasets. In: Proceedings of the 29th Annual IEEE International Conference on Local Computer Networks, Tampa, FL, USA, 16–18 November 2004. 2004. pp. 516-524

[35] Barth AT, Hanson MA, Powell HC Jr, Lach J. Tempo 3.1: A body area sensor network platform for continuous movement assessment. In: Proceedings of the 6th International Workshop on Wearable and Implantable Body Sensor Networks, Berkeley, CA, USA, June 2009. 2009. pp. 71-76

[36] Gietzelt M, Wolf KH, Marschollek M, Haux R. Automatic self-calibration of body worn triaxial-accelerometers for application in healthcare. In: Proceedings of the Second International Conference on Pervasive Computing Technologies for Healthcare, Tampere, Finland, January 2008. 2008. pp. 177-180

[37] Gao T, Greenspan D, Welsh M, Juang RR, Alm A. Vital signs monitoring and patient tracking over a wireless network. In: Proceedings of the 27th Annual International Conference of the IEEE EMBS, Shanghai, China, 1–4 September 2005. 2005. pp. 102-105

[38] Baker C, Armijo K, Belka S, Benhabib M, Bhargava V, Burkhart N, et al. Wireless sensor networks for home health care. In: Proceedings of the 21st International Conference on Advanced Information Networking and Applications Workshops, Niagara Falls, Canada, 21–23 May 2007. 2007. pp. 832-837

[39] Aziz O, Lo B, King R, Darzi A, Yang GZ. Pervasive body sensor network: An approach to monitoring the postoperative surgical patient. In: Proceedings of International Workshop on Wearable and Implantable Body Sensor Networks (BSN 2006), Cambridge, MA, USA; 2006. pp. 13-18

[40] Kwon DY, Gross M. Combining body sensors and visual sensors for motion training. In: Proceedings of the 2005 ACM SIGCHI International Conference on Advances in Computer Entertainment Technology, Valencia, Spain, 15–17 June 2005. 2005. pp. 94-101

[41] Boulgouris NK, Hatzinakos D, Plataniotis KN. Gait recognition: A challenging signal processing technology for biometric identification. IEEE Signal Processing Magazine. 2005; **22**:78-90

[42] Kimmeskamp S, Hennig EM. Heel to toe motion characteristics in Parkinson patients during free walking. Clinical Biomechanics. 2001;**16**:806-812

[43] Turcot K, Aissaoui R, Boivin K, Pelletier M, Hagemeister N, de Guise JA. New accelerometric method to discriminate between asymptomatic subjects and patients with medial knee osteoarthritis during 3-D gait. IEEE Transactions on Biomedical Engineering. 2008;**55**:1415-1422

[44] Bamberg SJM, Benbasat AY, Scarborough DM, Krebs DE, Paradiso JA. Gait analysis using a shoe-integrated wireless sensor system. IEEE Transactions on Information Technology in Biomedicine. 2008;**12**:413-423

[45] Choi JH, Cho J, Park JH, Eun JM, Kim MS. An efficient gait phase detection device based on magnetic

sensor array. In: Proceedings of the 4th Kuala Lumpur International Conference on Biomedical Engineering, Vol. 21, Kuala Lumpur, Malaysia, 25–28 June 2008. 2008. pp. 778-781

[46] Hidler J. Robotic assessment of walking in individuals with gait disorders. In: Proceedings of the 26th Annual International Conference of the IEEE Engineering in Medicine and Biology Society, Vol. 7, San Francisco, CA, USA, 1–5 September 2004. 2004. pp. 4829-4831

[47] Wahab Y, Bakar NA. Gait analysis measurement for sport application based on ultrasonic system. In: Proceedings of the 2011 IEEE 15th International Symposium on Consumer Electronics, Singapore, 14–17 June 2011. 2011. pp. 20-24

[48] ElSayed M, Alsebai A, Salaheldin A, El Gayar N, ElHelw M. Ambient and wearable sensing for gait classification in pervasive healthcare environments. In: Proceedings of the 12th IEEE International Conference on e-Health Networking Applications and Services (Healthcom), Lyon, France, 1–3 July 2010. 2010. pp. 240-245

[49] Purwar A, Jeong DU, Chung WY. Activity monitoring from realtime triaxial accelerometer data using sensor network. In: Proceedings of International Conference on Control, Automation and Systems, Hong Kong, 21–23 March 2007. 2007. pp. 2402-2406

[50] Software, Keysightt. Available from: http://www.keysight.com/en/pc-1297113/advanced-design-system-ads?cc=IL&lc=eng

[51] Pendry JB, Holden AJ, Stewart WJ, Youngs I. Extremely low frequency plasmons in metallic mesostructures. Physical Review Letters. 1996;**76**: 4773-4776

[52] Pendry JB, Holden AJ, Robbins DJ, Stewart WJ. Magnetism from conductors and enhanced nonlinear phenomena. IEEE Transactions on Microwave Theory and Techniques. 1999;**47**:2075-2084

[53] Marque's R, Mesa F, Martel J, Medina F. Comparative analysis of edge and broadside coupled split ring resonators for metamaterial design. Theory and experiment. IEEE Transactions on Antennas and Propagation. 2003;**51**:2572-2581

[54] Marque's R, Baena JD, Martel J, Medina F, Falcone F, Sorolla M, et al. Novel small resonant electromagnetic particles for metamaterial and filter design. In: Proceeding of the ICEAA'03, Torino, Italy. 2003. pp. 439-442

[55] Marque's R, Martel J, Mesa F, Medina F. Left-handed-media simulation and transmission of EM waves in resonator-subwavelength split-ring-loaded metallic waveguides. Physical Review Letters. 2002;**89**: 183901

[56] Baena JD, Marque's R, Martel J, Medina F. Experimental results on metamaterial simulation using SRR-loaded waveguides. In: Proceedings of the IEEE-AP/S International Symposium on Antennas and Propagation, June 2003, Ohio USA. 2003. pp. 106-109

[57] Marques R, Martel J, Mesa F, Medina F. A new 2-D isotropic left-handed metamaterial design: Theory and experiment. Microwave and Optical Technology Letters. 2002;**35**:405-408

[58] Shelby RA, Smith DR, Nemat-Nasser SC, Schultz S. Microwave transmission through a two-dimensional, isotropic, left-handed metamaterial. Applied Physics Letters. 2001;**78**:489-491

[59] Zhu J, Eleftheriades GV. A compact transmission-line Metamaterial antenna with extended bandwidth. IEEE Antennas and Wireless Propagation Letters. 2009;**8**:295-298

Chapter 3

Planar Antenna Design for Internet of Things Applications

Man Ho Tsoi and Steve W.Y. Mung

Abstract

Planar antenna plays an important role in Internet of Things (IoT) applications because of its small size, low profile and low cost. In IoT wireless module, antenna is typically occupied one-third size of overall circuit; therefore, planar antenna, i.e., integrated on printed circuit board (PCB) is one of attractive design. In this chapter, the fundamental of antenna is firstly discussed. Printed Inverted-F Antenna (PIFA) is taken as an example to explain the design process of simple planar antenna and a size-reduced 2.4 GHz ISM band PIFA is used for experimental explanation for the short-range IoT applications. Finally, a wideband antenna is shown as wideband planar antenna for short-range and long-range IoT applications.

Keywords: bluetooth low energy (BLE), industrial scientific and medical (ISM), printed circuit board (PCB), Internet of Things (IoT), planar antenna, printed inverted-F antenna (PIFA), wideband antenna

1. Introduction

Internet of Things (IoT) aims to connect existing sensors and devices to the internet in real time. Constantly collecting the information from surrounding help users making wiser decision, leading to higher quality in daily life and higher efficiency in industries [1–5]. Short-range and long-range wireless techniques are suitable used in different IoT applications. One of the wireless techniques for short-range IoT applications is bluetooth low energy (BLE) under 2.4 GHz industrial, scientific, and medical (ISM) band because of its low power consumption [6, 7]. On the other hand, cellular communications provide larger coverage in long-range IoT applications but they have high power consumption [8]. In RF/microwave modules, the size reduction and performance enhancement of the antenna are key design parameters, therefore, planar antenna, i.e., integrated on printed circuit board (PCB) is a suitable antenna for IoT applications.

In this chapter, the fundamental of antenna is firstly discussed. Printed Inverted-F Antenna (PIFA) is taken as an example to explain the design process of simple planar antenna and a size-reduced 2.4 GHz ISM band PIFA is used for experimental explanation for the short-range applications. Finally, a wideband antenna is shown as another approach on wideband planar antenna for short-range and long-range IoT applications.

2. Fundamental of antenna

2.1 Introduction of dipole antenna

Dipole antenna is one of the simple antenna that demonstrates the fundamental concept of antenna and it is a foundation of many practical antennas [9]. In **Figure 1**, dipole antenna is configured with two symmetric conductive arms carrying radio frequency current. Its length is required to be half wavelength (0.5λ) for maximum response and a half wavelength corresponds to approximately 6 cm (in air) in the 2.4 GHz ISM band. The current across the dipole generates the electromagnetic wave radiation propagating from the dipole arms.

2.2 Ground plane

A good conductive and reflective ground plane reduces the half wavelength dipole antenna to be quarter wavelength antenna [11]. This ground plane plays the same role of one of the arms and becomes a part of the antenna. This ground plane is considered as a mirror. In an optical mirror, if an object is placed in front of a mirror, a virtual image is generated with the same size and the same distance behind the mirror. In this case, if a signal source is placed above the ground plane, a virtual image of the source is generated with same current flowing direction and same phase shown in **Figure 2**, therefore, a quarter wavelength antenna and a ground plane form a half wavelength antenna. A well-designed ground plane should be very much larger in its dimensions than the half wavelength itself [11]. The active antenna measurement such as the over-the-air (OTA) measurement is a good method to indicate the overall antenna performance in the complete products compared to passive antenna measurement [12]. If the ground plane is significantly small, poor radiation performance is predicted in active antenna measurement.

2.3 Effect on length

An electric field with wavelength λ1 propagates toward to a dipole antenna with length L equals to 0.5λ1. This induces a sinusoidal current distribution shown in

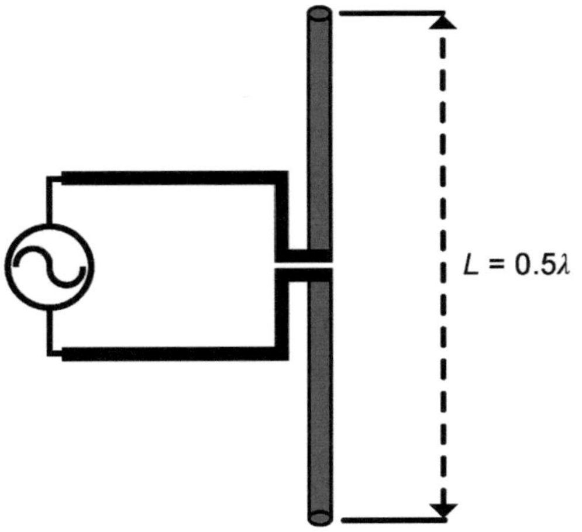

Figure 1.
Configuration of a half wavelength dipole antenna [10].

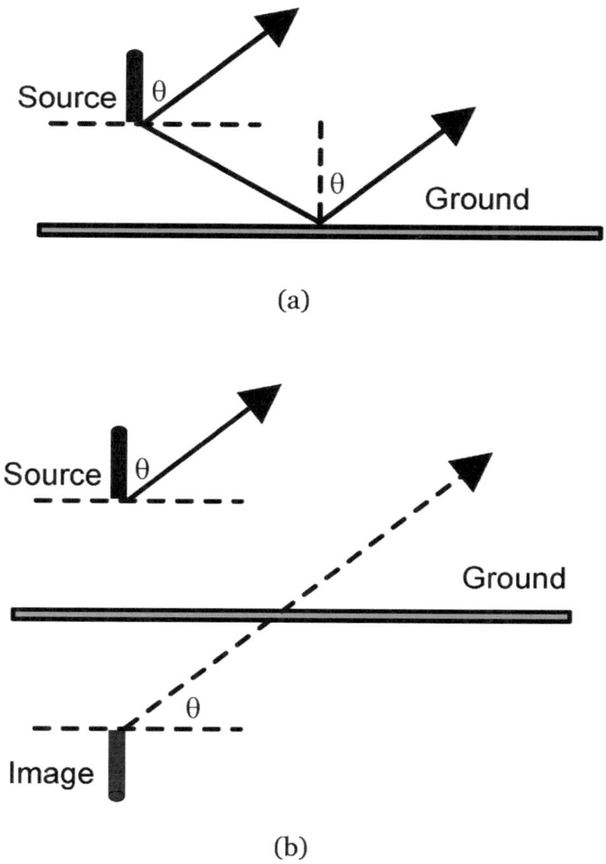

Figure 2.
Signal source and ground plane effect [11]. (a) Actual condition above a ground plane. (b) Equivalent condition with a virtual image under the ground plane.

Figure 3(a), given that the current distribution on the dipole is uniform. If the incident wave has wavelength λ2 which is much longer than the dipole length, for example L = 0.1λ2, current distribution is induced in triangular shape. The maximum current occurs at the center feed point and decreased linearly toward two ends to zero, as shown in **Figure 3(b)**.

For a dipole with uniform current distribution, the radiation resistance R_{rad} in free space is given by the following well-known equation [9, 10]:

$$R_{rad} = 80\,\pi^2 \left(\frac{L}{\lambda}\right)^2 \tag{1}$$

The radiation resistance of **Figure 3(b)** is much smaller than that of **Figure 3(a)** based on Eq. (1). Low radiation resistance is an indication of inefficient radiation. Most of the power is not radiated by the antenna when the length of antenna is not designed based on the wavelength of incident signal. This poor radiating condition occurs when the antenna operates not equal to its resonant frequency. Matching network is usually used to maximize the power transfer from the radio transceiver to the antenna [10]. Matching network sometimes is used to tune the operating frequency back to the desired value if the resonant frequency is little shifted to desired frequency.

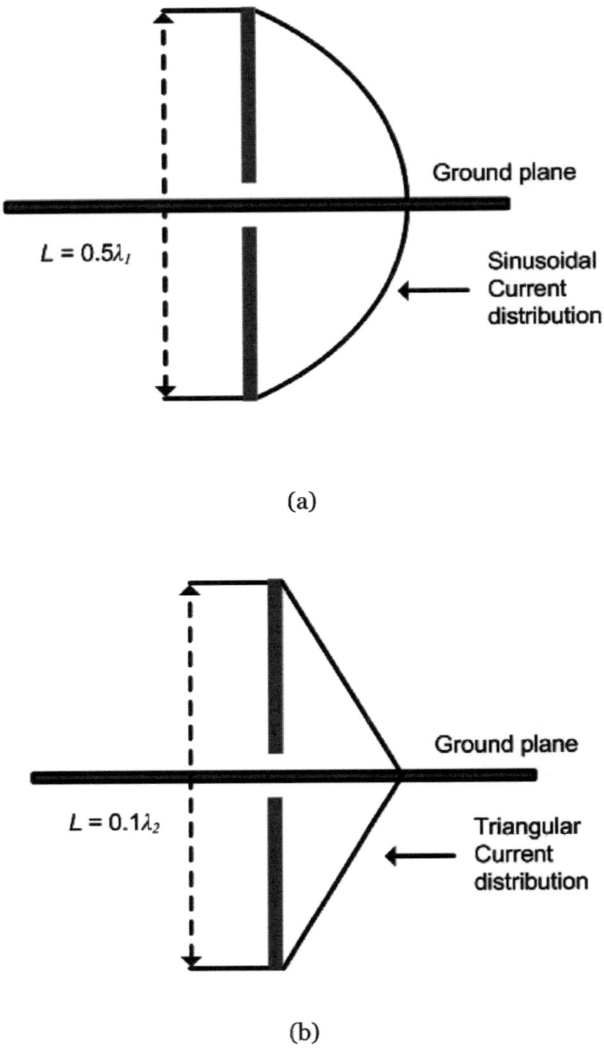

Figure 3.
(a) Sinusoidal current distribution for L = 0.5λ 1. (b) Triangular current distribution for L = 0.1λ 2 [9].

3. Design of general printed-F antenna (PIFA)

3.1 Printed inverted-F antenna (PIFA)

The printed inverted-F antenna (PIFA) is one of common planar antennas used in the commercial and medical devices because of its small size, low profile and low cost [13–20]. The typical PIFA structure is shown in **Figure 4**. Its working principle is same as monopole antenna with quarter-wave long along the main resonant line in **Figure 4**, therefore, the size of the ground plane is also an important part of antenna. It has a shorting feed point at the end of the main resonant line. This folded part introduces capacitance to the input impedance of the PIFA which is canceled by the shorting feed point. This foldable part, therefore, reduces the antenna size.

The matching network in **Figure 4** is used for maximum power transfer and, hence, efficient radiation [10]. Lump elements are normally used in matching network to minimize the size. In this section, PIFA is used as example of planar antenna since PIFA fulfills the requirements of IoT applications.

3.2 Meandered PIFA design

Figure 4 shows a typical PIFA structure on a printed circuit board (PCB) which is indicated with the dotted area at the PCB upper layer. Meandering line is commonly used to increase the total length in antenna design. The meandering line in **Figure 5** is used to replace the main resonant line in PIFA shown by combination of horizontal and vertical lines to form multiple turns.

The requirements of the PIFA are the operation frequency, power transmission efficiency, size and even cost. Simulation, fabrication and measurement are conducted until the antenna fulfills the defined application requirements. In general, a well-designed PIFA has the feature of having resonance at the operation frequency and good return loss, i.e., effective power transmission to antenna and compact in size.

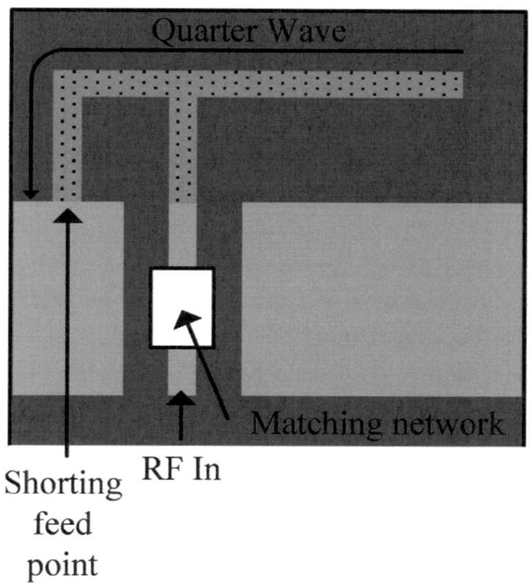

Figure 4.
Typical PIFA structure [13].

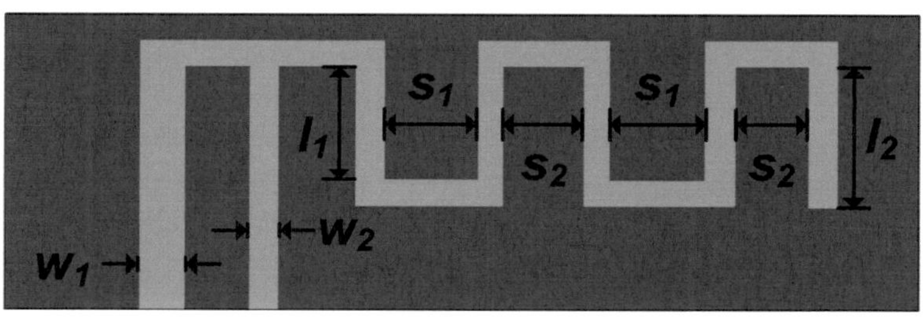

Figure 5.
PIFA with meandering line.

High directivity sometimes is considered in certain situation. However, it will not be discussed here since most of planar antennas are omni-directional transmission and reception instead of unidirectional antennas. The basic design rules and antenna performance characterization methods are addressed by case study in 2.4 GHz ISM band. The operation frequency of the antenna is governed by the basic dispersion relation $c = f\lambda$. The letter c represents the speed of electromagnetic wave in the air, which is a constant if only consider the wave traveling in single medium. In previous section, it shows that a dipole antenna resonates when the physical length of antenna equals to the quarter wavelength of incident signal and a sufficiently large ground plane form the mirror image under the plane. The length of the resonant line occupies a considerable area on the PCB, which is around one third of a wireless module. Proper selection of traces' length and width reduces the occupied area and impedance matching network for maximum power transfer [13].

3.3 Simulation and experimental results

Powerful computer simulation tools are used to drastically reduce the design time. Advanced Design System (from Keysight) is one of the electromagnetic (EM) simulators used to estimate the performance of certain designs in this chapter. Normally, there is small variation between the simulation and measurement results because of the fabrication variation, material variation and connectors mismatch, etc. There is limited effect on frequency below 6 GHz, however, this becomes significant when the frequencies are in millimeter wave. Two antennas are designed based on structure in **Figure 5**. The dimensions of these two examples (called Antenna PCB A and Antenna PCB B) of PIFA shown in **Table 1** and in **Figure 6**.

Antennas in **Figure 6** were simulated on FR4 substrate with a dielectric constant of 4.6 and 0.8 mm thickness. The PCB design mainly contains PIFA, ground plane, transmission line and 3.5 mm SMA connector. The size of PCB, 18.8 mm× 43.2 mm was chosen, which is the normal size of a 2.4 GHz ISM band wireless module. At end of meandering line is 4.2 mm for Antenna PCB A and 2.9 mm for Antenna PCB B. The resonance frequency of Antenna PCB A is expected to be lower due to the longer trace. Antenna PCB C was fabricated to calibrate the transmission line by the port extension measurement. The simulated results of Antenna PCB A and Antenna PCB B are shown in **Figure 7** and the return loss indicates the resonant frequency of the antennas [9]. The input feed point in the simulation is at Port A in **Figure 6(a)** which is without the transmission line and connector but others are the same as in **Figure 6(a)**.

In **Figure 7**, the resonance frequency of Antenna PCB A is lower than that of Antenna PCB B. Antenna PCB A and Antenna PCB B were fabricated with

Parameters	Length (mm)
w_1	0.9
w_2	0.5
s_1	2.0
s_2	1.7
l_1	2.5
l_2 (Antenna PCB A)	4.2
l_3 (Antenna PCB B)	2.9

Table 1.
Parameters used in the PIFA.

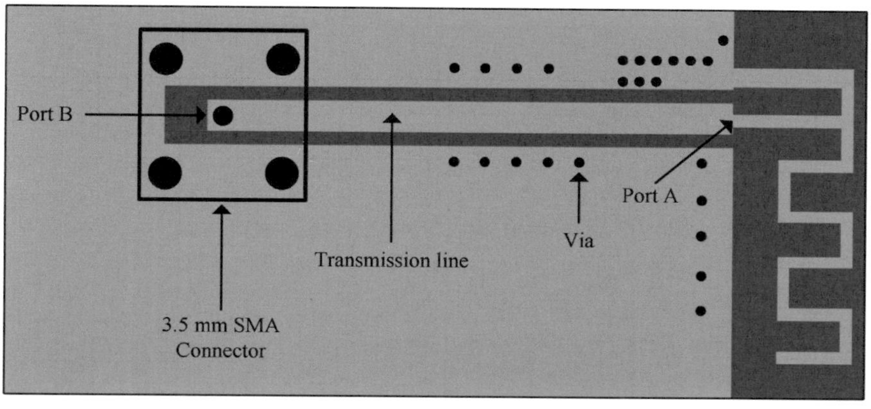

(a)

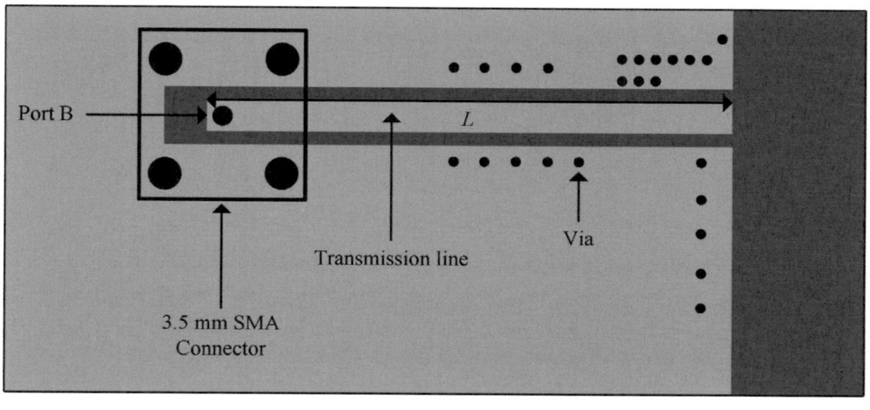

(b)

Figure 6.
PIFA: (a) Antenna PCB A (l_2 = 4.2 mm) and Antenna PCB B (l_3 = 2.9 mm) (b) and Antenna PCB C.

the transmission line and connector in **Figure 6(a)**. The width of the transmission line is 1.5 mm so that the characteristic impedance of the line is equal to 50 Ohm. The return losses of antenna were measured by vector network analyzer (VNA). If the return losses are not significant high enough, matching network is needed for maximum power transfer. Antenna PCB C is used for port extension by VNA so that the measurement reference plane is moved to Port A since the VNA can predict the open circuit at end of the transmission line from the connector in Antenna PCB C by the electrical length *L* of the transmission line. The simulated and measured results of Antenna PCB B are plotted in **Figure 8**. It is also shown that the overall performance of antenna at Port B is close to the same at Port A. The radiation patterns and gain measurement are carried out by passive antenna measurement system. The active antenna measurement sometimes is used to indicate the overall transmission and reception of the complete products. The maximum gain of the PIFA is normally around 3 dBi. Antenna PCB B, therefore, is suitable for 2.4 GHz ISM band applications. **Table 2** shows the comparison table of different antennas, which shows that the dielectric antennas have a little size smaller than the PIFA. However, PIFAs were only fabricated on PCB, which is approximately zero in thickness as well as zero cost of antenna and matching components.

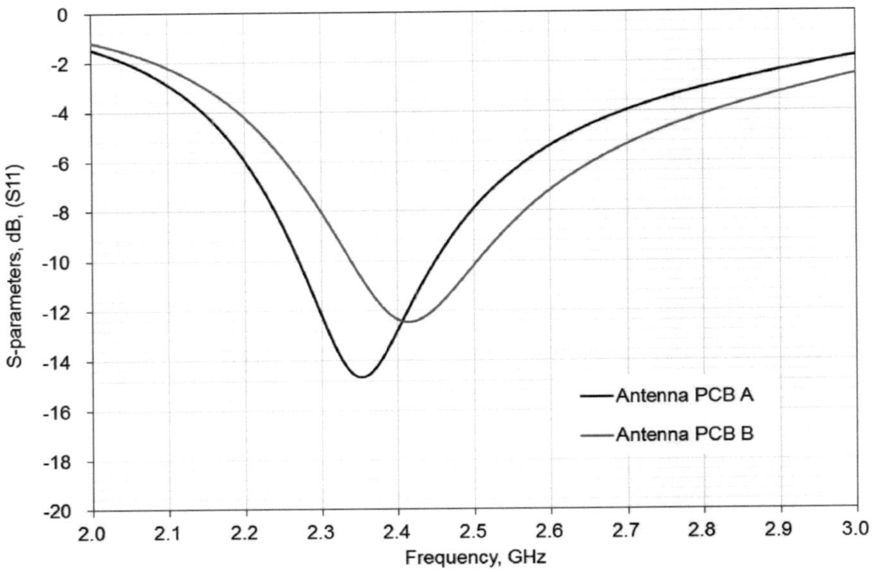

Figure 7.
Simulated S-parameter, S11 of Antenna PCB A and Antenna PCB B at Port A.

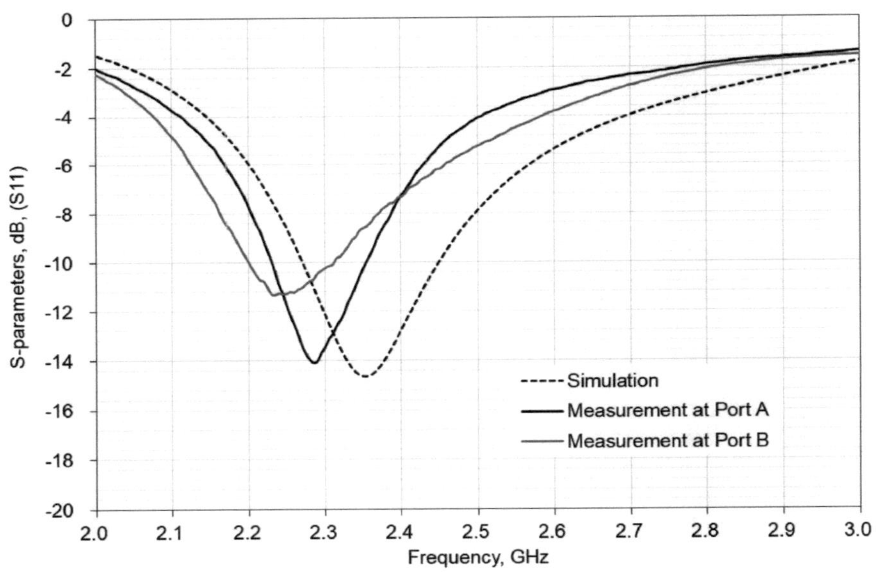

Figure 8.
Antenna PCB B: Simulated (at Port A) and measured (at Port A and B) S-parameter, S11.

4. Wideband planar antenna: foldable and non-foldable

4.1 Wideband planar antenna for IoT applications

Various IoT applications use different solutions to connect devices and sensors. Low power technologies such as Bluetooth and Zigbee are preferred for short-range applications because of their low power usage. Cellular communication technologies sometimes are required used for large coverage and high data rate applications

Antenna	Plane	Total average (dBi)	Area (mm^2)
PIFA (Antenna PCB B)	Y-Z	1.60	16 × 7.0
	X-Z	3.30	
	X-Y	1.10	
Miniaturized PIFA [13]	Y-Z	−0.70	15 × 6.0
	X-Z	−1.98	
	X-Y	−1.26	
Capacitive-loaded antenna	Y-Z	—	12 × 5.0
	X-Z	−1.76	
	X-Y	−3.32	
Dielectric antenna (3 mm length)	Y-Z	0.89	12 × 5.0
	X-Z	−1.85	
	X-Y	−2.56	
Dielectric antenna (5 mm length)	Y-Z	−3.22	18 × 11
	X-Z	−3.24	
	X-Y	−3.12	

Table 2.
Gain between dielectric and PCB antennas [10] (includes the area of matching network, but not the ground plane).

despite its large power consumption. Multiple narrowband antennas are needed when different technologies are used in IoT applications. One wideband antenna, therefore, is an attractive approach to replace multiple narrowband antennas. Different wideband structures were proposed to combine different frequency bands into one individual wideband antenna to serve different technologies in order to reduce the size and simplicity [21–23]. The wideband antenna is still large to be used in portable devices, therefore, foldable design [21] provides flexibility of IoT products as well as further size reduction.

4.2 Wideband planar antenna design

Dipole antenna in **Figure 1** could be extended to be a wideband antenna. Two conductive arms are replaced by thicker wire or even a plane to extend the bandwidth. One of example for wideband planar foldable and non-foldable antennas is shown in **Figure 9**.

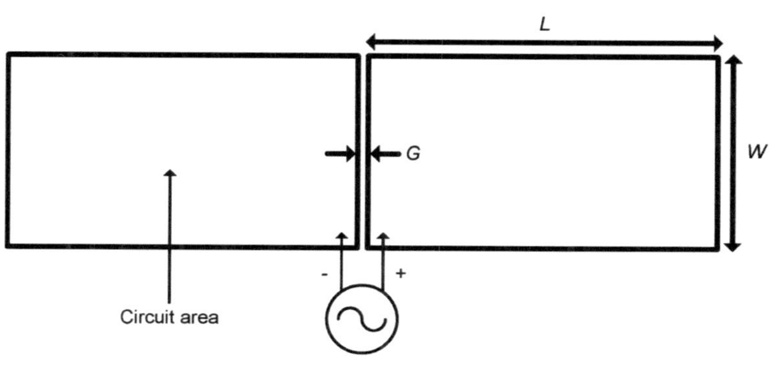

Figure 9.
Wideband planar foldable and non-foldable antennas [21].

Parameters	Length (mm)
G	0.9
W (foldable design)	29
W (non-foldable design)	25

Table 3.
Parameters used in the wideband planar foldable and non-foldable antennas.

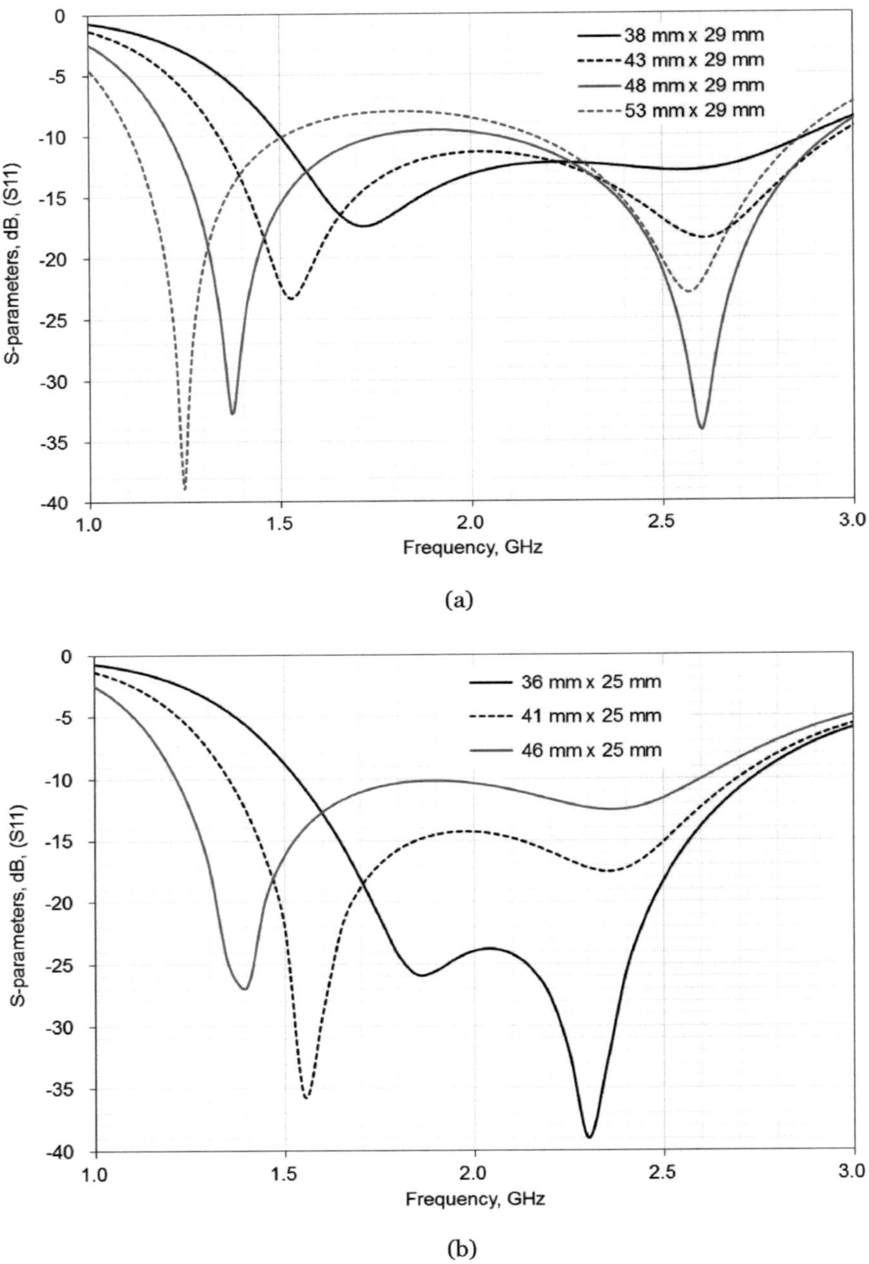

Figure 10.
Simulated results of wideband planar antennas [21]: (a) foldable and (b) non-foldable.

This planar antenna consists of two rectangular metal planes. The important parameters could be tuned in this design are width W, length L and gap G. The length is used for tuning in this example as it has significant impact on the performance of antenna. All parameters are fixed in **Table 3** except the length L which is the parameter chosen to be tuned for foldable and non-foldable antennas. The foldable design was fabricated in the metal sheet and the non-foldable design was fabricated on the FR4 substrate with a dielectric constant of 4.6 and thickness of 0.8 mm. The simulated results of return loss are shown in **Figure 10** with different lengths of sheet L. In **Figure 10**, the frequency range is shifted to the lower side with a longer length L because the length L is closer to the quarter-wavelength of a lower frequency.

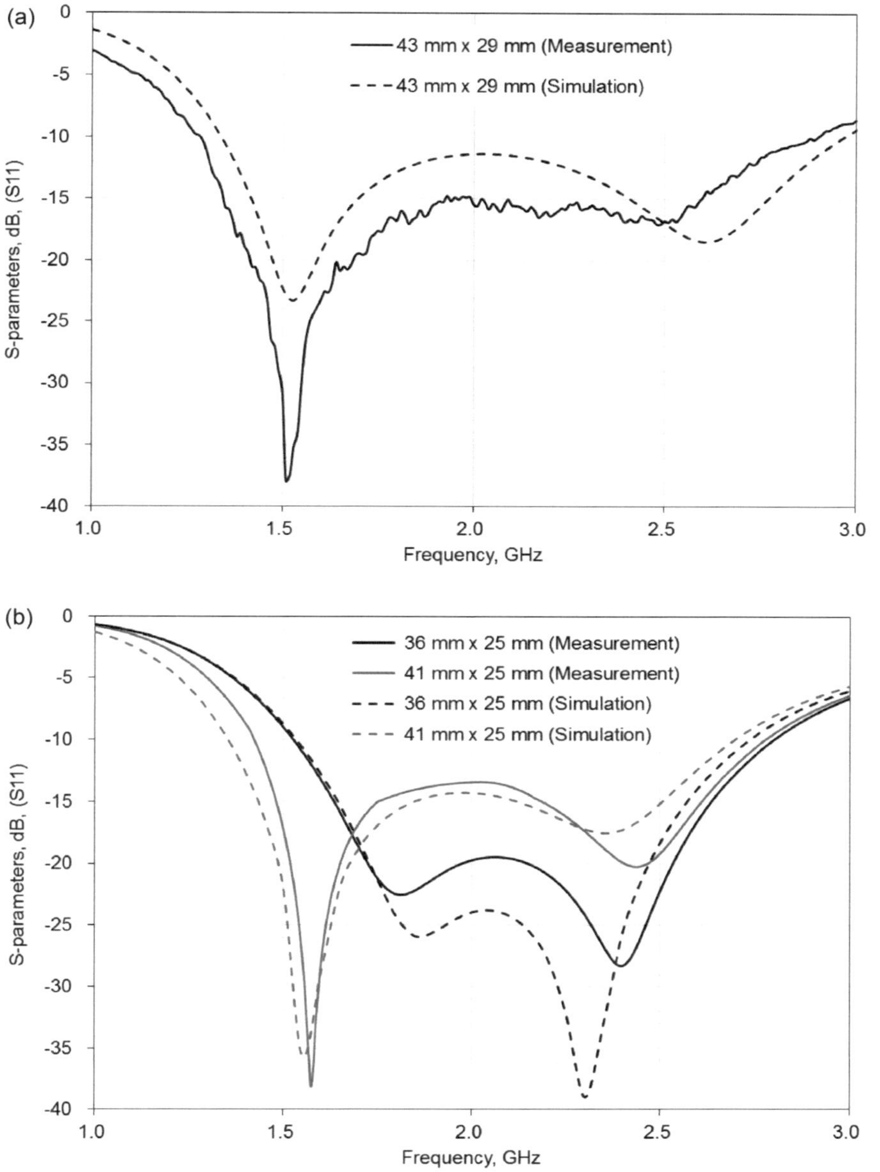

Figure 11.
Comparison between simulated and measured results of wideband planar antennas [21]: (a) foldable and (b) non-foldable.

Figure 11 shows the comparison between simulated and measured results of wideband planar foldable and non-foldable antennas. **Figure 11(a)** shows the simulated and measured results of the fabricated foldable antenna which shows that the simulated and measured results are close to each other with the bandwidth of 76% from 1.3 to 2.9 GHz. This range covers the applications in GPS, the 2.4 GHz ISM band, and the general 3GPP WCDMA bands and LTE bands. **Figure 11(b)** shows the simulated and measured results with L equal to 36 and 41 mm (same width of W = 25 mm). Simulated and measured results show that they are close to each other with the bandwidth of 76% from 1.35 to 2.75 GHz, which is little worse than the foldable design in **Figure 11(a)**. The maximum gain of the non-foldable is between 2.5 and 3.5 dBi.

5. Conclusion

The architecture of the PIFA on PCB with meandering line was shown. The measurement results of return loss and gain performances shown that it has better performances compared to the dielectric antennas as well as without any extra matching components. When only single communication technology is used in IoT product, PIFA is recommended. Using meandering line can reduce the antenna size as well as keeping the performance. PIFA design, therefore, is suitable for ISM band and other IoT applications. In the product utilizing numbers of communication technologies at same times, one wideband antenna integrated in the product is more suitable. Both foldable and non-foldable wideband structures, therefore, were proposed and fabricated for their different uses in IoT applications. Both measurement results of two structures show more than 65% in bandwidth. Their operating frequency covers IoT applications in GPS, the 2.4 GHz ISM band, and the common 3GPP WCDMA and LTE bands. And the foldable structure has advantage of wearable applications.

During the design process, the type of antenna is firstly confirmed and then the key parameters such as frequency and size need to be determined. Simulation software and measurement equipment are important tools to verify its performance and further design iterations may be required for fine-tuning the performance.

Abbreviations

IoT	Internet of Things
ISM	industrial, scientific, and medical
BLE	bluetooth low energy
PIFA	printed inverted-F antenna
PCB	printed circuit board
OTA	over-the-air
EM	electromagnetic
VNA	vector network analyzer
GPS	global positioning system
WCDMA	wideband code division multiple access
LTE	long-term evolution
3GPP	3rd generation partnership project

Author details

Man Ho Tsoi and Steve W.Y. Mung*
The Hong Kong Polytechnic University, Hong Kong

*Address all correspondence to: mungwaiyin@hotmail.com

References

[1] Ahlgren B, Hidell M, Ngai EC-H. Internet of things for smart cities: Interoperability and open data. IEEE Internet Computing. 2016;**20**:52-56. DOI: 10.1109/MIC.2016.124

[2] Sun Y, Song H, Jara AJ, Bie R. Internet of things and big data analytics for smart and connected communities. IEEE Access. 2016;**4**:766-773. DOI: 10.1109/ACCESS.2016.2529723

[3] He W, Yan G, Da Xu L. Developing vehicular data cloud services in the IoT environment. IEEE Transactions on Industrial Informatics. 2014;**10**(2):1587-1595. DOI: 10.1109/TII.2014.2299233

[4] Zanella A, Bui N, Castellani A, Vangelista L, Zorzi M. Internet of things for smart cities. IEEE Internet of Things Journal. 2014;**1**(1):22-32. DOI: 10.1109/JIOT.2014.2306328

[5] Satija U, Ramkumar B, Sabarimalai MM. Real-time signal quality-aware ECG telemetry system for IoT-based health care monitoring. IEEE Internet of Things Journal. 2017;**4**(3):815-823. DOI: 10.1109/JIOT.2017.2670022

[6] The Bluetooth Special Interest Group (Bluetooth SIG). Bluetooth Core Specification [Internet]. 2019. Available from: https://www.bluetooth.com/specifications/bluetooth-core-specification [Accessed: 25 February 2020]

[7] Tabish R, Ben Mnaouer A, Touati F, Ghaleb AM. A comparative analysis of BLE and 6LoWPAN for U-HealthCare applications. In: 2013 7th IEEE GCC Conference and Exhibition (GCC). 7-20 November, 2013; Doha. New York: IEEE; 2013. pp. 286-291

[8] The 3rd Generation Partnership Project (3GPP). Specifications. Available from: https://www.3gpp.org/specifications/specifications [Accessed: 25 February 2020]

[9] Kraus JD, Marhefka RJ. Antennas: For all Applications. 3rd ed. New York: McGraw Hill; 2003

[10] Mung SWY, Wu KM, Yuen JSM. Design trade-offs and challenges on miniaturized antenna for bluetooth in internet of things applications. Engineering Technology Open Access Journal. 2019;**3**(1):555602. DOI: 10.19080/ETOAJ.2019.03.555602

[11] Holzman E. Essentials of RF and Microwave Grounding. 1st ed. Boston: Artech House; 2006

[12] Huang Y, Loh TH, Foged LJ, Lu Y, Boyes S. Broadband antenna measurement comparisons. In: Proceedings of the Fourth European Conference on Antennas and Propagation. 12-16 April, 2010; Barcelona. New York: IEEE; 2010. pp. 1-5

[13] Cheung CY, Yuen JSM, Mung SWY. Miniaturized printed inverted-F antenna for internet of things: A design on PCB with a meandering line and shorting strip. International Journal of Antennas and Propagation. 2018;**2018**. DOI: 10.1155/2018/5172960

[14] Hristov HD, Carrasco H, Feick R. Bent inverted-F antenna for WLAN units. Microwave and Optical Technology Letters. 2008;**50**(6):1505-1510. DOI: 10.1002/mop.23405

[15] Ammann MJ, Doyle LE. A loaded inverted-f antenna for mobile handsets. Microwave and Optical Technology Letters. 2001;**28**(4):226-228. DOI: 10.1002/1098-2760(20010220)28:4<226::AID-MOP1002>3.0.CO;2-T

[16] Palukuru VK, Pekonen A, Pynttäri V, Mäkinen R, Hagberg J,

Jantunen H. An inkjet-printed inverted-F antenna for 2.4-Ghz wrist applications. Microwave and Optical Technology Letters. 2009;**51**(12):2936-2938. DOI: 10.1002/mop.24777

[17] Su SW. Linearly-polarized patch PIFA for GPS/GLONASS operation for tablet-computer applications. Microwave and Optical Technology Letters. 2015;**57**(1):149-153. DOI: 10.1002/mop.28800

[18] Soras C, Karaboikis M, Tsachtsiris G, Makios V. Analysis and design of an inverted-F antenna printed on a PCMCIA card for the 2.4 GHz ISM band. IEEE Antennas and Propagation Magazine. 2002;**44**(1):37-44. DOI: 10.1109/74.997891

[19] Yang HYD. Printed straight F antennas for WLAN and Bluetooth. In: IEEE Antennas and Propagation Society International Symposium, Vol. 2. 22-27 June, 2003; Columbus, OH, USA. New York: IEEE; 2003. pp. 918-921

[20] Ali M, Hayes GJ. Small printed integrated inverted-F antenna for Bluetooth application. Microwave and Optical Technology Letters. 2002;**33**(5):347-349. DOI: 10.1002/mop.10317

[21] Mung SWY, Cheung CY, Wu KM, Yuen JSM. Wideband rectangular foldable and non-foldable antenna for internet of things applications. International Journal of Antennas and Propagation. 2019;**2019**. DOI: 10.1155/2019/2125713

[22] Chen JH, Ho CJ, Wu CH, Chen SY, Hsu P. Dual band planar monopole antenna for multiband mobile systems. IEEE Antennas and Wireless Propagation Letters. 2008;**7**:769-772. DOI: 10.1109/LAWP.2008.2008110

[23] Li RL, Pan B, Laskar J, Tentzeris MM. A compact broadband planar antenna for GPS, DCS-1800, IMT-2000, and WLAN applications. IEEE Antennas and Wireless Propagation Letters. 2007;**6**:25-27. DOI: 10.1109/LAWP.2006.890754

Chapter 4

Phased Antenna Arrays toward 5G

Tran Cao Quyen

Abstract

An antenna array for 5G has to be able to create multibeam (approximately dozens or hundreds of beams), wide azimuthal coverage (approximately 360°), and high gain (>20 dB). The analyses of four kinds of multibeam phased arrays, namely a multibeam ULA analog phased array (MULA-analog-PA), multibeam ULA digital phased array (MULA-digital-PA), multibeam cylindrical digital phased array (MC-digital-PA), and multibeam cylindrical analog phased array (MC-analog-PA), are performed. The analyses show that all arrays could provide multibeam with different complexities and computations but MULA-analog-PA and MULA-digital-PA are with maximum 180° of azimuthal coverage; whereas MC-analog-PA and MC-digital-PA are with unlimited azimuthal angle. The simulations of the MC-analog-PA with 32 × 10 elements (10 rings with each ring of 32 elements) show that the array could provide 32 beams symmetrical over 3600 azimuthal coverage with the directivity of 27 dB. The obtained results proved the effectiveness of the phased array antennas for 5G applications.

Keywords: phased array antennas, multibeam, azimuthal coverage, cylindrical geometry

1. Introduction

5G will be at the heart of the future of communications in which the technologies such as new multiple accesses, massive MIMO, multiple beams, ultra-dense networking, etc. [1–4] are key technologies. 5G will bring new challenges for the designers of the physical infrastructure including antenna designers. Antennas for 5G have to be able to create multiple independent beams (approximately dozens or hundreds of beams), wide azimuthal coverage (approximately 360°), high gain (>20 dB), and acceptable complexity of feeding network for analog platform or powerful digital processing for digital one.

This chapter will present about multibeam phased array antennas toward 5G in terms of their principles of operation and theoretical limits. The analyses of four kinds of multibeam phased arrays, namely a multibeam ULA analog phased array (MULA-analog-PA), multibeam ULA digital phased array (MULA-digital-PA), multibeam cylindrical digital phased array (MC-digital-PA), and multibeam cylindrical analog phased array (MC-analog-PA), are performed, and some simulation results are given to demonstrate the performance of those arrays.

As is known, the phased array antennas were dating back from 60 to 80 decades with the main application of satellite communications and military radar [5–8]. In accompany with the use of higher carrier frequencies of 5G [9–10], phased array antenna could be smaller and more compact in size with civilization application.

In order to create a main beam pointing into Θ direction, it is necessary to make phase progressive of a uniform linear array (ULA) with the phase difference of kdcosΘ between two consecutive elements [11]. In other words, a MULA-analog-PA having M antenna elements required M phase shifters. If this array needs N independent beams, the array must have a matrix of M × N phase shifters [12].

Due to the complexity of the design of the MULA-analog-PA, the development of digital processing leads to the invention of digital beamforming of phased array. Performing the beamforming in a multibeam digital phased array antenna is a more flexible and versatile approach. For each antenna element, it has its own amplifying module but without any phase shifters or attenuators; but it required a strong central processing unit (CPU) in order to process beamforming algorithms [13, 14].

A MULA-digital-PA could create unlimited number of independent beams, but its azimuthal coverage could not be over 180°. It is intrinsic property of an ULA. To make a solution for this problem, we need to use the circular array [14] that has 360° beam coverage in azimuthal plane. Combining the advantage of multibeam and wide azimuthal angle leads to the construction of multibeam cylindrical phased array antenna in analog or digital beamforming. MC-digital-PA is preferred if a large numbers of beams and high computing performance are required, and MC-analog-PA is suggested if a moderate numbers of beams and low cost of computing performance are required.

2. Multibeam ULA analog phased array (MULA-analog-PA)

2.1 ULA analog phased array (ULA-analog-PA)

Let us introduce the subject of phase array antenna by considering the simplest situation, namely, uniform linear analog phased array (ULA-analog-PA). An array of identical elements (in this case, isotropic elements), all of identical magnitude and each with a progressive phase and arranging in a straight line, is referred to as a uniform linear array. A typical ULA-analog-PA in which each antenna element with equal spacing, d, is illustrated in **Figure 1**.

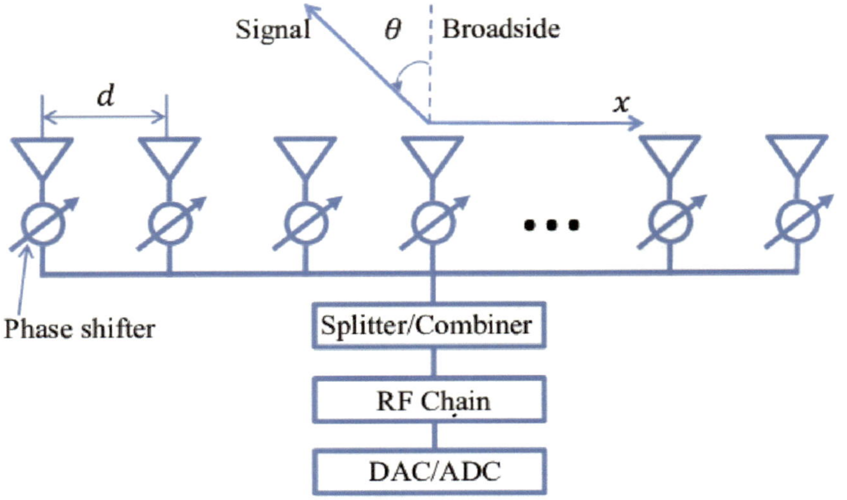

Figure 1.
A typical ULA-analog-PA.

The principle of phased array is that the phase of each antenna element is controlled by using an analog phase shifter. Assuming the array has M elements, the total field E at a large distance in the direction Θ is given by

$$E = 1 + e^{j\psi} + e^{j2\psi} + \cdots + e^{j(M-1)\psi} \tag{1}$$

where ψ is the total phase difference of the fields of adjacent elements as given by

$$\psi = \frac{2\pi d}{\lambda} \cos \phi + \alpha \tag{2}$$

where α is the phase difference of adjacent elements, that is, element 2 with respect to 1, 3 with respect to 2, etc.

After some manipulation, the total field E can be written as

$$E = \frac{\sin(M\psi/2)}{\sin(\psi/2)} \tag{3}$$

The array factor which is a ratio of total field E to its maximum is given by

$$AF = \frac{E}{E_{max}} = \frac{E}{E(\psi = 0)} \tag{4}$$

The array factor of ULA with equal amplitude, equal spacing in Z-axis, and $\alpha = 0$ (Broadside Array) is illustrated in Cartesian coordinate in **Figure 2** as follows.

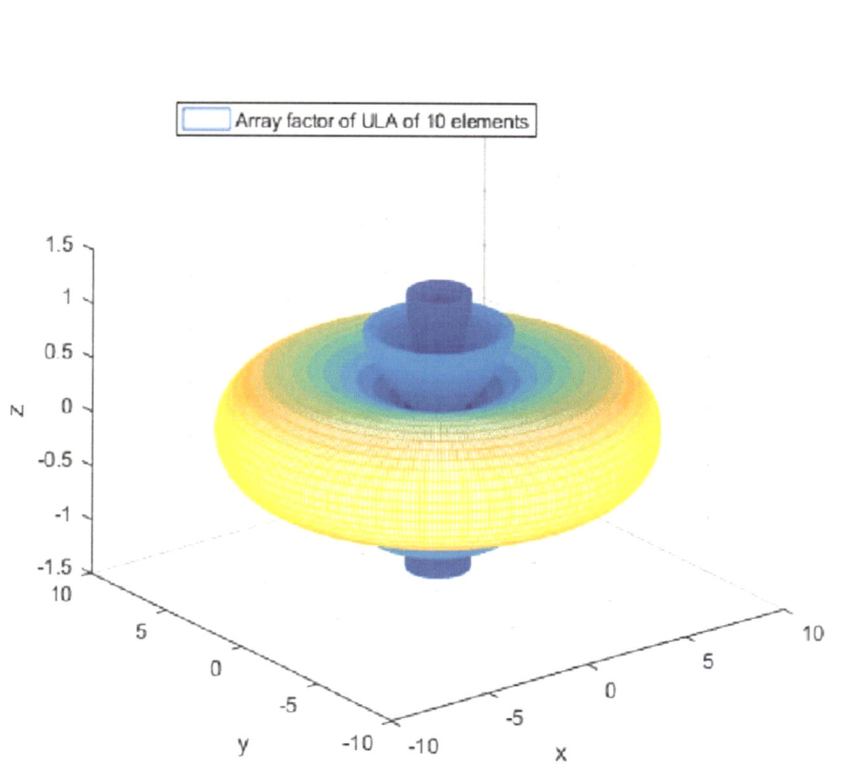

Figure 2.
The array factor of ULA-analog-PA.

2.2 The simulation result of the ULA phased array 1 × 4 elements using advanced design software (ADS)

When using ADS software to design a ULA-analog-PA of 1 × 4 elements in X-axis, we obtain the radiated field pattern E as in **Figure 3** as follows [15].

After the analysis and some illustrations of ULA-analog-PA, we may conclude that a ULA-analog-PA could provide only one main beam and some side lobes at a time.

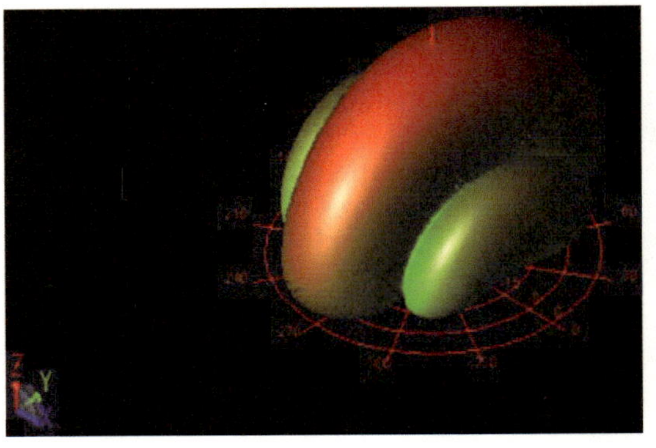

Figure 3.
The radiated field pattern of the ULA-analog-PA of 1 × 4 elements in X-axis.

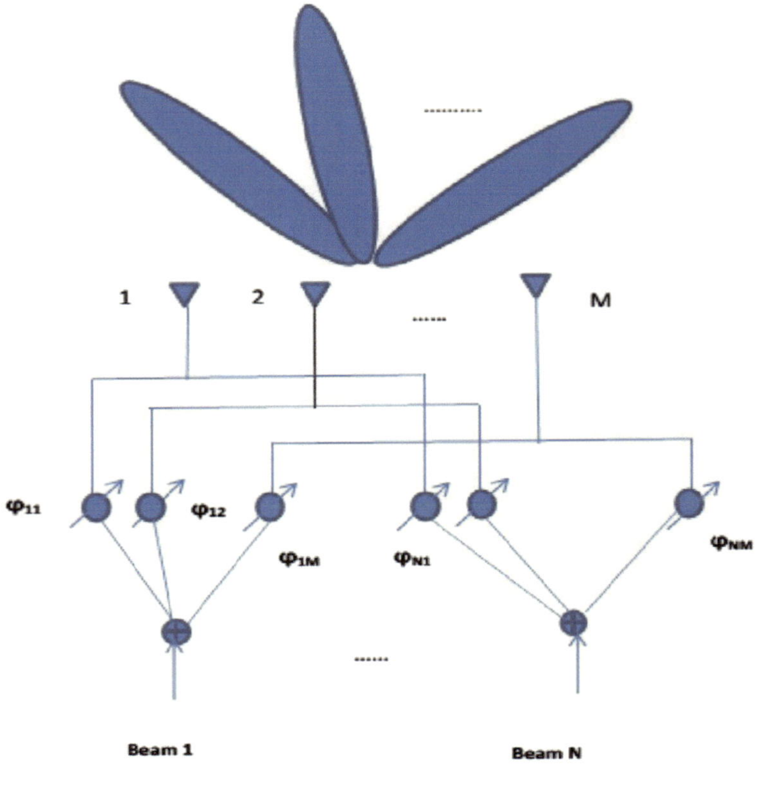

Figure 4.
A typical MULA-analog-PA.

2.3 Multibeam ULA analog phased array (MULA-analog-PA)

Now that the ULA-analog-PA has been introduced let us generalize to the construction of a MULA-analog-PA. A typical ULA-analog-PA is indicated as in **Figure 4** as follows.

In principle, a MULA-analog-PA could provide N independent beams in space. In order to drive the nth beam toward Θ direction, it is necessary to make phase progressive for the corresponding ULA-analog-PA.

In another statement, the phase difference between two consecutive elements of the nth beam is given by

$$\varphi_{nm} - \varphi_{n(m-1)} = -\frac{2\pi d}{\lambda}\cos\theta_n, \quad m = 2, \dots, M \tag{5}$$

To avoid grating lobe, the Nyquist condition for the distance between two consecutive elements, d, has to be satisfied. The Nyquist condition is

$$d < \frac{\lambda}{2} \tag{6}$$

The disadvantage of a MULA-analog-PA is the limit of number of independent beams due to the complexity of a matrix of N × M phase shifters or attenuators. Therefore, there is a little result for this kind of the array. Besides, the MULA-analog-PA has to deal with the problem of the limited azimuthal coverage which cannot be greater than 180°. It is one of the theoretical limits of ULA.

3. Multibeam ULA digital phased array (MULA-digital-PA)

In contrast to analog beamforming, the digital beamforming is performed without phase shifters or attenuators. MULA-digital-PA relies on a digital processing unit such as digital signal processing (DSP) or a strong computer in order to process the digital data that are the outputs of analog to digital converters (AD) and accompany with beamforming algorithms [13, 14]. The diagram of a typical MULA-digital-PA is shown in **Figure 5** as follows.

The principle of a MULA-digital-PA is that the collection of M antenna produces the received vector

$$X = [x_1, x_2, \cdots x_M], \quad m = 1, 2, \dots, M \tag{7}$$

Then, at the DSP unit, the received vector is multiplied by a matrix of weighting, W, which can be written as follows:

$$W = \begin{bmatrix} W_1 \\ W_2 \\ \vdots \\ W_M \end{bmatrix} = \begin{bmatrix} W_1 \cdots W_{1M} \\ W_2 \quad \vdots \\ \vdots \\ W_M \cdots W_{NM} \end{bmatrix} \tag{8}$$

where the component weighting is given by

$$W_n = \left[1, \exp\left(j\frac{2\pi}{\lambda}d\cos\theta\right), \cdots, \exp\left(j\frac{2\pi}{\lambda}nd\cos\theta\right)\right] \tag{9}$$

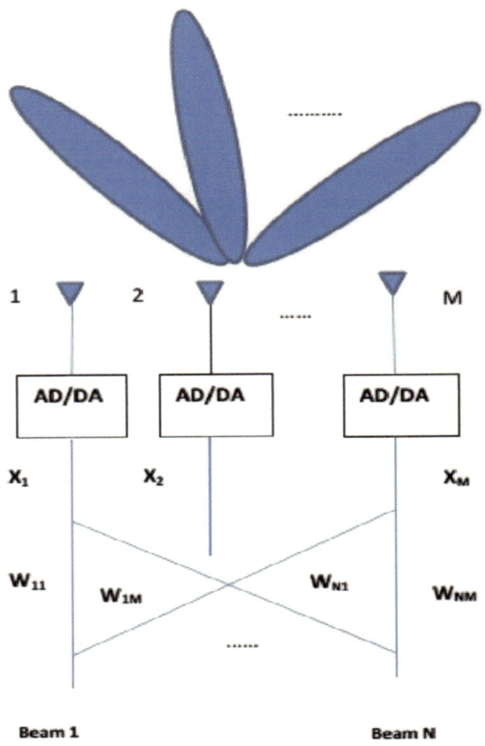

Figure 5.
A typical MULA-digital-PA.

Therefore, the output of DSP is

$$Y_n(\theta) = \mathrm{W}_n^H X \tag{10}$$

Finally, other constrains on the output of DSP is performed.

The advantage of MULA-digital-PA is that performing the beamforming is a more flexible and versatile approach. However, the array still has limited azimuthal coverage which cannot be greater than 180° since it is one of the theoretical limitations of ULA. In order to cover the azimuthal angle of 360°, we have to use at least three MULA-digital-PAs.

4. Multibeam cylindrical analog phased array (MC-analog-PA) of M × N elements

The objectives in many designs for 5G antenna are multiple independent beams (approximately dozens or hundreds of beams), wide azimuthal coverage (approximately 360°), and high gain (>20 dB). A MC-analog-PA of M × N elements has both the characteristics of multibeam and wide azimuthal coverage since it exploited the characteristics of not only of MULA-analog-PA but also of cylindrical geometry, that is, perfect symmetrical over Z-axis. However, since the array relies on analog technology, the number of beams is limited by the number of phase shifters. The geometry of a MC-analog-PA of M × N elements is shown in Cartesian coordinate in **Figure 6** as follows.

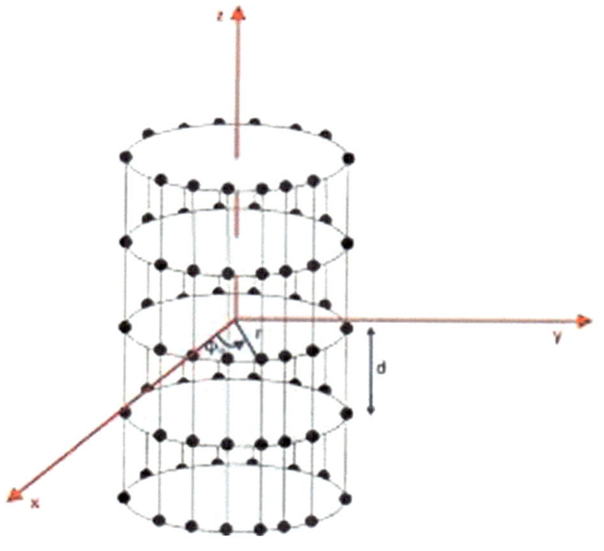

Figure 6.
The geometry of MC-analog-PA of M × N elements (each ring has M elements).

In this case, the array factor is given by [16].

$$AF_{cylinder}(\varphi, \theta) = AF_{ring}(\varphi, \theta) AF_{linear}(\varphi, \theta) \tag{11}$$

where $AF_{ring}(\varphi, \theta)$ is the array factor of the circular array [17] in XOY plane and $AF_{linear}(\varphi, \theta)$ is the array factor of ULA-analog-PA in Z-axis.

Generally, the array factor of an array having M elements in space is given by

$$AF(\varphi, \theta) = I_1 e^{-j\beta\Delta r_1} + \cdots + I_M e^{-j\beta\Delta r_M} = \sum_{k=1}^{M} I_k e^{-j\beta\Delta r_k} \tag{12}$$

where $\Delta r_k = r_k.a_r = x_k \cos\theta \sin\varphi + y_k \sin\theta \sin\varphi + z_k \cos\varphi$ is the phase difference of the Kth element to the reference element, I_k is the excited current of the Kth element, r_k is position vector of the Kth element, and α_r is directional unit vector.

The directivity of an antenna can be approximated as [16].

$$D_{\max} = \frac{4\pi}{HPBW_\theta.HPBW_\varphi} \approx \frac{41253}{HPBW_\theta.HPBW_\varphi} \tag{13}$$

where half power beam width in Θ plane, $HPBW_\theta$, is perpendicular to the half power beam width in ϕ plane, $HPBW_\varphi$.

4.1 The simulation result of the multibeam circular analog phased array of 32 elements

To illustrate the performance of MCr-analog-PA, we do some simulations. First, isotropic antenna elements are arranged symmetrical in a circular with the reservation of the Nyquist condition of the distance between two consecutive elements.

Second, uniform currents are excited for all elements then the array factor of MCr-analog-PA of 32 elements is calculated (Eq. (12)) and depicted in Cartesian coordinate in **Figure 7** as follows.

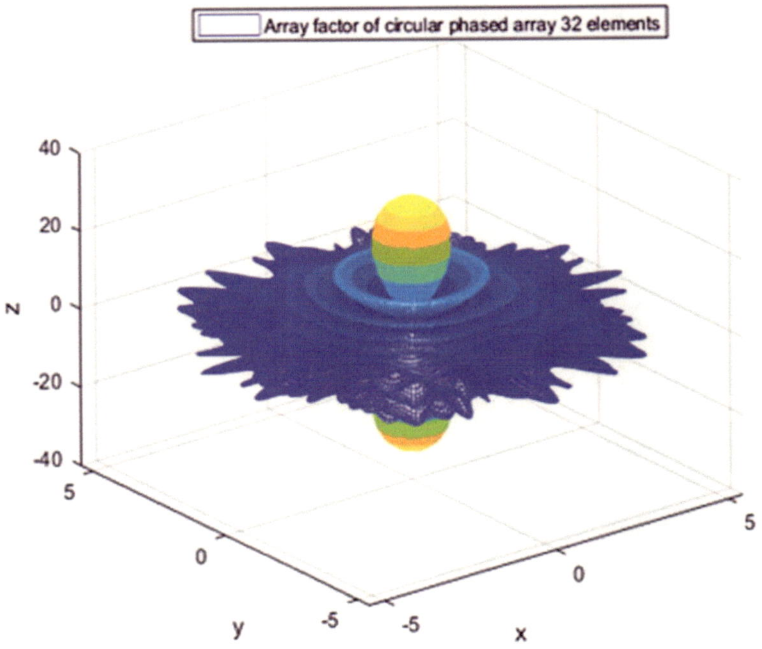

Figure 7.
The array factor of MCr-analog-PA of 32 elements.

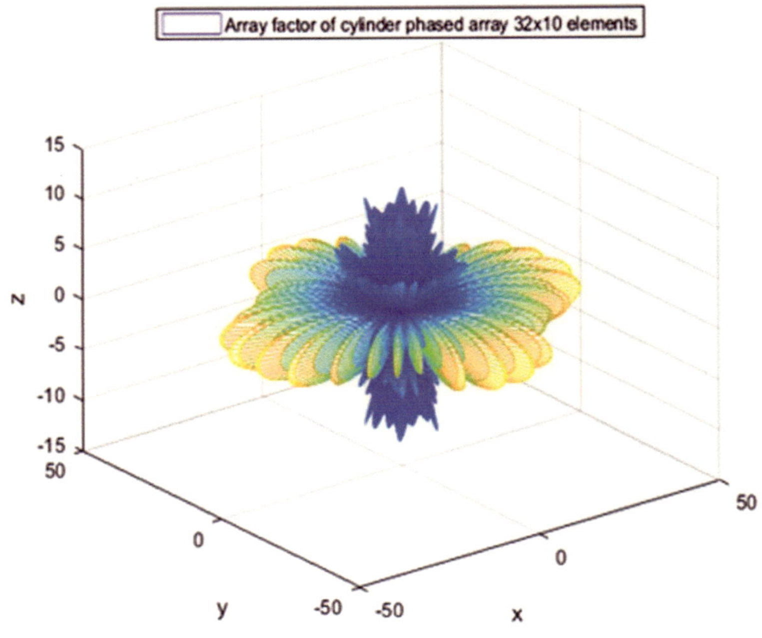

Figure 8.
The array factor of MC-analog-PA of 32 × 10 elements.

4.2 The simulation result of the multibeam cylindrical analog phased array of 32 × 10 elements

When combining 10 rings of Section 4.1 in Z-axis with each ring spaced by λ/2 (half of wave length), we obtain the MC-analog-PA of 32 × 10 elements. The array factor of the MC-analog-PA of 32 × 10 elements is calculated [Eqs. (11) and (12)] and depicted in Cartesian coordinate as in **Figure 8** as follows.

From the result in **Figure 8**, we can see that the array can produce 32 independent beams symmetrical in azimuthal plane.

Since $HPBW_\theta = HPBW\varphi \approx 9^o$, the directivity of the MC-analog-PA of 32 × 10 elements can be approximated as

$$D_{cylinder} = \frac{41253}{(9).(9)} \approx 509 \approx 27dB \tag{14}$$

From the obtained results of the multibeam over azimuthal angle of 360° and directivity of the array, we may conclude that the MC-analog-PA of 32 × 10 elements can meet the requirements of multibeam, wide azimuthal coverage and high gain of 5G applications.

5. Multibeam cylindrical digital phased array (MC-digital-PA)

Let us introduce the last array of this chapter, namely, multibeam cylindrical digital phased array (MC-digital-PA). The MC-digital-PA of M × N elements is the expanding of a MULA-digital-PA combing with a cylindrical geometry. Therefore, it has both the characteristics of multibeam and wide azimuthal coverage. Especially, not only the array structure follows the cylindrical geometry as described in Section 4 but also the digital beamforming is performed using DSP units or a strong computer accompany with beamforming algorithms.

Generally, if huge beams and 3600 azimuthal angle are required, the MC-digital-PA will become a promising candidate. The only shortcoming of the MC-digital-PA is the cost of intensive computations.

6. Conclusions

The analyses of MULA-analog-PA, MULA-digital-PA, MC-analog-PA, and MC-digital-PA show that all arrays could provide multibeam with different complexities and computations but MULA-analog-PA and MULA-digital-PA are with maximum 180° of azimuthal coverage; whereas MC-analog-PA and MC-digital-PA are with unlimited azimuthal angle. The simulations of the MC-analog-PA with 32 × 10 elements (10 rings with each ring of 32 elements) show that the array could provide 32 beams symmetrical over 360 azimuthal coverage with the directivity of 27 dB. In view of antenna design, those arrays are possible candidates for 5G applications.

Acknowledgements

This work is partially supported by the Vietnam National University, Hanoi (VNUH).

Abbreviations

ADS	advanced design software
DA/AD	digital to analog converter/analog to digital converter
DSP	digital signal processing
MIMO	multiple output multiple input
MULA-analog-PA	multibeam uniform linear phased array of analog technology
MULA-digital-PA	multibeam uniform linear phased array of digital technology
MC-analog-PA	multibeam cylindrical phased array of analog technology
MC-digital-PA	multibeam cylindrical phased array of digital technology
MCr-analog-PA	multibeam circular phased array of analog technology
ULA	uniform linear array
ULA-analog-PA	uniform linear array of analog technology

Author details

Tran Cao Quyen
University of Engineering and Technology, Hanoi, Vietnam

*Address all correspondence to: quyentc@vnu.edu.vn

References

[1] Chin WH, Fan Z, Haines R. Emerging technologies and research challenges for 5G wireless networks. IEEE Wireless Communications Magazine. 2014;**21**(2):106-112. DOI: 10.1109/MWC.2014.6812298

[2] Shaf M, Molisch AF, Smith PJ, et al. 5G: A tutorial overview of standards, trials, challenges, deployment, and practice. IEEE Journal on Selected Areas in Communications. 2017;**35**(6):1201-1221. DOI: 10.1109/JSAC.2017.2692307

[3] Asteley D et al. LTE: The evolution of mobile broadband. IEEE Communications Magazine. 2009;**47**(4):44-51. DOI: 10.1109/MCOM.2009.4907406

[4] Larsson EG, Edfors O, Tufvesson F, Marzetta TL. Massive MIMO for next generation wireless systems. IEEE Communications Magazine. 2014;**52**(2):186-195. DOI: 10.1109/MCOM.2014.6736761

[5] Zaghloul AI, Hwang Y, Sorbello RM, Assal FT. Advances in multibeam communications satellite antennas. Proceedings of the IEEE. 1990;**78**(7):1214-1232. DOI: 10.1109/5.56934

[6] Jain A. Multibeam synthetic aperture radar for global oceanography. IEEE Transactions on Antennas and Propagation. 1979;**27**(4):535-538

[7] Brautigam B, Schwerdt M, Bachmann M. An efficient method for performance monitoring of active phased array antennas. IEEE Transactions on Geoscience and Remote Sensing. 2009;**47**(4):1236-1243. DOI: 10/1109/TGRS.2008.2008719

[8] Agrawal A, Kopp B, Luese MH, Haver KW. Active phased array development for modern shipboard radar systems. Johns Hopkins APL Technical Digest. 2001;**22**(4):600-613

[9] Carey E, Lidholm S. Millimeter-Wave Integrated Circuits. Berlin, Gemany: Springer; 2005. DOI: 10.1007/b102136

[10] Marcus MJ. 5G and 'IMT for 2020 and beyond. [spectrum policy and regulatory issues]. IEEE Wireless Communications. 2015;**22**(4):2-3. DOI: 10.1109/WMC.2015.7224717

[11] Hansen RC. Phased Array Antennas. New York: Wiley Interscience; 2009. DOI: 10.1002/9780470529188

[12] Hong W, Jiang ZH, Yu C, et al. Multibeam antenna technologies for 5G wireless communications. IEEE Transactions on Antennas and Propagation. 2017;**65**(12):6231-6249. DOI: 10.1109/TAP.2017.2712819

[13] Litva J, Lo TKY. Digital Beamforming in Wireless Communication. Boston, London: Artech House; 1996. ISBN: 0-89006-712-0

[14] Balanis C. Introduction to Smart Antenna. California, USA: Morgan and Claypool Publishers; 2007. ISBN: 1598291769

[15] Quyen TC. Developing a symmetrical phased array antenna with low complexity. IEEE: 7th International Conference on Integrated Circuit, Design and Verification (ICDV). Hanoi, Vietnam; 2017. pp. 133-137. DOI: 10.1109/ICDV.2017.8188653

[16] Kraus J. Antennas. 2nd ed. New Delhi: Tata McGraw-Hill. ISBN: 0-07-463219-1

[17] Jackson BR, Rajan S, Liao BJ, Wang S. DOA estimation using directive antennas in uniform circular array. IEEE Transactions on Antennas and Propagation. 2015;**63**(2). DOI: 10.1109/TAP.2014.2384044

Chapter 5

Design of a UWB Coplanar Fed Antenna and Circular Miniature Printed Antenna for Medical Applications

Adnane Latif

Abstract

Breast cancer is the second deadly cancer for women; for more efficiency and an early detection, the biomedical field need new systems that should be safe, comfortable, and sensible. The medical field already has its methods to detect breast cancer. In this chapter, a new ultra-wide band (UWB) planar antenna is presented for microwave imaging, the antenna is designed to operate in a frequency band from 2.9-10.8GHz. The antenna was designed to be adaptable for multi-viewing imaging due to it simple form, low cost and ease to be manufactured. The simulation results of the new ultra large band antenna and a performance comparison with other UWB antennas. We also offer a circular miniature printed antenna that satisfies the UWB characteristics in terms of bandwidth and reflection coefficient. This antenna is intended for a system for detecting malignant tumors by microwave imaging. The antenna made has a patch on a FR-4 type substrate with ε_r = 4,3, thickness h =1.575 mm and dimensions l_s = 25 mm, and w_s = 25 mm. A rectangular slot is inserted on the radiating element ensuring its miniaturization. The latter is powered by a microstrip line width wm with matching impedance at 50 Ohm.

Keywords: ultra wide band, coplanar antenna, microwave imaging, CPW (coplanar waveguide) antenna, wireless body area network (WBAN) antennas, BAN communications, UWB antenna, printed antennas, circular patch antennas, wireless communications, radars, microwaves, medical applications

1. Introduction

Since the FCC allows the use of the UWB spectrum, the development of new systems that will adapt to this band has been growing. Much attention has been given to developing new systems that fit with UWB applications as radar detection, biomedical imaging. The proposed UWB antenna is more likely to be used in the biomedical imaging applications as a system to detect breast cancer.

Breast cancer is the second deadly cancer for women, for more efficiency and an early detection, the biomedical field need new systems that should be safe, comfortable and sensible [1, 2]. The medical field already has its methods to detect breast cancer like X-ray mammography, magnetic resonance imaging (MRI), as efficient as those techniques are they are missing between 10 and 30% [2] of very early breast cancer stages.

The UWB microwave imaging is one of the techniques that has been developed, in the Hunt for the next techniques that will detect breast cancer in its early stages. The basis for microwave detection is to compare between the dielectric properties of a normal tissue and cancerous tissue, the result of the comparison gives a prediction of malignant tissue characteristics as it size, shape, placement, ... [3].

The second part of this chapter, I present a microstrip antenna that we offer is circular in shape and, a priori, has a low gain and a narrow bandwidth. To meet the requirements of our specifications would be to expand the bandwidth. For this reason, we have developed a design methodology, which has enabled us to develop an antenna capable of meeting these requirements.

The reduction in size is also a consideration to take into account when designing this antenna, which would allow it to be more easily integrated into the system and reduce the size. For this, the certain techniques are used. Among these techniques, we will use the slot insertion at the level of the radiating element.

The design procedure followed consists of three steps:

- The first step is to develop a structure with ULB characteristics.
- The second step is used to optimize the performance of the structure developed in terms of adapting the impedance
- The third step is to develop an antenna network.

2. Design of a UWB coplanar fed antenna

2.1 Ultra wide band

UWB is defined as a system with a very large band, this large spectrum comes usually with some advantages as a low power, high debit of data, high time resolution, low-cost and an ease of implementation, resistance to interference and so on. Those advantages opened a wide range of UWB application to radar detection, biomedical imaging, and HD communication.

The definition of UWB is not a special one, the FCC (Federal Communication Commission) [4] defines it as a system with a bandwidth larger than 500 MHz or larger than 20% if we are working whit the relative bandwidth W/fc (W is the width of the band and fc is the carrier frequency). The UWB is being defined by its very large band that is 7.5 GHz between 3.1 GHz and 10.6 GHz for a limitation of the power emission level −41.3 dBm/MHz [4].

2.1.1 Matching and efficiency

The UWB communication systems uses a very short pulse duration of tens hundreds of nanoseconds, and sense the pulse and the bandwidth are inversely proportional, the shorter pulses is the wider the spectrum is going to be.

Antennas would match the UWB requirements if it have a bandwidth greater than 500 MHz defined at −10 dB according to the FCC, or to have a relative bandwidth more than 20% [4, 5].

Where f_l and f_h are the low and the high frequency respectively.

The efficiency of an UWB antenna can be evaluated by the specter efficiency, the evaluation matching has to be over the whole range of frequencies.

2.1.2 Signal dispertion ans distortion

The signal passing through the UWB antennas is a very short pulse, the shorter the pulse is more likely the UWB antenna's response is going to be distorted and delayed due to the ripple after the pulse called the rippel effect [5].

The rippling effect is caused by the geometry of the antenna, and it causes frequency translation, dispersion or delay on the transition reducing the speed of data transmission [6].

2.2 Antenna design

This section is dedicated to the design and the performance of the proposed antenna. The design proposed is a notched coplanar antenna. The geometry of the antenna consisted of a 25 × 30 mm^2, Rogers RT5850 substrate with ε_r = 2.2 and thickness h = 3.175 mm. The top part of the antenna is a circle with two cutouts of 90° each. The structure of the antenna is shown in **Figure 1**.

The two main characteristics that effect the performance of the antenna was the inner radius of the circle R and the laminate thickness. The gap between the transmission line and the ground structure was optimized so that the antenna's impedance can match a 50 Ω SMA port.

2.3 Performances of the antenna

2.3.1 The return loss S11

Figure 2 shows the return loss or more known as the parameter S11 of the antenna, the spectrum of **Figure 2** contains the UWB frequency band spectrum. The spectrum's antenna range is form 2.8 GHz to 10.9 GHz and contains tow resonate frequencies

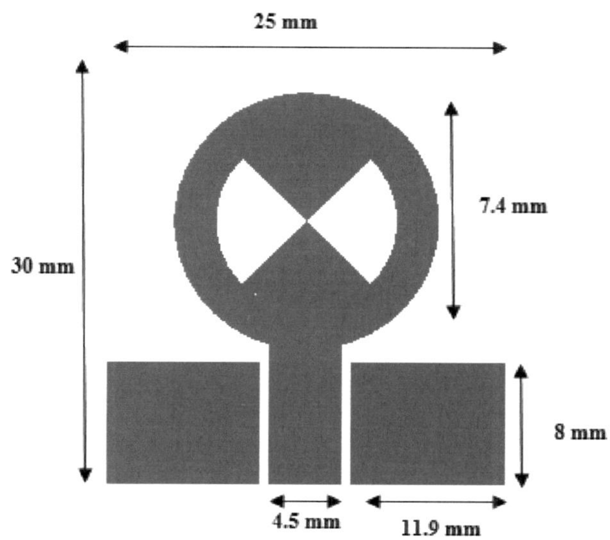

Figure 1.
Geometry of the antenna proposed.

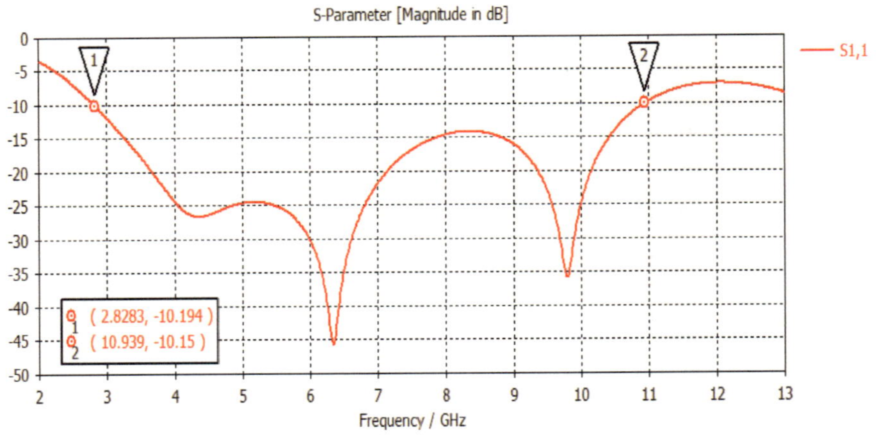

Figure 2.
The return loss S11 as a function of frequency.

6.1 GHz and 9.8 GHz with their S11 parameters -46 dB and -35 dB respectively. The antenna can be easily used in the UWB (3.1 GHz – 10.6 GHz) applications [4].

2.3.2 Voltage standing wave ratio (VSWR)

VSWR is a parameter that describes the power that is reflected by the antenna. It is a function of the reflection parameter.

Figure 3 shows the graph of the VSWR of the antenna proposed, we can tell from the graph that the VSWR is under tow in all the bandwidth so the VSWR can be considered as good.

2.3.3 Radiation pattern

The main focus of the chapter is to design a UWB antenna that can be easily implemented in UWB application with a small dimension and a larger frequency spectrum. **Figure 4** shows the radiation pattern of the antenna proposed for the frequency of 9.8 GHz, the color red defined the higher range of the gain and the green refers to the lower part [7].

2.4 Paramertic study

The parametric study is done in two parts the first part was to choose the laminate and then the value of the radius of the patch.

2.5 Comparison with others UWB antennas

The antennas that was compared to the antenna proposed in the chapter is: a coplanar microstrip antenna with defected ground structure for UWB applications [7]. A printed UWB antenna with full ground plane also for WBAN applications and CPW-fed slot patch antenna [8–10]. The study is to compare the proposed geometry on tree different Rogers laminates. **Table 1** shows the characteristics of the tree laminates.

Figure 5 presents the return loss of the antenna with the different laminates, according to the results of **Figure 5**, the laminate Rogers with a thickness of 3.175 mm was chosen to complete the parametric study. The variation of the radius is

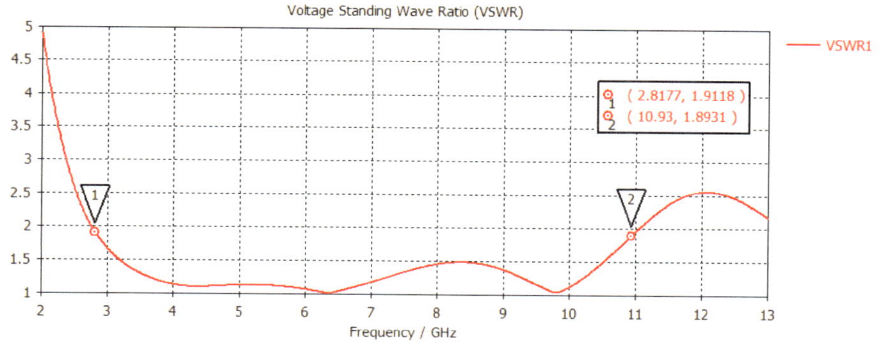

Figure 3.
VSWR as a function of frequency.

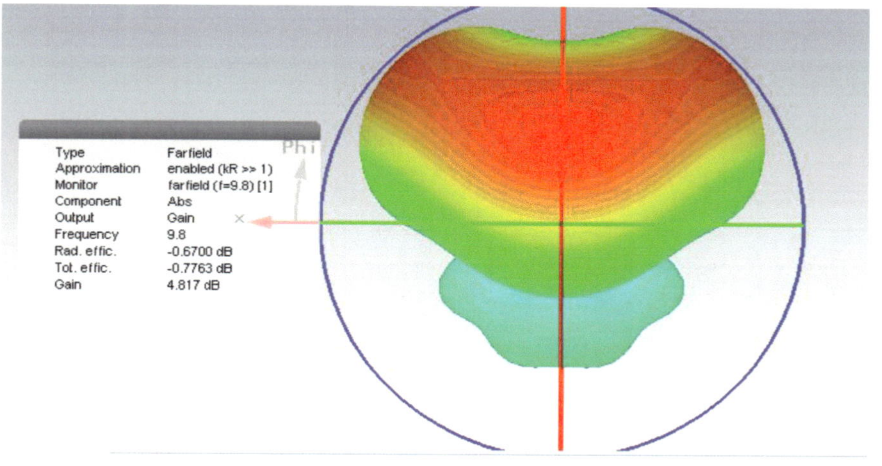

Figure 4.
Radiation pattern of the proposed antenna.

Laminates	Laminate characteristics	
	Standard thickness	Standard copper cladding
Laminate 1	0.787 mm	35 μm
Laminate 2	1.57 mm	35 μm
Laminate 3	3.175 mm	35 μm

Table 1.
Standard laminales characteristics.

altered around the theoretical radius to optimize the performance of the antenna. The patch of the antenna is separated into two parts, first where the cutouts are and the second part is the ring that contour the cutouts, this parametric study consisted on varying the inner radius of the patch, the radius of the circle that contain the cutouts. **Figure 6** shows a relation between the radius and the depth of the return loss, R = 4.8 mm shows a better S11 parameter.

Table 2 shows the difference between four UWB antennas, the proposed antenna has a wider band, includes all UWB frequencies, the highest gain is around 5.6 dBi.

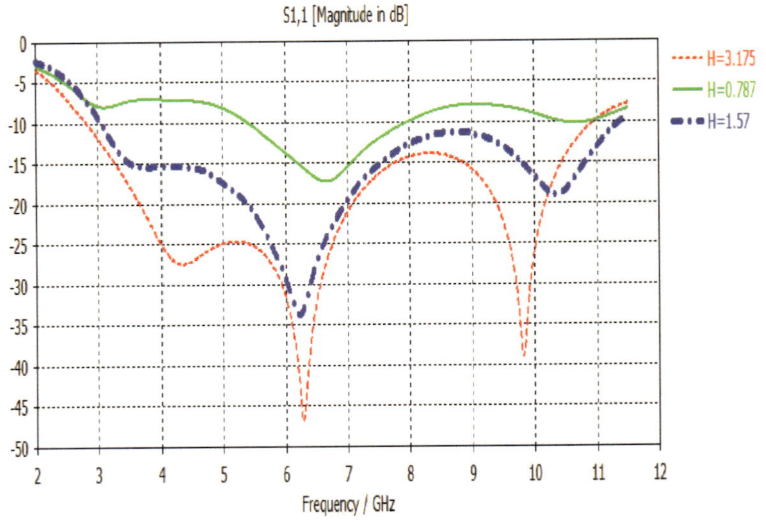

Figure 5.
Effect of the thickness of the substrate.

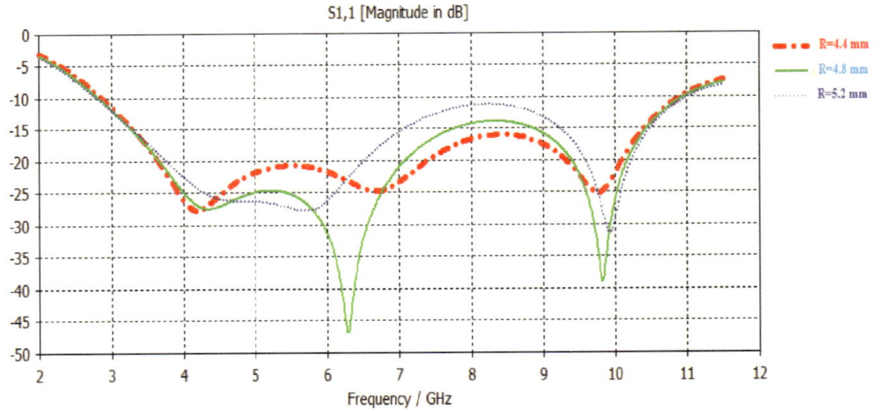

Figure 6.
Effect of the inner radius on the antenna.

Antennas	Characteristic of the antennas		
	Dimension (w × L)	**Bandwidth**	**Gain**
Antenna proposed	25 mm × 30 mm	2.7–10.98 GHz	3.5 dBi
Antenna [7]	30 mm × 32 mm	3.1–9.9 GHz	—
Antenna [8]	75 mm × 85 mm	4–9.5 GHz	—
Antenna [9]	24 mm × 30 mm	4.82–8.87 GHz	3.9 dBi

Table 2.
Performance comparison.

3. Design of a UWB circular miniature printed antenna

3.1 Design of circular miniature printed antenna

3.1.1 Antenna geometry

The patch of radius (r) is produced on a substrate of the FR-4 type (dielectric permittivity ε_r = 4.3, thickness h = 1.575 mm) and of dimensions ls = 25 mm and w_s = 25 mm. A rectangular slot is inserted on the radiating element (u × s) ensuring its miniaturization. The latter is supplied by microstrip line of w_m east width in order to adapt it to a 50 Ω supply [11, 12]. **Figure 7** shows the proposed antenna.

From **Table 3** values, we perform a simulation on the CST software [13].

3.1.2 Simulation results

3.1.2.1 S_{11} parameter

The parameter S_{11} is the coefficient that most concerns designers of printed antennas because it represents the reflection coefficient which plays the role of disturbance on data transmission [14]:

We see that the coefficient S is around −22.35 dB for a resonant frequency of 3.8 GHz, the bandwidth is 3.17–15 GHz, as shown in **Figure 8** [15].

Figure 7.
Antenna geometry.

Parameters	r	l_m	w_m	t_m	lg	u	s	v	g	k
mm	8	9	2.25	0.035	8	1	10	4	0.27	5

Table 3.
The dimensions of the circular miniature printed antenna.

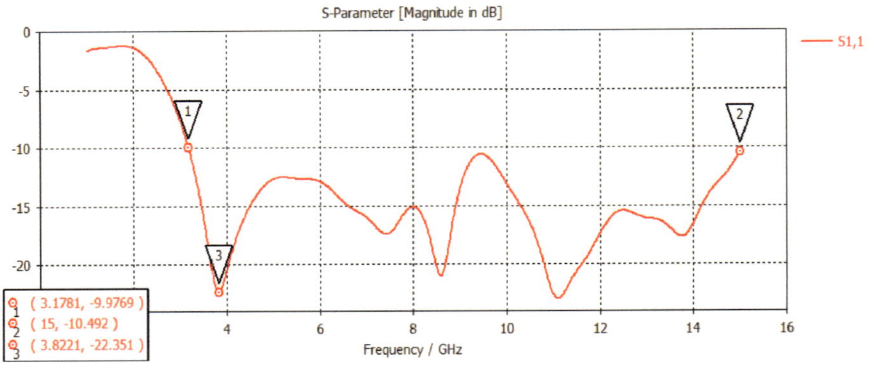

Figure 8.
Return loss of circular miniature printed antenna.

3.1.2.2 Antenna gain

Figure 9 gives us the variation of the gain of our antenna as a function of the frequency. It happens to be between 2 and 4.87 dBi on the frequency band which interests us, which is 3.1–10.6 GHz. This limitation of the gain could be improved by a possible networking of antennas [16].

3.1.2.3 Radiation diagram

The 3D radiation pattern at the 8 GHz frequency. We can say that the radiation is focused on both sides of the antenna. At the frequency f = 8 GHz the gain of antenna is 4.28 dBi (**Figure 10**) and a directivity of antenna is 4.87 dBi (**Figure 11**).

Figures 12 and **13** show the far-field gain and far-field directivity in plane E. For both the opening of the main lobe at -3 dB for the frequency 8 GHz is 66.9° [16, 17].

3.1.3 Parametric study

We are going to play on three parameters, to see its effect on the behavior of this antenna.

- The parameter effect r (RO) (the radius of the radiating part) on coefficient S_{11} (**Figure 14**) and gain (**Figure 15**):

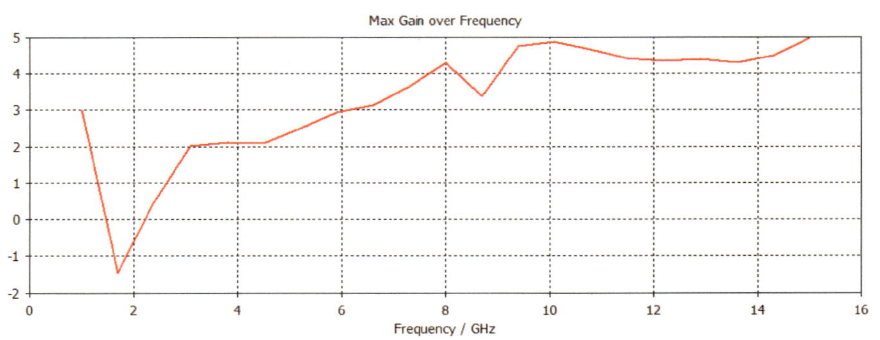

Figure 9.
Gain diagram of circular patch antenna.

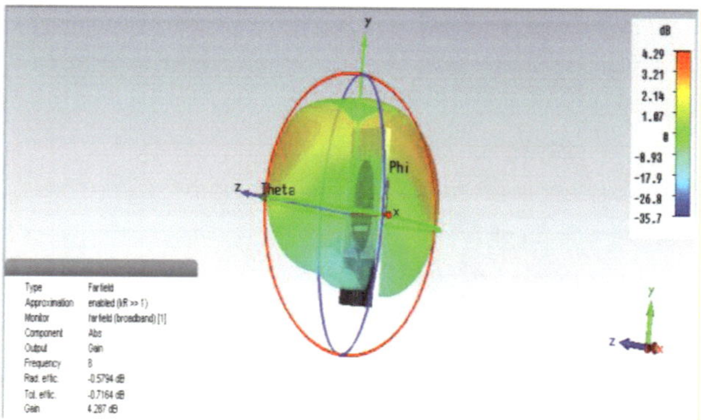

Figure 10.
Gain diagram (3D) of circular patch antenna.

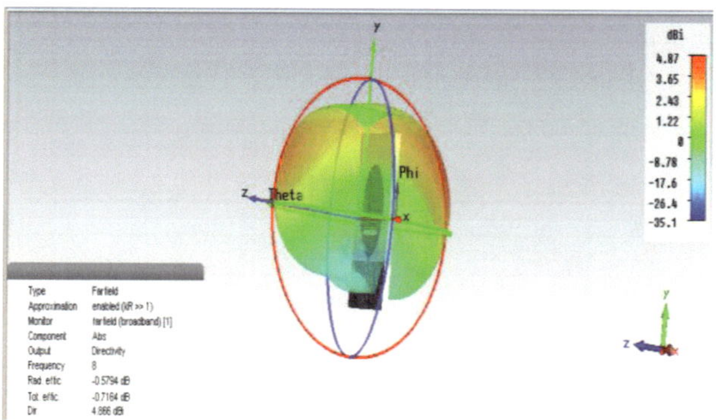

Figure 11.
Directivity diagram (3D).

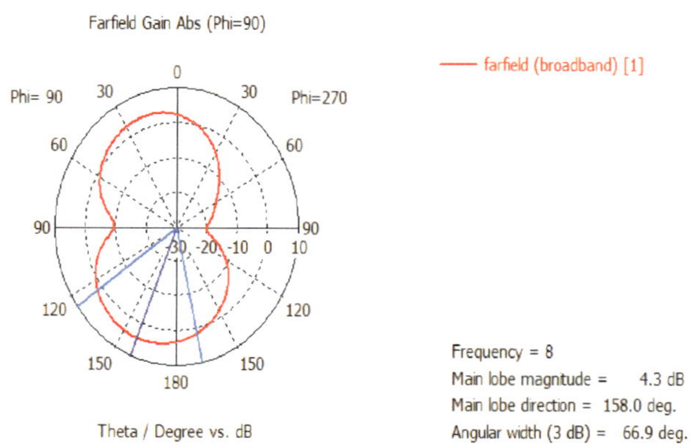

Figure 12.
Far-field gain (2D) of circular miniature printed antenna.

S_{11} parameter:
Gain:

- The parameter effect l_g (length of the ground plane) on coefficient S_{11} (**Figure 16**) and gain (**Figure 17**):

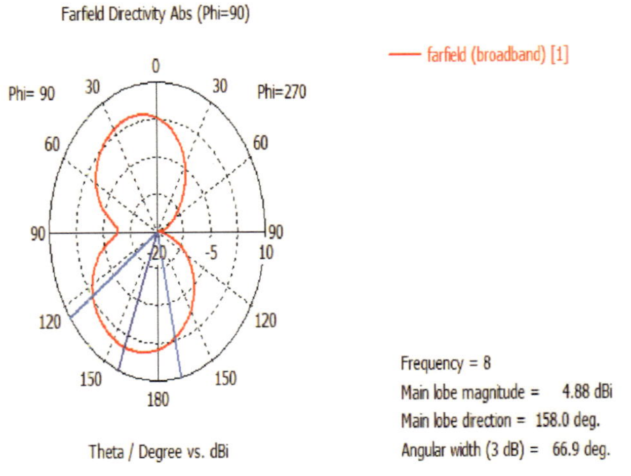

Figure 13.
Far-field directionality (2D).

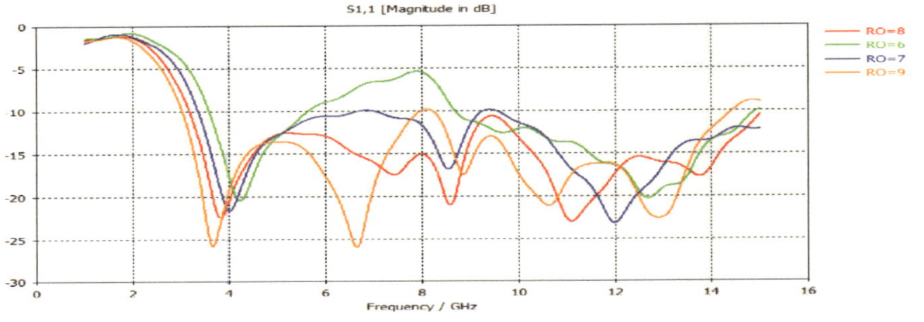

Figure 14.
Return loss with effect of the antenna radius.

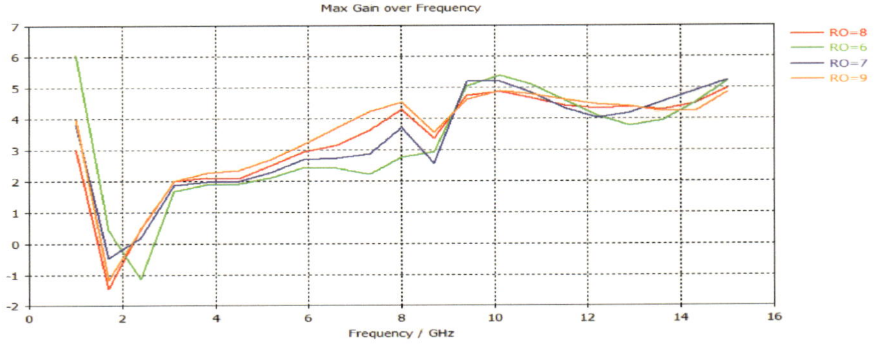

Figure 15.
Gain diagram with effect of the antenna radius.

S_{11} parameter:
Gain:

- The parameter effect v (slot location) on coefficient S_{11} (**Figure 18**) and gain (**Figure 19**):

S_{11} parameter:
Gain:

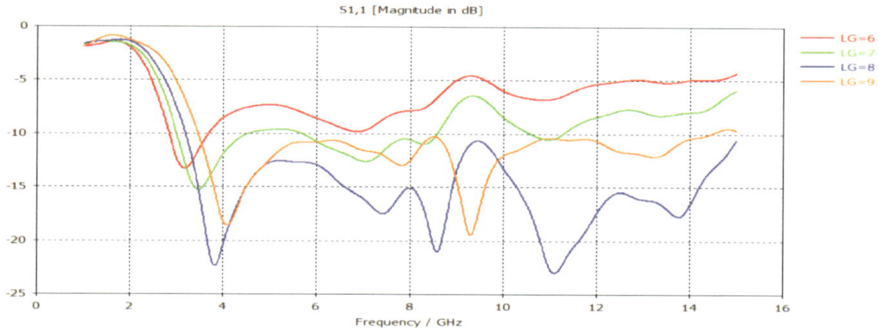

Figure 16.
Return loss with effect of the length of the antenna ground plane.

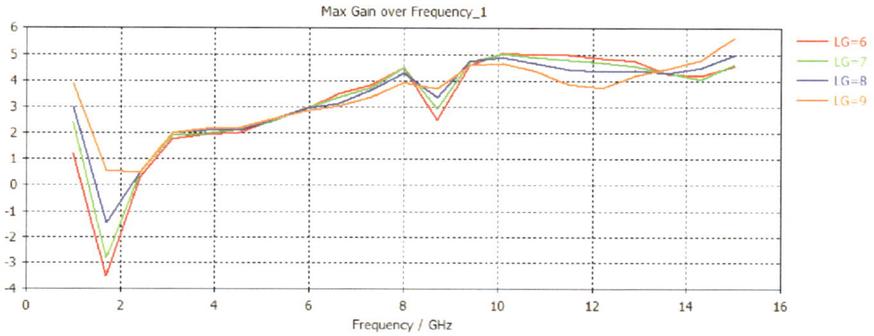

Figure 17.
Gain diagram with effect of length of the antenna ground plane.

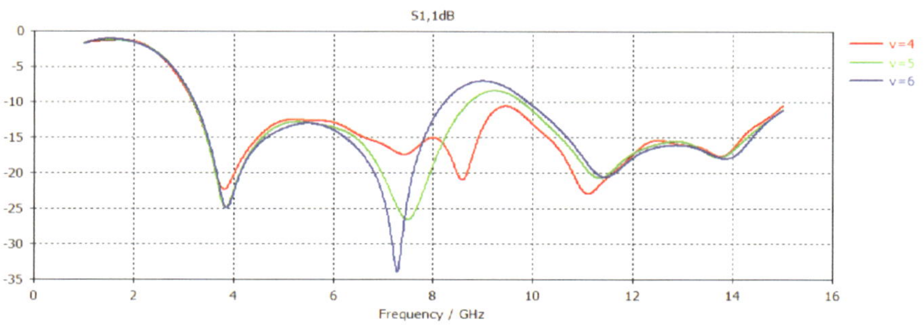

Figure 18.
Return loss with effect of the antenna slot location.

3.2 Design of array of circular miniature printed antenna

3.2.1 Antenna array geometry

We are going to make our antenna network (**Figure 20**) from the previous antenna with some modifications in terms of the parameters and its values in order to have the desired results (**Table 4**).

3.2.2 Simulation results

S11 parameter:

In **Figure 21,** we see that the coefficient S_{11} is of the order of −42 dB for a resonant frequency of 6.08 GHz, we obtained exactly the bandwidth that interests us, which is [2.7 10.6 GHz] [18].

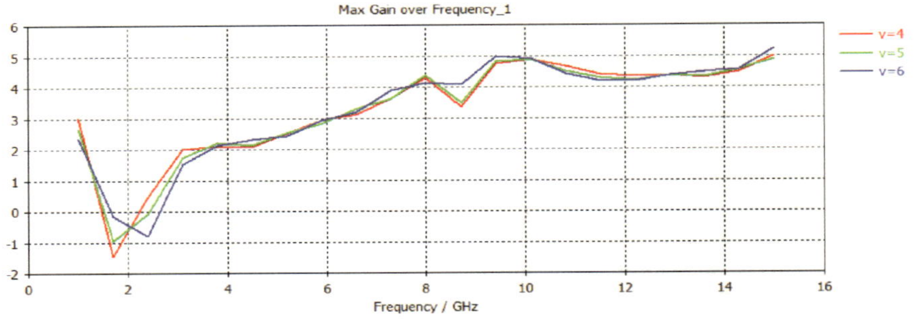

Figure 19.
Gain diagram with effect of the antenna slot location.

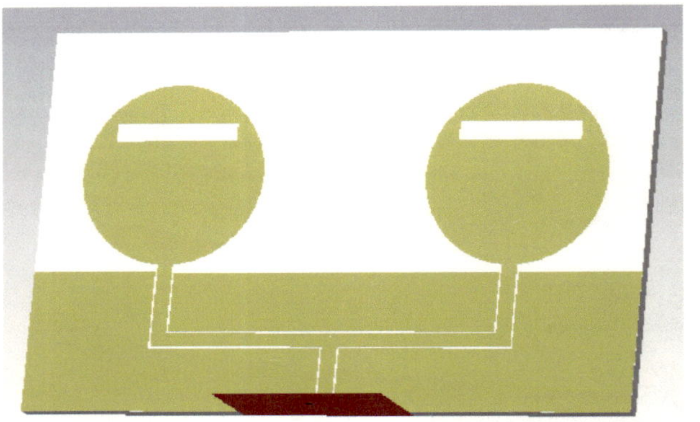

Figure 20.
Antenna array structure.

Parameter	r	L_m	w_m	t_m	lg	w_p	lp	s	u	v	g	k
mm	9	7	1.5	0.035	6	4	9	12	1.8	13	0.18	5

Table 4.
Dimensions of the array antenna.

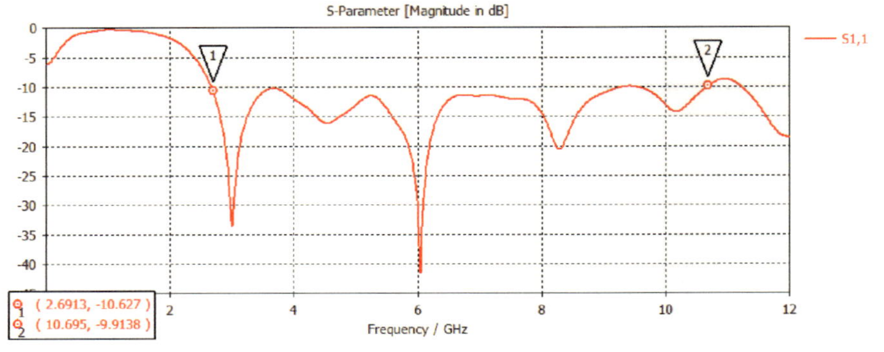

Figure 21.
Return loss of array antenna.

Gain:

Figure 22 gives us the variation of the gain of our antenna as a function of the frequency. Note that the gain has increased to 7.6 dBi on the frequency band [3.1–10.6 GHz]. This networking of antennas has allowed us to increase the gain [19].

At the frequency of 6.14 Hz in **Figure 23** (Gain Diagram) and **Figure 24** (Directivity Diagram), we can say that the radiation is focused on both sides of the

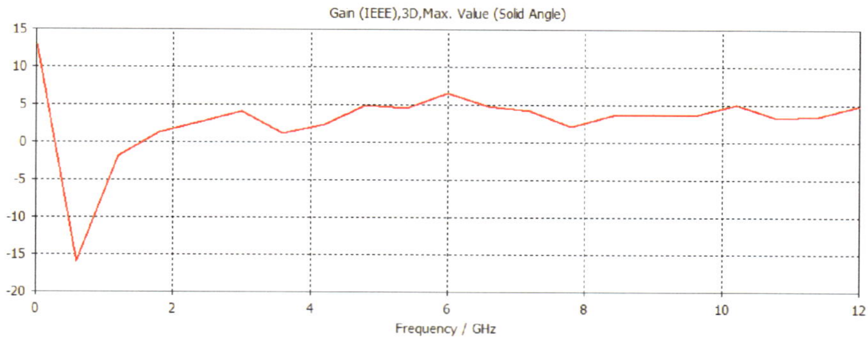

Figure 22.
Gain diagram of array antenna.

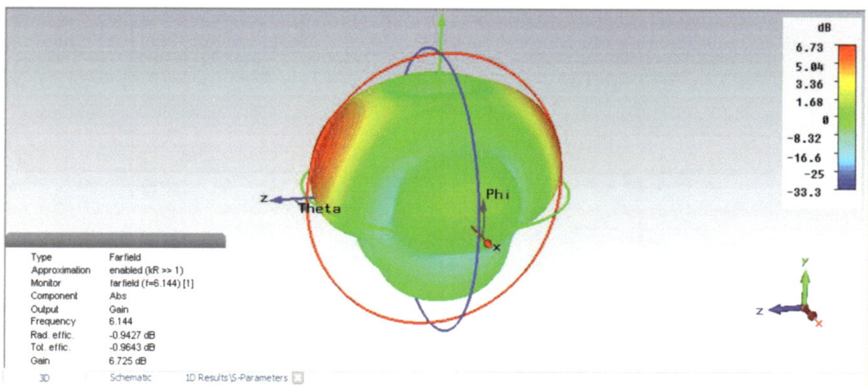

Figure 23.
Gain diagram of array antenna.

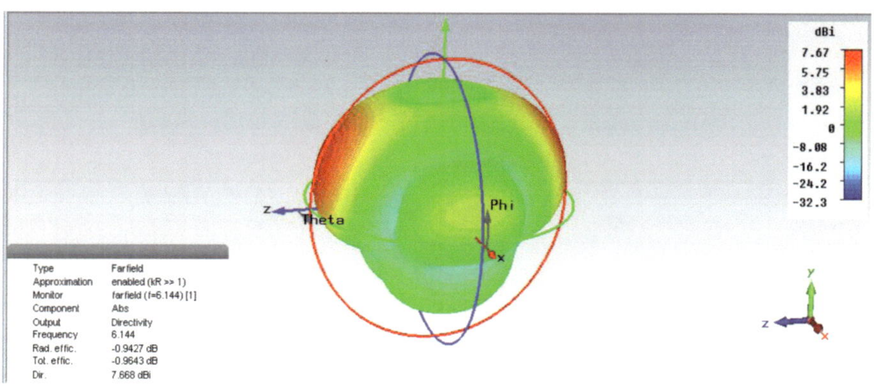

Figure 24.
Directivity diagram.

antenna but this time we have side lobes. With a gain of 6.7 dBi and a directivity of 7.6 dBi at f = 6.14 [20, 21].

Radiation diagram:

Figures 25 and **26** show respectively the far-field gain and far-field directivity in plane E. The opening of the main lobe at -3 dB for the frequency 6.14 GHz is 32.8° [22–25].

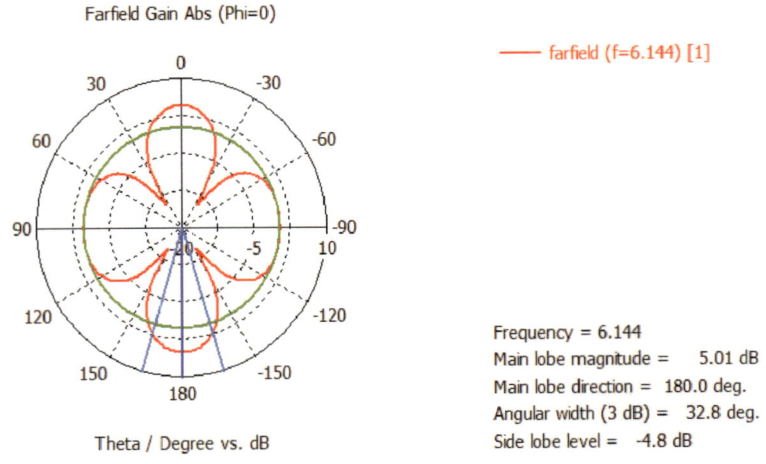

Figure 25.
Farfield gain.

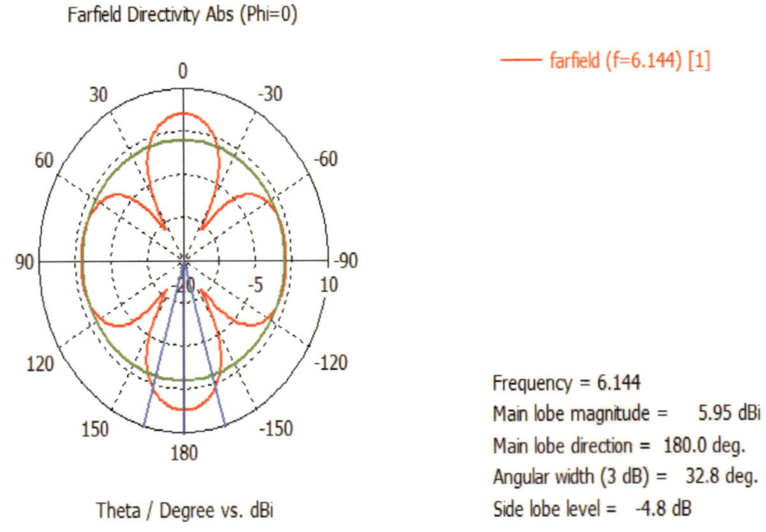

Figure 26.
Farfield directivity.

4. Conclusion

In the first part of this chapter, the new coplanar antenna with cutouts on a circular patch demonstrate advantages: larger frequency spectrum, a good VSWR, improved parameter S11. The geometry of the patch was calculated and design with CST microwave studio. The antenna was design for UWB application more likely the microwave imaging in the range from 3.1 GHz to 10.6 GHz. Moreover the antennas are smaller than traditional antennas, and as a coplanar it can be easy to implement. Also, enhancing gain will be a good improvement by making this antenna in an array due to its small size.

In the second part of this chapter, we proposed a miniature circular patch antenna intended for an application in medical imaging.

The antenna network satisfies the imposed requirements satisfactorily and exhibits ultra broadband (ULB) behavior. In fact, the simulations under CST resulted in a reflection coefficient at −10 dB between 2.6 GHz and 10.7 GHz.

We have shown by this study that the dimensions and shape of the ground plane could have a significant impact on the bandwidth of the structure.

Future work includes the fabrication of a prototype of the antenna, as well as measurements in anechoic chamber to verify the agreement between the measured and simulated results.

Due to the complexity of each human tissue, our antenna should be tested in different parts of the body to determine is the optimum distance of 20 mm is valid wherever the antenna is placed within the body. The implementation of an array of antennas will be also evaluated in future studies.

Author details

Adnane Latif
Cadi Ayyad University, Marrakesh, Morocco

*Address all correspondence to: a.latif@uca.ac.ma

References

[1] Shao W, Zhou B. UWB microwave imaging for breast tumor detection in inhomogeneous tissue. In: Proceedings of the 27th Annual Conference of 2005 IEEE Engineering in Medicine and Biology, Shanghai, China. 2005. pp. 1496-1499

[2] Xiao X, Kikkawa T. Early breast cancer detection by ultrawide band imaging with dispersion consideration. Japanese Journal of Applied Physics. 2008;**47**(4 Part 2):3209-3213

[3] Hagness SC, Taflove A, Bridges JE. Three-dimensional FDTD analysis of a pulsed microwave confocal system for breast cancer detection: Design of an antenna-array element. IEEE Transactions on Antennas and Propagation. 1999;**47**(5):783-791

[4] Ultra-wideband operation FCC report and order. Tech. Rep. US 47 CFR Part 15. 2002

[5] Valderas D, Sancho JI, Puente D. Ultrawideband Antennas: Design and Applications. Broche. 210 pp. ISBN-13: 978-1848164918, Editeur: Icp (15 Octobre 2010)

[6] Balanis CA. Handbook of Microstrip Antennas. New York: John Wiley and Sons; 1982

[7] Cheema L, Sherdia K. Design of microstrip antenna with defected ground structure for UWB applications. International Journal of Advanced Research. 2013;**2**(7):2525-2528

[8] Simorangkir RBVB, Abbas SM, Esselle KP. A Printed UWB Antenna with Full Ground Plane for WBAN Applications. 2016. pp. 127-130

[9] Niranjan V, Saxena AK, Srivastava KV. CPW-fed slot patch antenna for 5.2/5.8 GHz WLAN application. In: Progress Electromagn. Res. Symp. PIERS 2012, Kuala Lumpur. 2012. pp. 1350-1352

[10] Dikmen CM, Çimen S, Çakir G. Planar octagonal-shaped UWB antenna with reduced radar cross section. IEEE Transactions on Antennas and Propagation. 2014;**62**(6):2946-2953

[11] Lu JH, Yeh CH. Planar broadband arc-shaped monopole antenna for UWB system. IEEE Transactions on Antennas and Propagation. 2012;**60**(7):3091-3095

[12] Majumdar B, Esselle KP. A simple reconfigurable patch antenna for mobile applications. In: IEEE International Symposium on Antennas and Propagation & USNC/URSI National Radio Science Meeting. 2015. pp. 2259-2260

[13] Mahlaoui Z, Antonino-daviu E, Ferrando-bataller M, Benchakroun H, Latif A. Frequency reconfigurable patch antenna with defected ground structure using varactor diodes. In: 2017 11th European Conference on Antennas and Propagation Frequency. Vol. 11. 2017. pp. 2217-2220

[14] Li M, Behdad N. A compact, capacitively-fed UWB antenna with monopole-like radiation characteristics. IEEE Transactions on Antennas and Propagation. 2016;**65**(3):1-1

[15] Alipour A, Hassani HR. A novel omni-directional UWB monopole antenna. IEEE Transactions on Antennas and Propagation. 2008;**56**(12):3854-3857

[16] Miralles E, Andreu C, Cabedo-Fabrés M, Ferrando-Bataller M, Monserrat JF. UWB on-body slotted patch antennas for in-body communications. In: 2017 11th

European Conference on Antennas & Propagation. 2017. pp. 1-4

[17] Jianying L et al. Bending effects on a flexible Yagi-Uda antenna for wireless body area network. In: 2016 Asia-Pacific International Symposium on Electromagnetic Compatability, APEMC 2016. 2016. pp. 1001-1003

[18] Lin JC. Frequency optimization for microwave imaging of biological tissues. Proceedings of the IEEE. 1985;**73**(2):374-375

[19] Nouri A, Dadashzadeh GR. A compact UWB band-notched printed monopole antenna with defected ground structure. IEEE Antennas and Wireless Propagation Letters. 2011;**10**:1178-1181

[20] Weiland T, Timm M, Munteanu I. A practical guide to 3D simulation. IEEE Microwave Magazine. 2008;**6**:62-75

[21] Haq MA, Koziel S, Cheng QS. EM-driven size reduction of UWB antennas with ground plane modifications. In: Engineering 2017 International Applied Computational Electromagnetics Society Symposium (ACES). 2017. pp. 7-8

[22] Buller W, Wilson B. Measurement and modeling mutual capacitance of electrical wiring and humans. IEEE Transactions on Instrumentation and Measurement. 2006;**55**(5):1519-1522

[23] Benchakroun H, Cabedo-Fabrés M, Ferrando-Bataller M, Mahlaoui Z, Latif A. New approach for the design of UWB monopoles with directive radiation for BAN applications. In: 12th European Conference on Antennas and Propagation (EuCAP 2018). 2018. p. 5

[24] Belguessoum A. Conception et réalisation d'une antenne multibandes alimentée par un guide d'onde coplanaire CPW, Ottawa, Canada; 2008

[25] Fortino N. Conception et caractérisation d'antennes imprimées pour systèmes Ultra-Large Bande impulsionnels. Université Nice Sophia Antipolis; 2006 (Français)

Chapter 6

Beamforming Phased Array Antenna toward Indoor Positioning Applications

Nguyen Thanh Huong

Abstract

Indoor positioning systems based on radio wave have drawn a lot of research interest over the last few decades. One of the positioning methods, Angle of Arrival, locating object based on the relative angle of object to the reference points, requires a directional antenna in order to improve the accuracy of indoor positioning system. From this situation, an eight-port phased array antenna using reflection type phase shifter is designed. The input power is split to each antenna through eight-port Wilkinson power divider with insertion loss of about 11 dB and isolation of about 20 dB. To extract more accurate position, the main beam of phased array antenna can be steered smoothly by a design 18 of a continuous and full 360° reflection type phase shifter with low insertion loss variation. Microstrip patch antennas are used as elements in phased array antenna. The steering of main beam is presented by radiation patterns of phased array antenna, measured in anechoic chamber from −45 to 45° with step 5°. The measurement result shows that the main beam direction is quite close to the desired direction in simulation. In most case, the side lobe level is less than the main lobe about 10 dB.

Keywords: angle of arrival (AoA), antenna array, beamforming, phased array, low loss, phase shifter, steering beam

1. Introduction

Nowadays, positioning has played an important part in daily life and is the foundation for many other applications such as navigation, tracking, location-based services, and games. While outdoor positioning has become widespread and popular with satellite-based navigation systems, that is, GPS, GLONASS, and Galileo, indoor positioning systems have attracted a lot of research interest over the last decade. Indoor localization promises to create a lot of new services such as guiding users in museum, preventing thefts from valuable assets, locating products in supermarkets, navigating in mall, saving power consumption of electronic devices, and so on.

Due to scattering and attenuation of microwave on roofs, walls, and other objects, the systems for outdoor positioning are infeasible solutions when applying to indoor positioning. Instead, indoor positioning systems have been implemented based on several technologies: infrared (IR) [1], bluetooth [2, 3], radio-frequency identification (RFID) [4], wireless local area networks (WLAN) [5], ultra-wideband [6], ultra-sound [7], magnetic positioning [8], and audible sounds [9]. Among them, WLAN-based approach appears to be among the most suitable solutions owing to its wide range and the popularity of equipment. Nowadays, WLAN devices can be found in almost all civil and commercial properties due to their ease of installation and usage. The number of WLAN devices has reached billions and is continuously increasing day by day, as a result, the study in WLAN-based indoor positioning promise to be applied and spread in the near future. Besides, with the current 5G supports, this indoor positioning technology can even attain to longer reach, greater capacity, high accuracy, and reliability.

For WLAN-based approach, indoor localization techniques are classified into RSS scene analysis, ToA, TDoA, RToF, and AoA [1]. While the highly unstable feature of RSS in indoor environments is the major challenge of RSS scene analysis technique and ToA, TDoA, and RToF are based on a precise clock synchronization of devices, the AoA technique requires a directional antenna design to estimate the relative angle of object to reference points.

There have been several antenna designs for AoA-based indoor localization presented in past few years. Giorgetti and Cidronali [10] introduced a switched-beam directional antenna, including six circular antennas to cover six areas in a room. Rzymowski et al. [11] also introduced an antenna design using 12 passive elements electrically steerable parasitic array radiator antenna with one active monopole in the center of the ground plane. By controlling single-pole, single-throw switches, parasitic elements connect to the ground and act as reflectors, which change the main lobe's directions. With 12 passive elements, this antenna has 12 different directions. Kamarudin et al. [12] proposed reconfigurable antennas using PIN diodes to switch lumped components such as inductors and capacitors in order to change the structure of antennas and switch between four beams. Bui et al. [13] built a switch-beam array antenna based on 4 × 4 Butler matrix to create four beams toward four angles. It is found that these designs are based on switching between the limited number of predefined beams, which limits the resolution in determining the location of object and leads to significant errors in positioning.

For indoor localization, the system needs to determine the location of object with high accuracy. Therefore, an antenna design with high-resolution steering beam ability is very essential in the AoA technique. As discussed above, the current antenna designs for indoor localization system mainly focus on switched-beam antenna structures. With these structures, the resolution of angle of arrival primarily depends on the number of antenna elements. Therefore, to create a system with high resolution of beam scanning, the number of antennas as well as switching elements such as PIN diode, FET also increases, which makes the system more cumbersome and complex. In order to improve the accuracy for AoA-based indoor localization, the aim of this chapter is to study and develop a phased array antenna at Wi-Fi band (2.4 to 2.484 GHz), capable of finely steering beam without increasing the number of antennas.

This chapter addresses the subject of a beamforming phased array antenna, which plays the vital role for AoA-based indoor positioning system. The array antenna is designed to steer the main beam in high resolution without the increase of antenna elements as well as switching elements. The design of this antenna array has focused on developing the controllable 360° continuous reflection type phase shifter. This chapter is organized as follows. In Section 2, we explain the principles of forming beam and several methods to feed antennas and to control the phase differences for phased array antennas. In Section 3, we discuss the structure of phased array antennas used for practical indoor localization, advancing from antenna elements, power dividers, feed network, and phase shifters. In Section 4, we verify the performance the printed antenna array by the Vector Network Analyzer and in anechoic chamber.

2. Design factors of beamforming phased array antenna

Phased array antenna is a multiple antenna system in which antenna elements are fed coherently with variable phase or amplitude control to provide for pattern shaping [14]. The concept of phased antenna array was introduced in military applications in the 1940s and first used for military applications for several decades. It improved the reception and transmission patterns of antennas and enabled the antenna system to be electronically steered to receive or transmit information from a particular direction without mechanically moving the structure, which cannot be obtained by any single antenna type. Furthermore, in applications for long distance communication, an antenna design with very high directive characteristics is required. As a single antenna usually provide low directivity and gain with wide radiation pattern [15], an array of antennas with high directivity gain is usually used to meet that demand. Recent growth in civilian radar-based sensors and communication systems has drawn increasing interest in utilizing phased array technology for commercial applications. In most cases, the antenna elements in an array are identical to conveniently adjust the directivity and other parameters of the array. Besides, the antenna elements can be arranged in different geometries to create different beams. The complexity of system increases according to the array geometry. The total beam of the array is a combination of the fields radiated by each antenna elements. The fields of elements will interfere constructively to reinforce in desired directions and interfere destructively to cancel in undesired directions in order to provide directive patterns. The influence of element fields on total beam is considered as array factor and the beam steering principle is based on the change of array factor. In addition, feeding techniques also contribute to the array performance and feed network is the most important component since it supplies the signals to the whole antenna structure and determines the amplitude and phase of electromagnetic waves in order to create the desired beam.

2.1 Array geometry

As mentioned in [15], the geometrical configuration of the overall array may be linear, planar, circular, spherical, etc. For each case, the effective field distribution and mutual coupling will be different from one another. A linear phased array is where the antenna elements are aligned along a straight line, called the array axis. Normally, identical antennas with equal distances are selected in order to simplify calculation as well as beam control. An array of identical elements all of identical magnitude and each with a progressive phased is referred to as a uniform array [15]. The distance between two antenna elements is called element spacing d in **Figure 1a**. For linear phased array, the main beam can only scan along the x-axis, and the angle θ represents the "angle of arrival" of the radio waves. A planar phased array antenna is a set of antennas located in a plane with equal spacing between N elements (d_y) in each column and M elements (d_x) in each row, as shown in **Figure 1b**. By adjusting the magnitude and phase of incoming wave for each antenna, the main beam of this structure can be steered along two x- and y-axis. Therefore, it can provide a 2D angular scan, both horizontal ϕ and vertical θ scans. With this advantage, planar phased array antennas are being used in smart antenna

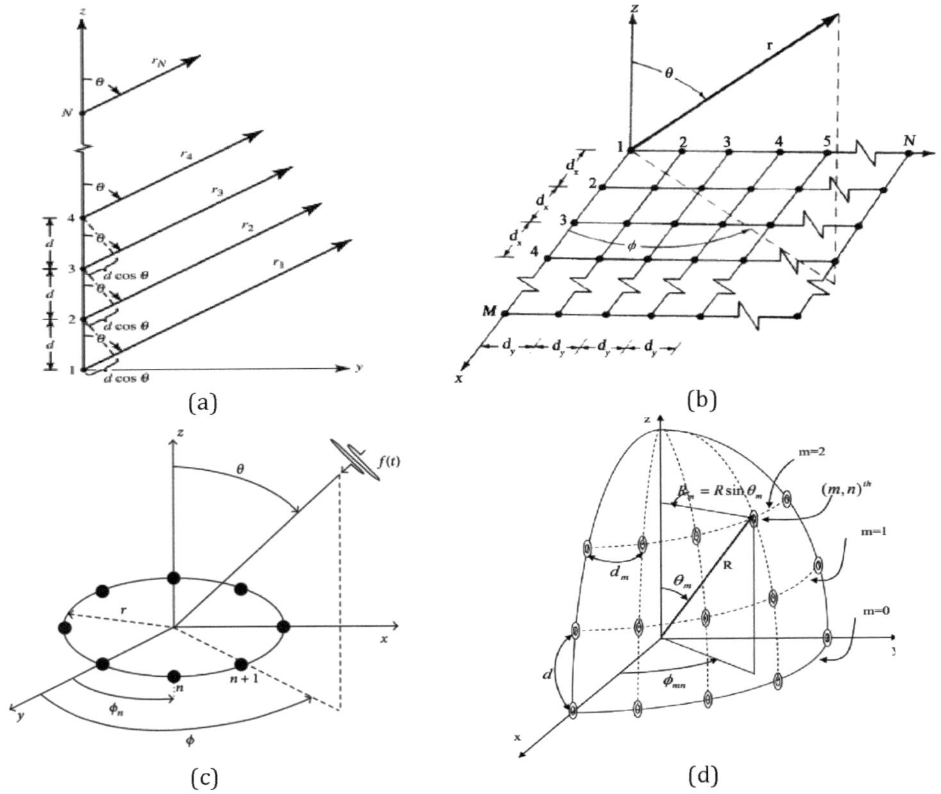

Figure 1.
Phased array geometry: (a) linear, (b) planar, (c) circular, and (d) spherical [15].

and beamforming antenna system. In circular phased array, antennas also lie on the same plane but are arranged on a circular ring of radius r with N equally spaced elements (**Figure 1c**). Circular arrays are basically 1D linear arrays but in circular form and can steer beam in 2D. A spherical phased array consists of antenna elements arranged on the surface of sphere with uniform or non-uniformly spacing (**Figure 1d**). The element spacing is defined as the distance between two adjacent elements along the curved surface. With such geometry, spherical phased array may radiate in any direction and achieve omnidirectional or isotropic coverage. However, due to difficulties in model, design as well as fabrication, this geometry seems to not receive much less attention than above geometries.

2.2 Array factor

Another important factor – Array Factor (AF) depends on some parameters such as the number of antennas in array, geometrical arrangements, relative magnitude, phase difference, and element spacing. It is a mathematical factor representing the relationship between total field of array and the field of single element. If E_s is the field of a single antenna and AF is the array factor, the total field E_{total} at the far-field of the array can be calculated as [15]:

$$E_{total} = E_s \times AF \quad (1)$$

In case the array of two dipole antennas with element spacing $d = \frac{\lambda}{4}$, the total fields are different corresponding to different phase excitation β shown in **Figure 2**.

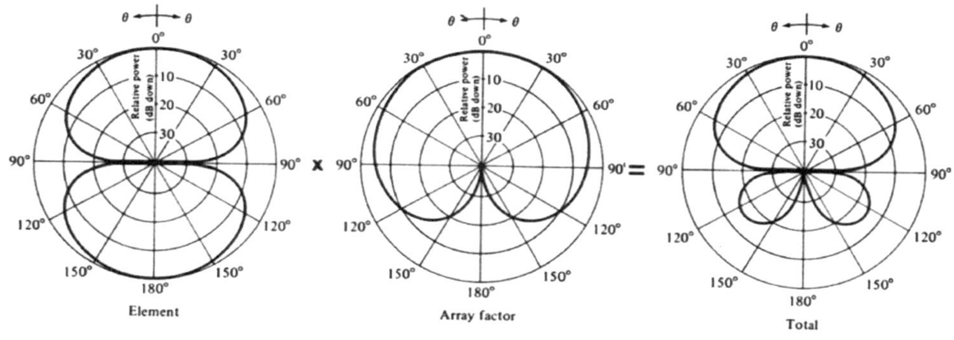

Figure 2.
Total field patterns of two dipole antenna array with λ/4 element spacing and different phase excitation β = −90° [15].

If a linear array has uniform amplitude, distances between two adjacent elements are identical with identical powers traveling to antennas. The AF is given as

$$AF = \sum_{n=1}^{N} e^{\,j(n-1)(kd\ cos\theta+\beta)} \tag{2}$$

where N is the number of antenna elements, d is the element spacing, β is the phase difference of incident waves to antennas, and θ changes from 0 to 2π. The maximum value of array factor is at an angle of $\theta_m = cos^{-1}\left(\frac{\lambda\beta}{2\pi d}\right)$, which enables to control the maximum value of AF through controlling the phase difference β. That is the method to steer the main beam in the phased array antenna system. For AoA-based indoor localization, the use of uniform amplitude and spacing linear array is enough. For non-uniform amplitude linear array, planar array, circular array, and spherical array, AF is quite complicated and not really necessary but already introduced in [10, 11].

2.3 Grating lobe

When steering main beam toward the desired direction, some side lobes happen to be substantially larger in amplitude and reach the level of the main lobe. The lobe of maximum radiation toward unintended direction is known as a grating lobe which is undesirable in phased array antenna applications. During the transmission, if the energy does not focus in the desired direction, that means it is allocated in the direction of the grating lobes, the transmission distance is significantly reduced. In smart antenna systems where the user's direction is determined, the presence of grating lobes can cause the system to misidentify the user that means while transmitting, unwanted users are also treated as expected user, thus it affects information security or interferes other users. In a whole, to avoid the grating lobes, element spacing d is required to meet the following condition:

$$\frac{d}{\lambda} \le \frac{1}{1 + |cos\theta|} \tag{3}$$

where λ is the wavelength of operating frequency and θ is the angle between main beam direction and z-axis **Figure 1a**. However, in this chapter, the angle θ = 90° is used as original angel of main beam direction, so angle of main beam will be (θ – 90°). The scanning main beam angle range is designed from −60 to 60°.

Expression (3) indicates that element spacing d in this case is smaller than 0.53λ. Therefore, to avoid grating lobes in scanning angle sector from −60 to 60°, an element spacing not more than 0.53λ is chosen.

2.4 Mutual coupling

When signals are transmitted to antenna arrays, the antenna elements will interfere each other, which is the so-called mutual coupling effect. The amount of coupling depends on the radiation characteristic, relative orientation of each antenna and spacing between elements. As discussed in [15], even if both antennas are transmitting, partial energy radiated will be received by other antennas because of the directional characteristics of practical antennas. Part of the incident energy on antenna elements may be backscattered in different directions, thereby allowing them to behave as secondary transmitters. In many cases, it is very complex to analyze and difficult to predict this effect but the coupling must be taken into account because of its significant contribution.

When two radiating elements are positioned along the E-plane, very small spacing exhibits the smallest coupling isolation, while the H-plane exhibits the small coupling for large spacing [15]. By selecting the correct distance, these fields can be decomposed to surface waves, and the spacing at which on plane coupling overtakes the other one also depends on the electrical properties and the geometrical dimensions of the microstrip antenna. In general, the element spacing should be designed to reduce the adverse effects of mutual coupling. According to [16], the spacing is recommended to be between 0.33λ and 0.5λ.

2.5 Feed network

The angle of radiation beams mainly depends on the range of phase difference at the feed lines to the antenna array. Meanwhile, the phase difference is directly generated from the feed network. In electronically scanned arrays or phased arrays, the feed networks for phase difference generation are typically realized using microwave circuit types such as Hybrid Coupler, Delay Line, Crossover, Power Divider, Phase Shifter, ... Generally, all types of feed networks can be classified into the categories: constrained feeds, space feeds, and hybrid feeds. In an example for the space feed network, a lens array is fed by a single horn antenna located at an expected distance from the array [14] with the phase control at every element in the lens. The main advantage of this configuration is to reduce the cost and weight of the system compared with using hybrid feed, therefore it is applicable to lower cost ground-based arrays as well as very large space-based radar and communication system. However, this configuration is quite complex and requires the precision mechanical system to use the phase control at the objective aperture. Therefore, we select constrained feed which is commonly used for antenna array. The constrained feed can then be categorized into two basic types: series feed and parallel feed.

In series feed, antenna elements are placed in series along the feed line, and phase shifters can be inserted series either antennas or feed line, as shown in **Figure 3a**. The input signal is fed from one end of the feed network and then coupled serially to each antenna element. The compactness and low loss are two main advantages that make series feed more attractive than parallel feed. Additionally, the number of required phase shifters is also less than ones in parallel feed. However, bandwidth limitation is the main disadvantage of series feed. As the feed line is also treated as a delay line, the phase shifts on feed line are different at different frequencies. Therefore, the series feed only operates at designed frequency. Moreover, when phase shifter is placed on the feed line, the loss through

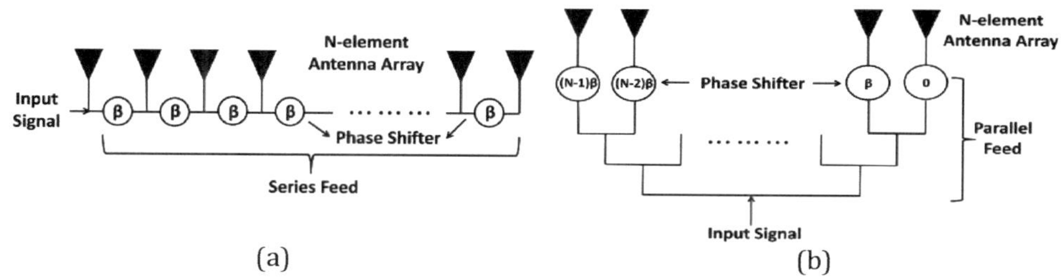

Figure 3.
(a) Series feed network for phased array antenna; (b) parallel feed network for phased array antenna [14].

phase shifters is also cumulative, which can be an issue in the design of array with a large number of elements. Calculating and controlling the phase shift value of the phase shifters become elaborate as they depend on the length of the feed lines.

In parallel feed, which are often called corporate feeds, the input signal is divided in a parallel tree network to all the antenna elements as shown in **Figure 3b**. The parallel feed is composed of power dividers and phase shifters followed by antenna elements. Power dividers split input signal to *N* signals with same amplitudes and same phases at output ports, then phase shifters control phase shift to each antenna. Thanks to the parallel structure, lengths of transmission lines do not affect phase shift. Therefore, it eliminates the major disadvantage of series feed, operating at only one frequency. Furthermore, because of parallel structure, the phase shifts and losses of phase shifters do not affect each other, which lead to simpler control circuit. However, the disadvantage of parallel feed is cumbersome as discussed above. Moreover, in applications requiring long transmission, transceiver modules including power amplifier, low noise power amplifier, and switch, are used. Sometimes, in array antenna system forming the beam based on amplitude control, power amplifiers with variable gain ability are needed. In practice, variable

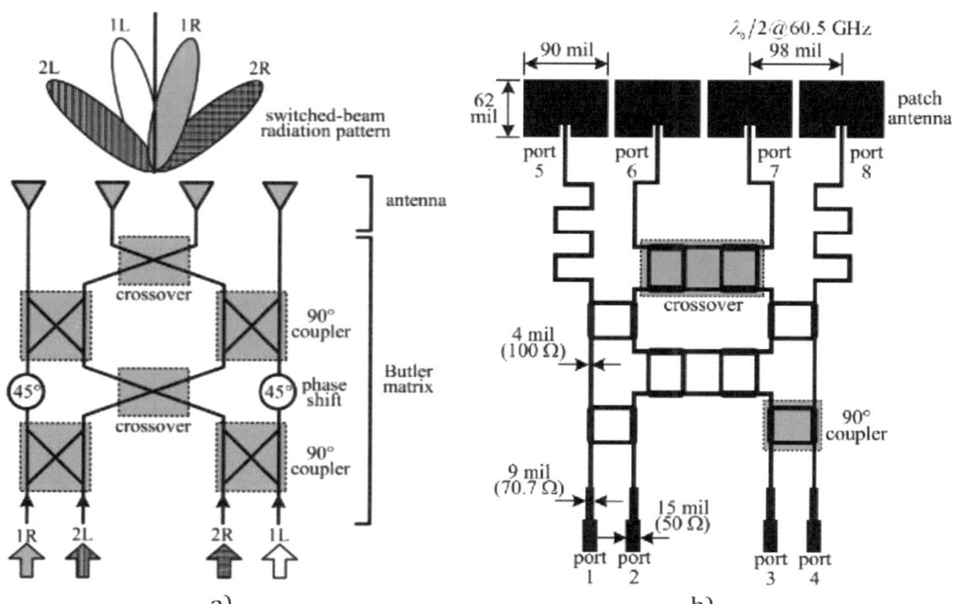

Figure 4.
4 × 4 Butler matrix network [20].

gain also enables to compensate the loss on phase shifter, of which loss value is not small.

Butler matrix is also an approach usually studied in phased array antennas [17–19]. The Butler matrix is a type of beamforming network and first described by Jesse Butler and Ralph Lowe in [20]. It has N inputs and N outputs; with N is usually 4, 8, and 16. One input signal fed from one of input ports goes through components of Butler matrix like couplers, phase shifters, crossovers in order to create phase difference of wave at N output ports, which combines with antenna elements to create beams with different predefined directions, as shown in **Figure 4**. Because beam direction is predefined, this system is also considered as a switched beam system. Nevertheless, there are two main drawbacks when using Butler matrix. First of all, the complexity and dimension significantly increase when N increases. That is why Butler matrix network is hardly designed with the number of input ports of 16 or more. Secondly, the number of beams is limited. With N inputs, this system can only create N beams with different directions, hence it is not suitable to provide beams with high resolution scan angle.

3. Practical phased array antenna design for indoor positioning

The indoor localization and positioning system depend considerably on the phased array antenna structure. The structure as well as parameters of phased array antenna such as array geometry, element spacing, and feed network needs to be carefully defined. For indoor positioning, we only care about the location of users and ignore altitude information, hence a linear array structure with 2D scanning can fully meet this requirement. Other architectures such as planar array, circular array, and spherical array are also options, but they make the system much more complex and cumbersome. With a straightforward beam steering principle, the uniform amplitude and spacing linear array is chosen.

Regarding the element spacing of linear array antenna system, it should be a suitable value in order to limit the mutual coupling, avoid grating lobes, and increase the directivity. As mentioned in Section 2, to reduce the adverse effects of mutual coupling, element spacing should be from 0.33λ to 0.5λ. With respect to the usage, the antenna system is aimed to be located at the corner of the room and the user will move inside the room, yielding the main beam angle from -45 to $45°$. Thus, from Eq. (3), the element spacing is chosen to be less than 0.58λ to prevent the appearance of grating lobes when scanning main beam angle from -45 to $45°$. In [21], Rabinovich and Alexandrov show the relationship between directivity and element spacing of array antenna (**Figure 5**). When element spacing d is less than wavelength λ, the directivity is proportional to d/λ. To sum up, the element spacing equals to half wavelength in my antenna system. Finally, with advantages compared to serial feed and butler matrix, the parallel feed network is selected for this beamforming phased array.

3.1 Power divider

A parallel feed network is composed of power dividers and phase shifters. The incident wave is split to multiple ports through power dividers. With design of uniform amplitude and spacing linear array, the power should be equally split to each output port. If the phases of outgoing waves are identical, the compensation of phase will not be required. In microwave, reducing the loss enables more power to feed into antennas, which increases the transmission range. Finally, the mismatch in microwave circuits causes the reflection waves, which can play another role as

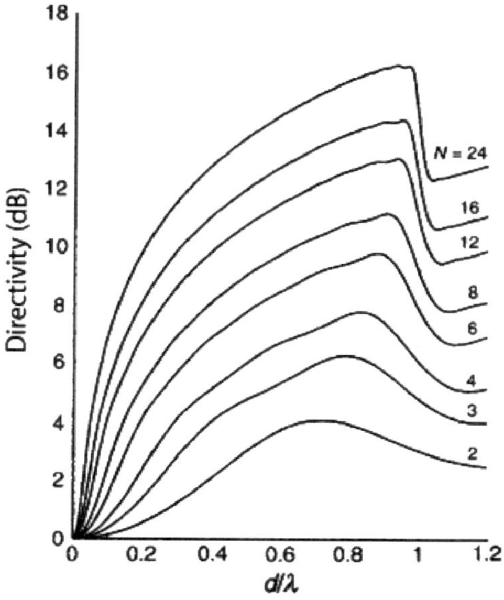

Figure 5.
Directivity as a function of the element spacing of linear array antenna [21].

incoming waves at output ports. Therefore, if output ports of power divider are highly isolated to each other, the reflected waves will not affect other ports.

Power divider is a passive component to split the input signal into two or more lower power output signals. Power dividers usually provide in-phase output signals with both equal power division ratio (3 dB) and unequal power division ratios. As described in [22], T-junction and Wilkinson are two kinds of most popular power dividers. While the T-junction dividers have some disadvantages related to impedance matching, power loss or isolation, the Wilkinson Power Divider (WPD) can simultaneously solve these problems. The WPD appears lossless when the output ports are matched. The structure of WPD, shown in **Figure 6**, comprises two quarter-wave transmission lines with impedance of $\sqrt{2}Z_0$ and a $2Z_0$ lumped-element resistor at the end. To analyze this circuit, the "even-odd" mode analysis

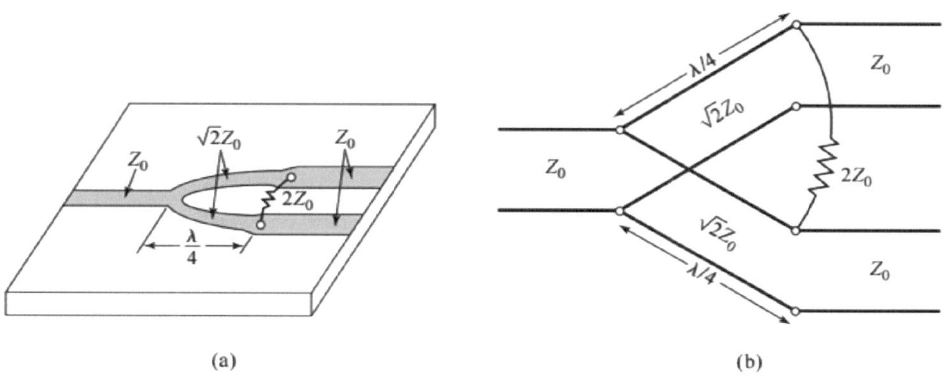

Figure 6.
The Wilkinson power divider: (a) microstrip line configuration, (b) equivalent transmission line circuit [22].

technique is often used, and the scattering matrix calculation is represented in [22]. Finally, the scattering matrix of WPD is:

$$S = \frac{-1}{\sqrt{2}} \begin{bmatrix} 0 & j & j \\ j & 0 & 0 \\ j & 0 & 0 \end{bmatrix} \tag{4}$$

We can see that the power from input port is split into two output ports without power loss on transmission line, there is no reflection wave from output ports and output ports are completely isolated. With advantages of WPD over T-junction dividers, the WPD is utilized as the power divider for feed network of antenna array. In [22], professor Pozar introduced the N-way WPD (**Figure 7**) in which the transmission line has impedance depending on the number of output ports. This circuit can be matched at all ports, with isolation between all ports. The drawback of this structure is to use crossovers for the resistors for $N \geq 3$, which makes fabrication difficult in planar form. The WPD is arranged in cascade structure for low loss and better isolation, shown in **Figure 8**. Two-way WPD is connected together by 50 Ω transmission

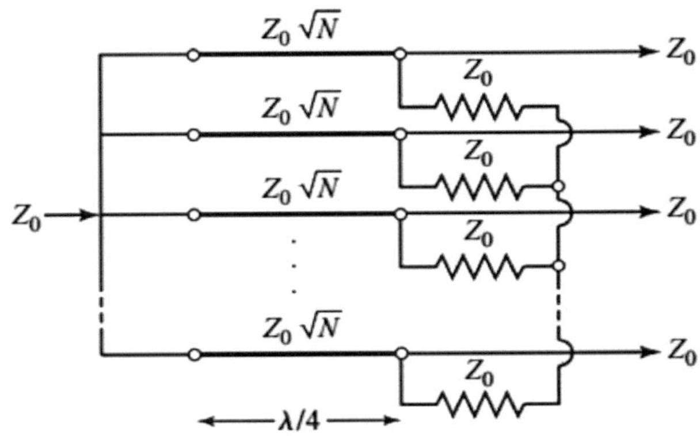

Figure 7.
An N-way, equal-split Wilkinson power divider [22].

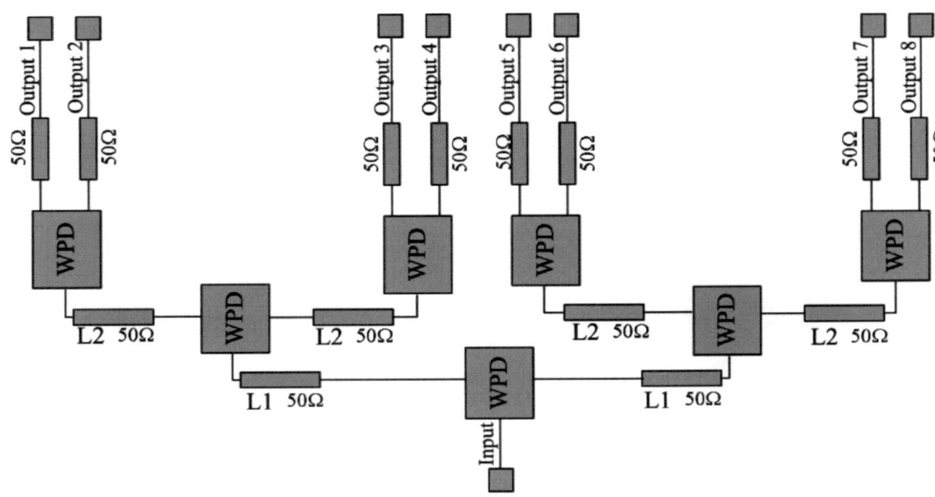

Figure 8.
The eight-way equal-split Wilkinson power divider [23].

lines in a three-stage structure. The 50 Ω transmission lines allow to arbitrarily adjust distances between output ports in order to satisfy the dimension of phase shifters and element spacing between antennas in phased array antenna system.

3.2 Phase shifter

3.2.1 Phase shifter theoretical calculation

In parallel feed network, the main role of phase shifter is to create the phase difference of waves coming to the antennas in order to scan a beam or to reconfigure a beam shape. Three characteristics are required for this phase shifter. Firstly, it must be capable of shifting phase of waves in full 360° range in order to meet all of the demands of the phase difference in phased array antenna system. Secondly, continuously shifting makes it possible to create any phase, which enables steering main beam with high resolution. Finally, with a uniform amplitude and spacing linear array, when power divider equally splits power to output ports, a phase shifter with low insertion loss variation is very important. Without this property, our linear array should be treated as a non-uniform amplitude linear array, which is more complex in beam steering principle.

Depending on the required output, phase shifters are classified into analog phase shifter and digital phase shifter. Digital phase shifters with semiconductor components such as PIN diode and FET switch predefined states to provide predetermined phase shift. The phase shift, generated by this type, has high accuracy. However, structure of this type will become cumbersome to meet high-resolution demands. With high resolution, the number of predefined states increases, resulting in the expansion of the number of switching elements as well as size of controller for them. Meanwhile, analog phase shifter with the use of varactors or Schottky diodes can continuously change the phase shift. These diodes have capacitance depending on the bias voltage, for this reason, they can be used as electrically variable capacitor in tuned circuit accordingly. By shifting the phase continuously, this type can provide high-resolution phase shift without changing its hardware structure. However, less accuracy and relatively narrow bandwidth are drawbacks of this type. On implementing a microwave phase shifter, these disadvantages must be enhanced for better performance.

There exist four types of phase shifters, as shown in **Figure 9**. One of the digital phase shifter, the switched-line one, adopts delay lines and switching elements to generate time delay differences.

The phase shift depends only on transmission line length; therefore, it is very stable over time and temperature. A basic schematic of switched line phase shifter is shown in **Figure 9a**. The second type, switched network phase shifter (**Figure 9b**), is similar to switched line phase shifter but delay lines are replaced by networks composed of inductors and capacitors. The dimension of this type does not change as much as switched line phase shifter, besides this type is suitable for low frequency design. Loaded line phase shifters (**Figure 9c**) are loaded with a shunt reactance that is electrically shortened or lengthen by PIN diode or FET in order to get the desired phase shift. This type has advantages of simplicity and low insertion loss for phase shift less than 45°. However, for larger values of phase shift, high sensitivity is required in order to increase insertion loss. Therefore, this type is only suitable for phase shift less than 45°. Reflection type phase shifter comprises a 3-dB hybrid couple and two tunable loads, as shown in **Figure 9d**. By selecting the appropriate load, this type can shift more than 360° continuously and has low insertion loss, like in [24–26]. From requirement of phase shifter, it is obvious that reflection type phase shifter with full 360° and continuous phase shift is the most appropriate choice for this phased array antenna system.

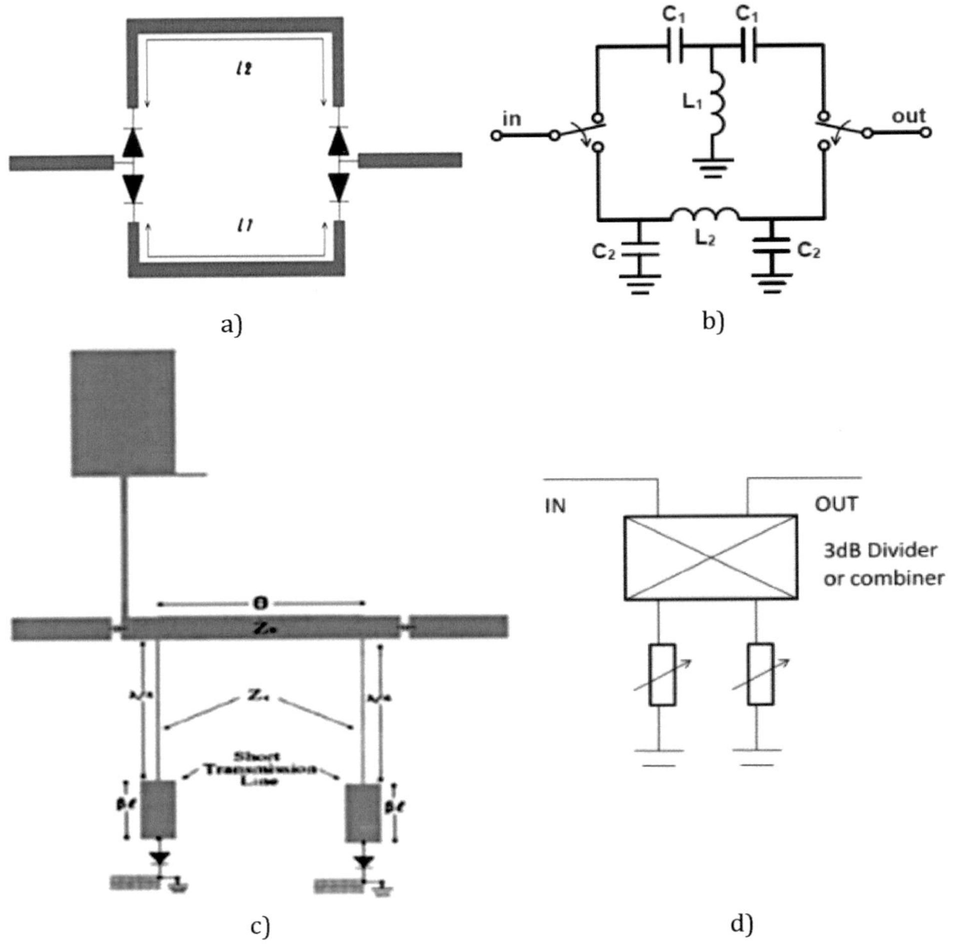

Figure 9.
Types of phase shifter: (a) switched line; (b) switched network; (c) loaded line; (d) reflection type [24–26].

The Reflection Type Phase Shifter (RTPS), an electrically adjustable phase shift, is made up of a 3-dB hybrid coupler combined with some components such as capacitors, resistors, transmission line, and varactors to eliminate the reflected wave at input port and shift phase of wave at output port. The 3-dB hybrid coupler is a 4-port network with all ports matched to the reference impedance of 50 Ω. When stimulating at port 1, incoming wave transmits to port 3, port 4, and is

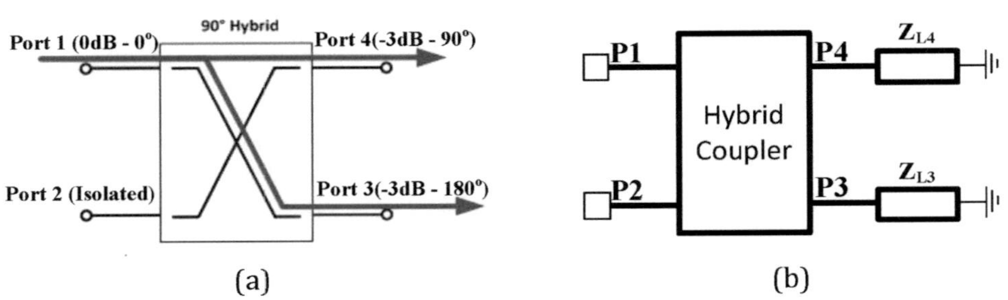

Figure 10.
(a) 3-dB hybrid coupler; (b) structure of RTPS.

suppressed at port 2. The amplitude of waves at ports 3 and 4 equal to $\frac{1}{\sqrt{2}}$ one of incoming wave at port 1 and phase of waves at port 3 and 4 shift 90° and 180° compared with one at port 1, respectively (**Figure 10a**).

In mathematical model, 3-dB hybrid coupler is characterized by a scattering matrix that represents voltage relationship between incoming (V+) and reflected (V) waves at different ports.

$$[V^-] = [S][V^+] \tag{5}$$

$$[S] = \frac{1}{\sqrt{2}}\begin{bmatrix} 0 & 0 & -j & -1 \\ 0 & 0 & -1 & -j \\ -j & -1 & 0 & 0 \\ -1 & -j & 0 & 0 \end{bmatrix} \tag{6}$$

Here, S_{ij} is ratio of the reflected wave at port i to the incoming wave at port j when there is only input wave from port j.

$$[S_{ij}] = \left.\frac{V_i^-}{V_j^+}\right|_{V_k^+=0, k\neq j} \tag{7}$$

To be more specific, we have:

$$\begin{bmatrix} V_1^- \\ V_2^- \\ V_3^- \\ V_4^- \end{bmatrix} = \begin{bmatrix} \left(\frac{-j}{\sqrt{2}}V_3^+\right) + \left(\frac{-1}{\sqrt{2}}V_4^+\right) \\ \left(\frac{-j}{\sqrt{2}}V_3^+\right) + \left(\frac{-1}{\sqrt{2}}V_4^+\right) \\ \left(\frac{-j}{\sqrt{2}}V_1^+\right) + \left(\frac{-1}{\sqrt{2}}V_3^+\right) \\ \left(\frac{-1}{\sqrt{2}}V_1^+\right) + \left(\frac{-j}{\sqrt{2}}V_2^+\right) \end{bmatrix} \tag{8}$$

Considering the reflection coefficient (Γ) generated by the load, the reflection coefficient is determined from the load impedance and the source impedance.

$$\Gamma = \frac{V^-}{V^+} = |\Gamma|e^{j\phi} \tag{9}$$

From the two components constituting the RTPS, the 3-dB hybrid coupler and the reflective load mentioned above, the model of RTPS is shown as in **Figure 10b**, with port 1 as incoming port, port 2 as outgoing port, and two loads Z_{L3} and Z_{L4} at port 3 and 4, respectively. The phase shift of RTPS is the phase difference between outgoing wave at port 2 compared with incoming wave at port 1. Suppose there is only an incoming wave at port 1. When passing through the 3-dB hybrid coupler, from (8), waves at ports 3 and 4 are:

$$V_3^+ = -\frac{j}{\sqrt{2}}V_1^+; V_4^+ = -\frac{1}{\sqrt{2}}V_1^+ \tag{10}$$

Because of appearance of ZL3 load at port 3, there is wave reflection with reflection coefficient Γ3. Therefore, from (9) and (10), the reflection wave at port 3 is:

$$V_3^- = \Gamma_3 V_3^+ = -\frac{j}{\sqrt{2}} \Gamma_3 V_1^+. \tag{11}$$

Similarly, we have the reflected wave at port 4:

$$V_4^- = \Gamma_4 V_4^+ = -\frac{j}{\sqrt{2}} \Gamma_4 V_1^+ \tag{12}$$

The reflected waves at ports 3 and 4 can be also regarded as incoming waves from ports 3, 4 transmitting to ports 1 and 2. We set $V_3'^+$ and $V_4'^+$ as incoming wave at ports 3 and 4, respectively. We have:

$$V_3'^+ = V_3^- = -\frac{j}{\sqrt{2}} \Gamma_3 V_1^+ \tag{13}$$

$$V_4'^+ = V_4^- = -\frac{j}{\sqrt{2}} \Gamma_4 V_1^+ \tag{14}$$

These two waves transmit to ports 1 and 2 through 3 dB hybrid coupler, and outgoing waves from ports 1 and 2 are calculated from (6), (11), and (12) as follows:

$$\begin{aligned} V_1^- &= -\frac{j}{\sqrt{2}} V_3'^+ - \frac{1}{\sqrt{2}} V_4'^+ = -\frac{1}{2}\left(\Gamma_3 V_1^+ - \Gamma_4 V_1^+\right) \\ V_2^- &= -\frac{1}{\sqrt{2}} V_3'^+ - \frac{j}{\sqrt{2}} V_4'^+ = \frac{j}{2}\left(\Gamma_3 V_1^+ + \Gamma_4 V_1^+\right) \end{aligned} \tag{15}$$

For the phase shifter, we wish that there is no reflected wave at port 1. Therefore, $\Gamma_3 = \Gamma_4$ enables us to eliminate the reflected wave at port 1. From (9) and (13), choosing $Z_{L3} = Z_{L4}$ makes $\Gamma_3 = \Gamma_4 = \Gamma$ and outgoing wave at port 2 as $V_2^- = j\Gamma V_1^+$. Finally, the forward voltage gain S_{21} of RTPS will be as follow:

$$\begin{aligned} S_{21} &= \frac{V_2^-}{V_1^+} = j\Gamma = j|\Gamma| e^{j\phi} = |\Gamma| e^{j\left(\frac{\pi}{2}+\phi\right)} \\ \angle S_{21} &= \frac{\pi}{2} + \phi \end{aligned} \tag{16}$$

3.2.2 Design of reflection type phase shifter

The reflection type phase shifter (RTPS) is a phase shifter, empowering continuous change of the phase of wave without the need to change hardware structure, based on the change in capacitance of varactors. **Figure 11** shows the principle diagram of the RTPS, which consists of a 3-dB hybrid coupled and reflection loads X_L putting at ports 3 and 4. Waves enter port 1 and go out port 2.

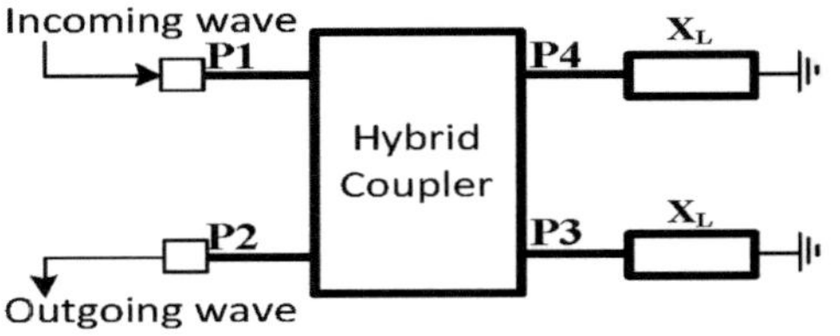

Figure 11.
Schematic diagram of RTPS.

When putting reactive loads to ports 3 and 4, the reflection coefficient is:

$$\begin{aligned} \Gamma &= \frac{jX_L - Z_0}{jX_L + Z_0} = \frac{(jX_L - Z_0)^2}{(jX_L + Z_0)(jX_L - Z_0)} = \frac{1}{Z_0^2 + X_L^2}(jX_L - Z_0)^2 \\ &= \frac{1}{Z_0^2 + X_L^2}\left(\sqrt{Z_0^2 + X_L^2}\, e^{j\arctan\left(-\frac{X}{Z_0}\right)}\right)^2 = e^{j2\arctan\left(-\frac{X}{Z_0}\right)} \end{aligned} \tag{17}$$

From (14) and (15), the phase shift of RTPS with reactive loads is:

$$\angle S_{21} = \frac{\pi}{2} - 2\arctan\left(\frac{X_L}{Z_0}\right) \tag{18}$$

From (16), we found that the range of phase shift values $\angle S_{21}$ entirely depends on the capacitance range of varactors, which limits the phase shift range. To be more specific, in order to achieve the 360° phase shift, theoretically, the capacitance of varactors must reach infinity, which is impossible. Therefore, RTPS cannot get the full 360° with only one varactor on reflection loads. To overcome the above problem, the reflection load structure shown in **Figure 12** is employed. This endows us to choose the appropriate capacitance range, thus selecting the suitable varactors and its voltage as follows:

Based on the transmission line impedance theory [27], because Z_{T2} is a quarter-wavelength transmission line, the equivalent impedance of Z_{T2} and D_2 is:

$$Z_2 = \frac{Z_{T2}^2}{Z_d} \tag{19}$$

The equivalent circuit becomes (**Figure 13**):

Similarly, Z_{T1} is also a quarter-wavelength, so the equivalent impedance is:

$$Z_L = Z_{T1}^2\left(\frac{1}{Z_d} + \frac{1}{Z_2}\right) \tag{20}$$

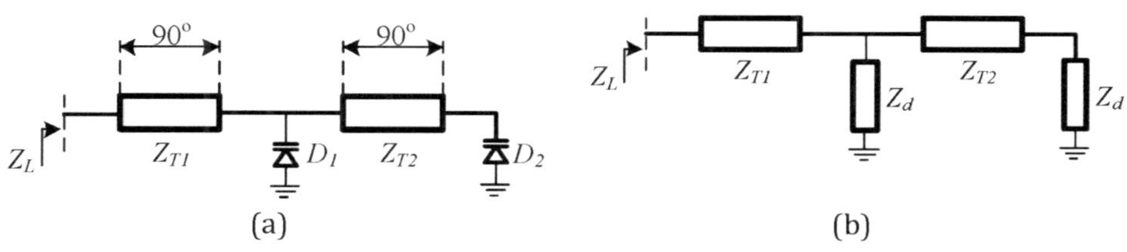

Figure 12.
(a) Reflection load of RTPS; (b) the equivalent circuit of reflection load.

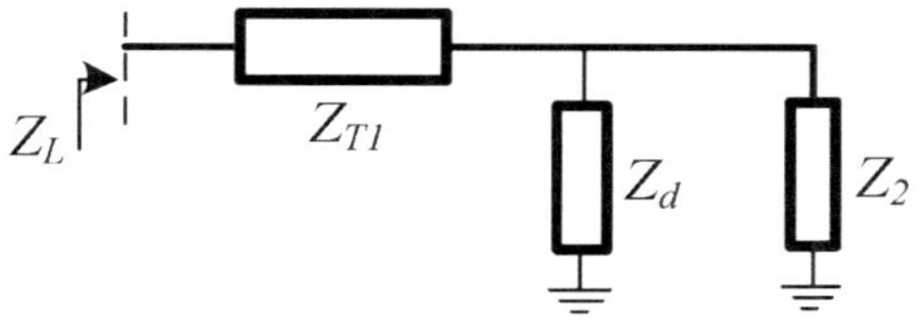

Figure 13.
The equivalent circuit of final RTPS.

Substitute (17) to (18):

$$Z_L = Z_{T1}^2\left(\frac{1}{Z_d}+\frac{1}{Z_2}\right) = Z_{T1}^2\left(\frac{1}{Z_d}+\frac{Z_d}{Z_{T2}^2}\right) = \frac{Z_{T1}^2}{Z_{T2}^2}\frac{Z_{T2}^2+Z_d^2}{Z_d} \tag{21}$$

On the other hand, the impedance of the capacitance diodes is:

$$Z_d = j\omega L_s + \frac{1}{j\omega C_d} = j\left(\omega L_s - \frac{1}{\omega C_d}\right) = jX_d \tag{22}$$

Substitute (20) to (19):

$$Z_L = jX_L = \frac{Z_{T1}^2(Z_{T2}^2 - X_d^2)}{Z_{T2}^2 jX_d} = j\frac{Z_{T1}^2(X_d^2 - Z_{T2}^2)}{Z_{T2}^2 X_d} \tag{23}$$

Substitute (21) to (16):

$$\angle S_{21} = \frac{\pi}{2} - 2arctan\left(\frac{X_L}{Z_0}\right) = \frac{\pi}{2} - 2arctan\left(\frac{Z_{T1}^2(X_d^2 - Z_{T2}^2)}{Z_{T2}^2 X_d Z_0}\right) \tag{24}$$

In order to get 360° phase shift, we have:

$$\angle S_{21}|_{X_{dmin}} = \angle S_{21}|\,X_{dmax}$$

$$\Leftrightarrow \frac{\pi}{2} - 2arctan\left(\frac{Z_{T1}^2(X_{dmin}^2 - Z_{T2}^2)}{Z_{T2}^2 X_{dmin} Z_0}\right) = \frac{\pi}{2} - 2arctan\left(\frac{Z_{T1}^2(X_{dmax}^2 - Z_{T2}^2)}{Z_{T2}^2 X_{dmax} Z_0}\right)$$

$$\Leftrightarrow \frac{Z_{T1}^2(X_{dmin}^2 - Z_{T2}^2)}{Z_{T2}^2 X_{dmin} Z_0} = \frac{Z_{T1}^2(X_{dmax}^2 - Z_{T2}^2)}{Z_{T2}^2 X_{dmax} Z_0}$$

$$\Leftrightarrow X_{dmin} - \frac{Z_{T2}^2}{X_{dmin}} = X_{dmax} - \frac{Z_{T2}^2}{X_{dmax}}$$

$$\Leftrightarrow (X_{dmin} - X_{dmax})\left(1 + \frac{Z_{T2}^2}{X_{dmin}X_{dmax}}\right) = 0$$

Since $X_{dmin} \neq X_{dmax}$, we have:

$$Z_{T2}^2 = -X_{dmin}X_{dmax} \tag{25}$$

Thus, by calculation of Z_{T2} satisfying (23), we will have 360° phase shift.
From (23), we have: $X_{dmin}X_{dmax} < 0$

$$\Leftrightarrow \left(\omega L_s - \frac{1}{\omega C_{dmin}}\right)\left(\omega L_s - \frac{1}{\omega C_{dmax}}\right) < 0$$

$$\Leftrightarrow \begin{cases} C_{dmin} < \dfrac{1}{\omega^2 L_s} \\ C_{dmax} > \dfrac{1}{\omega^2 L_s} \end{cases}$$

$$\Rightarrow \frac{1}{\omega^2 L_s} \in [C_{dmin}; C_{dmax}] \tag{26}$$

At 2.45 GHz, for impedance matching, the varactor diode SMV1247, package type SC-79 of Skyworks is chosen with L_s = 0.7nH. To simplify the voltage controller of the varactors, the model capacitance range of SMV1247 from 8.5 pF to 0.7 pF is selected corresponding to the voltage range from 0 V to 5 V. From (23), Z_{T2} = 15 Ω.

For the impedance of Z_{T1}, besides calculating the entire phase shift, the change of phase shift needs to be taken into account. It is desired that the relationship between the control voltage and the phase shift becomes as close to linear as possible, so that controlling phase shift with small resolution is easier. Thus, it is important to examine the expression $\frac{d\angle S_{21}}{dV_R}$ where V_R is the bias voltage of varactors and $\angle S_{21}$ is the phase shift. As for SMV1247, the relationship between bias voltage and capacitive value as follows:

$$C_d = \frac{C_{J0}}{\left(1+\frac{V_R}{V_J}\right)^M} + C_P \tag{27}$$

where C_{J0} = 8.47 pF, V_J = 80 V, M = 70, C_P = 0.54 pF are package parameters from varactor SMV1247's datasheet.

Substitute (25) to (22), the relational expression between $\angle S_{21}$ and V_R is:

$$\begin{cases} \angle S_{21} = \frac{\pi}{2} - 2\,arctan\left(\frac{Z_{T1}^2\left(X_d^2 - Z_{T2}^2\right)}{Z_{T2}^2 X_d Z_0}\right) \\ X_d = \omega L_s - \frac{1}{\omega C_d} \\ C_d = \frac{C_{J0}}{\left(1+\frac{V_R}{V_J}\right)^M} + C_P \end{cases} \tag{28}$$

Derivative $\angle S_{21}$ in VR:

$$\frac{d\angle S_{21}}{dV_R} = \frac{2Z_{T1}^2 Z_{T2}^2 Z_0 M C_{J0}\left(X_d^2 + Z_{T2}^2\right)}{\omega V_J C_d^2 \left(1+\frac{V_R}{V_J}\right)^{M+1}\left[Z_{T2}^4 Z_0^2 X_d^2 + Z_{T1}^4\left(X_d^2 - Z_{T2}^2\right)^2\right]} \tag{29}$$

where ω, V_J, C_d, Z_{T2}, M, C_{J0}, Z_0, L_s, C_P are known, Z_{T1} is unknown, and V_R and $\angle S_{21}$ are variables.

From expression (27), to see the effect of Z_{T1} on the phase shift, different values of Z_{T1} are investigated by MATLAB to find the $\frac{d\angle S_{21}}{dV_R}$ with smallest maximal point. Firstly, Z_{T1} is roughly experimented with different values from 10 to 80 Ω with the spacing of 10 Ω. From results in **Figure 14**, the value of Z_{T1} can be found around 20 Ω.

Because the value of Z_{T1} is found around the value of 20 Ω, the value of Z_{T1} is examined at a narrower range from 15 to 22 Ω with the spacing of 1 Ω (**Figure 15**).

Among the surveyed values of Z_{T1}, the value of 19 Ω gives the best result and that is the best-chosen value. Thus, the RTPS will be implemented with the load topology, where the impedance of the two microstrip circuits is Z_{T1} = 19 Ω and Z_{T2} = 15 Ω, respectively. The varactors are Skyworks SMV1247 with capacitance from 8.5 to 0.7 pF corresponding to bias voltage from 0 to 5 V.

3.2.3 Direct current (DC) feed and DC block

As mentioned above, varactors are the elements in microwave circuit, and their capacitance value is controlled by a DC bias from controllers. With respect to the

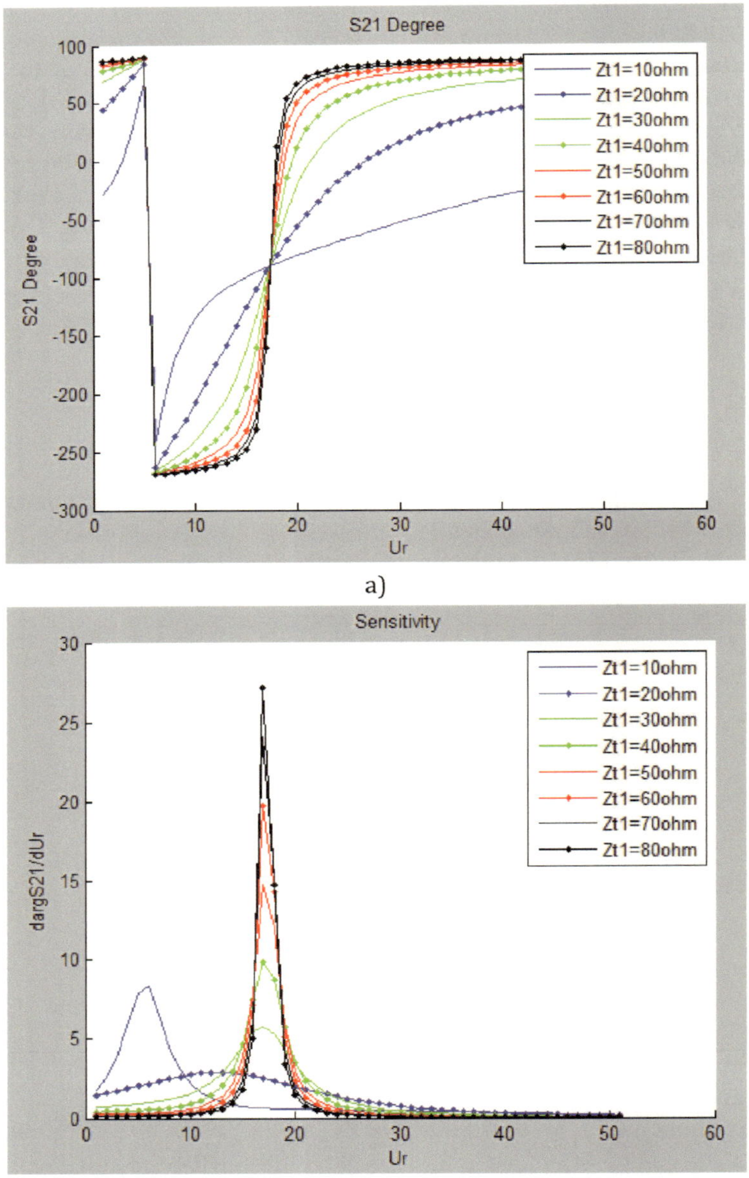

a)

(b)

Figure 14.
The results from the first survey of Z_{T1} *value: (a) phase shift, (b)* $\frac{d\angle S_{21}}{dV_R}$.

principle of microwave circuit, the varactors directly connect to transmission lines. From controller perspective, the DC bias is put to at two polar, anode and cathode, of varactors. Therefore, it is easily found that the microwave circuit and the controller are physically connected to each other, which leads to that high frequency waves can propagate to controller, and DC current can go to microwave circuit. When propagating to controller, controller can be considered as a new part in microwave circuit, so characteristics of our design might change unpredictably. For microwave circuit with reference impedance of 50 Ω, the maximal DC bias for varactors is 20 V, is corresponding to the power of approximately +36 dBm. This

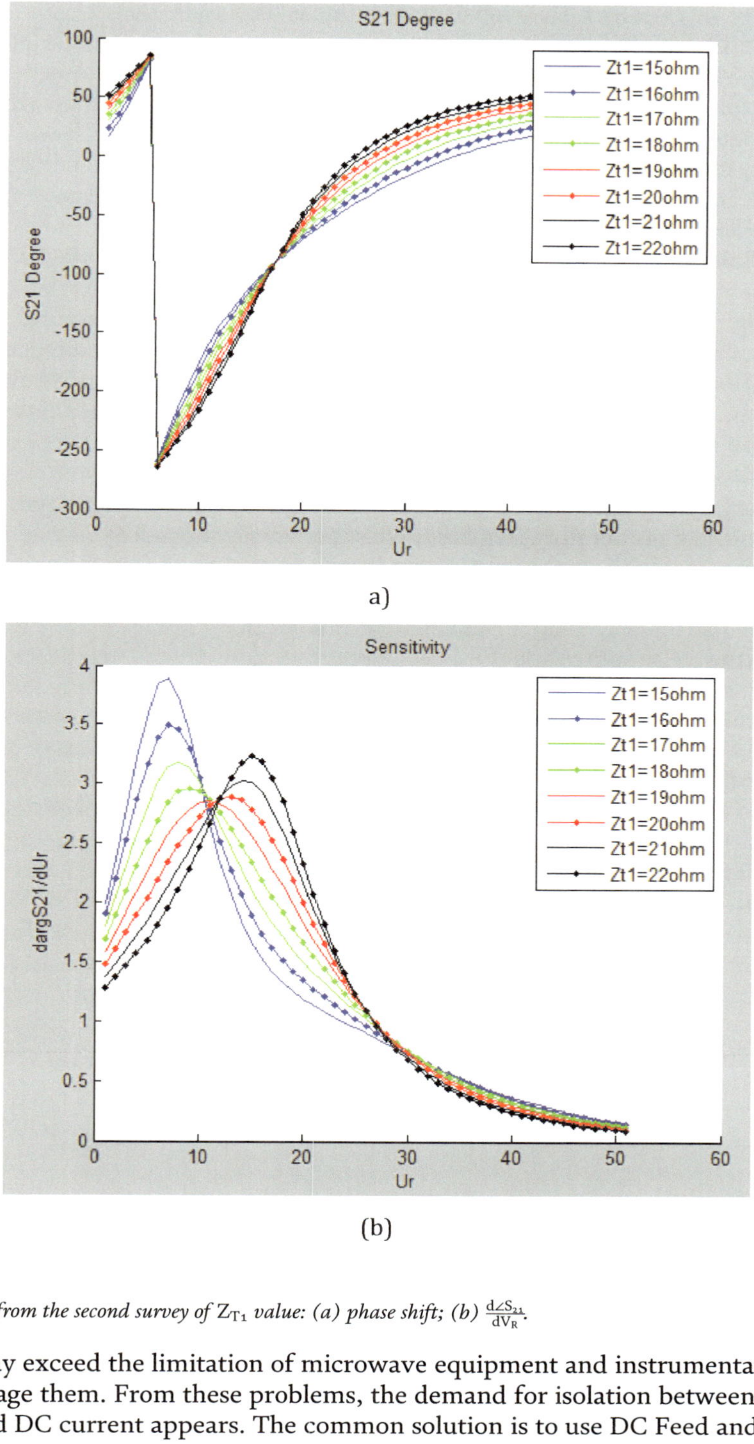

Figure 15.
The results from the second survey of Z_{T1} *value: (a) phase shift; (b)* $\frac{d\angle S_{21}}{dV_R}$.

value may exceed the limitation of microwave equipment and instrumentation and can damage them. From these problems, the demand for isolation between microwave and DC current appears. The common solution is to use DC Feed and DC Block components.

DC Block is usually a capacitor, which prevents DC current, but permits microwave circuit to pass easily. The impedance of DC Block must be very low (<5 Ω) compared with reference impedance (50 Ω) at operating frequencies (2.45 GHz). To deploy capacitors on microstrip, gap and microstrip interdigital capacitors may be options. However, with impedance less than 5 Ω, the capacitance must be larger than 13.2 pF at 2.45GHz. That capacitance value is hard to be met by gaps or microstrip interdigital capacitors. For capacitor components, in practice, at microwave frequency, a capacitor component is equivalent to a series of inductor, resistor, and capacitor, so the impedance of DC Block changes in frequency. Above a certain frequency threshold, that DC Block will act as an inductor. Thus, DC Blocks are usually capacitors with its self-resonant frequency near operating frequency, 2.45 GHz. Finally, the DC Block is VJ0603D8R2CXP capacitor of Vishay/Vitramon with impedance about 3.8 Ω at 2.45 GHz, as shown in **Figure 16**.

In contrast, the DC feed is usually an inductive element, which blocks high frequency waves and passes DC current. The impedance of DC Feed element must be very high compared with the reference impedance (50 Ω) at operating frequency, 2.45 GHz. Similar to DC Block, the DC Feed may be an inductance element with a self-resonant frequency near 2.45 GHz operating frequency. However, in practice, due to cost constraint and the availability of inductors, these inductors are replaced by another option, taking advantage of some special case in length of transmission lines. To minimize the whole structure, the quarter-wavelength transmission line used is shorted at the end, of which the equivalent impedance is positive infinity at operating frequency. In other words, a quarter-wavelength transmission line with shorted at the end can be treated as an open circuit at the beginning of that transmission line. Consequently, the DC bias can be replaced at the end of quarter-wavelength transmission line, so that high frequency waves do not reach the DC circuit while the DC currents can flow directly to varactors.

There exist two ways to short a circuit. The first way is to use via holes to create physical connection from electric lines to the ground; however, this method also shorts the DC bias. The other way is to employ open quarter-wavelength straight stubs or radial stubs. These two kinds of stubs do not physically connect to the ground, so they do not short the DC bias. In [28], Gardner and Wickert mentioned that radial stubs, possibly realized as shunt stubs with low characteristic impedance, may avoid the problems of transverse resonance and poorly defined point of attachment associated with straight stubs. In our DC Feed, the radial stubs are the suitable choice. Some research groups have studied about the equations for the formulation of microstrip radial stubs [27, 29–31]. Based on these studies combined with optimization tool in Advanced Design System software, the optimized

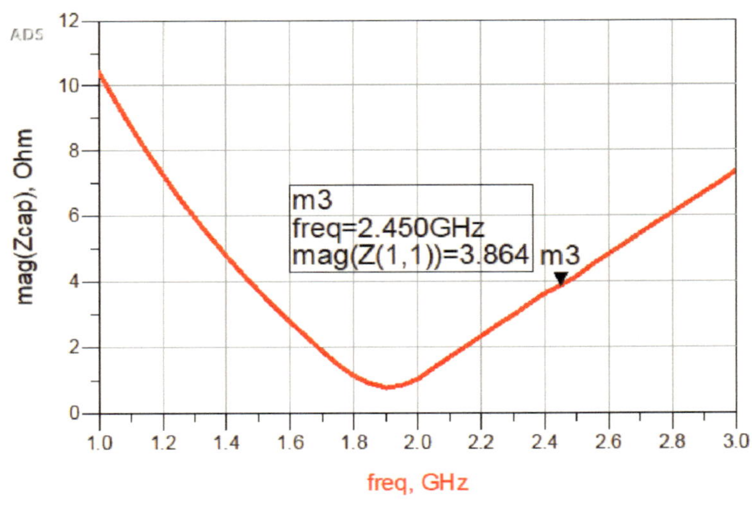

Figure 16.
Impedance of DC block VJ0603D8R2CXP.

parameters calculated for the Roger4003c substrate radial stubs are: the outer radius R_o = 540.621 mil and the inner radius R_i = 56.688 mil and angle α = 30°.

3.2.4 Design of controller for reflection type phase shifter

In phase array antennas, in order to be able to steer the antenna's main lobe in different directions, the phase-shift system must be placed in front of the antenna elements to generate the phase difference of the wave to each antenna so as to create the angle of beam. The phase shift entirely depends on the capacitance value of the varactors whereas every varactor always has its own C – V characteristic curve, representing the relationship between the input applied voltage and output capacitance. In other words, the phase shift is controlled by the voltage applied to the anode and cathode pins of varactors. For SMV1247 varactor, the voltage range of 0–5 V is applied. Thus, the controller is required to generate adjustable voltage in the range of 0–5 V, providing eight different voltages for eight RTPS and each voltage channel must meet the total power of four varactors in a RTPS.

To generate a DC voltage controlled by the microcontroller, popular methods are to modulate pulse width or control open angle of power semiconductor components such as Triac or Thysistor to convert AC to DC. These two methods are commonly used in power circuits, so we usually only care about the average power and voltage in a cycle, but in essence, the voltage generated is not flat over time but it is ripple. Obviously, these two methods will not be able to generate the control voltage for the varactors, which constantly changes the capacitance and results in undesired phase shifts. From the datasheet of SMV1247 varactor, the reverse current is very small, just a few nA to μA, so it is possible to use Digital to Analog Converter (DAC) units. Output current of these elements is about 20 mA. It is enough to satisfy the reverse currents of four varactors in a RTPS. In addition, the DACs have a variety of resolution options ranging from 8 bits to 24 bits. This facilitates flexible adjustment of the RTPS resolution. Therefore, the controller will be designed based on the eight DAC elements to generate bias voltages for eight RTPS. The controller model is described in **Figure 17**. The main components are DACs, and in order to increase the accuracy of the output voltage, the DACs are powered by a different source with a higher accuracy than the source for the microcontroller or can be adjusted for backup case when higher voltage range of varactors is demanded.

3.2.5 Design of antenna element

Because the positioning system is executed indoor, the WiFi or Wireless LAN should be the most suitable protocol to use in order to radiate power to indoor mobile devices; therefore the 2.4 GHz to 2.484 bands is chosen as the operating frequency. On the other hand, in order to communicate with the mobile devices for localizing the position of the object, the angel of main beam of phased array antenna should change from −45 to 45° to scan the desired object, hence antenna element must have half power beam width [15] greater than 90°. These requirements demonstrate that the radio signal strength comparison completely depends on the array factor. Omnidirectional antennas can satisfy both requirements; however, some omnidirectional antennas such as dipoles, ... [15] still have some undesired effects, that is, radiating to the back side of the antenna array, which results in some high side back lobes due to the reflection. The microstrip patch antenna can both meet the above condition and have small back lobes. In our array, the microstrip patch antenna was designed to operate at Wi-Fi band frequency, 2.4 to 2.484 GHz with an input impedance of 50 Ω using a low cost FR4 substrate with dielectric constant ε = 4.3, loss tangent tanδ = 0.02 and thickness h = 62 mil = 1.58 mm. The antenna

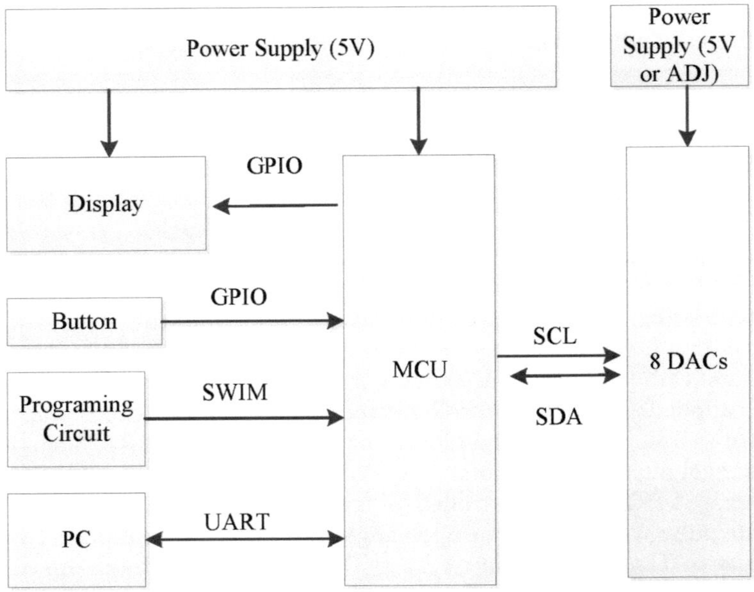

Figure 17.
Block diagram of varactor diode controller.

width W and length L are computed with the following Equation [15, 32, 33] where f is center frequency (2.45 GHz):

$$W = \frac{3 \times 10^{11}}{2 \times f} \sqrt{\frac{2}{\varepsilon + 1}} = 36.60\ mm \tag{30}$$

$$L = L_{eff} - 2 \times \Delta L = 28.07\ mm \tag{31}$$

$$\text{where} \begin{cases} L_{eff} = \dfrac{3 \times 10^{11}}{2 \times f \times \sqrt{\varepsilon_{eff}}} = 29.53\ mm \\ \Delta L = 0.412 \times h \times \dfrac{\varepsilon_{eff} + 0.3}{\varepsilon_{eff} - 0.258} \times \dfrac{\frac{W}{h} + 0.264}{\frac{W}{h} + 0.8} = 0.73\ mm \\ \varepsilon_{eff} = \dfrac{\varepsilon + 1}{2} + \dfrac{\varepsilon - 1}{2} \left[1 + 12 \dfrac{h}{W} \right]^{-\frac{1}{2}} = 3.9953 \end{cases}$$

with L_{eff} is the effective length, ΔL is the length adjustment, ε_{eff} is the effective permittivity, and h is the substrate thickness.

4. Evaluation of phased array antenna performance

To validate the design of the antenna, every component in the phased array structure is evaluated and measured. All the simulations are performed on Keysight Technologies's Advanced Design System (ADS) software and CST Microwave Studio. All the components are fabricated on the FR4 substrate with dielectric constant ε of 4.3, loss tangent tanδ of 0.02, substrate thickness h of 62 mil (1.58 mm), and the conductive copper thickness t of 1.4 mil (1 oz).

4.1 Array component design and evaluation

4.1.1 Wilkinson power divider (WPD)

In the two-way WPD, theoretically, the total input power is equally divided into two output ports. Assume that the input port is port 1 and the output ports are port 2 and 3, the transmission coefficient S21 = S31 = − 3dB as shown in **Figure 18b**.

However, microstrip circuits on any substrate are affected by the loss tangent coefficient of the substrate and the microstrip discontinuities. The loss tangent represents the loss in the dielectric. The higher the loss tangent is, the more the loss is. For FR4 substrate, the loss tangent is about 0.025, so the forward gain of WPD on FR4 is less than theoretical one. Furthermore, the microstrip discontinuity phenomenon, as well as [22, 34, 35], appears in practice and generates parasitic

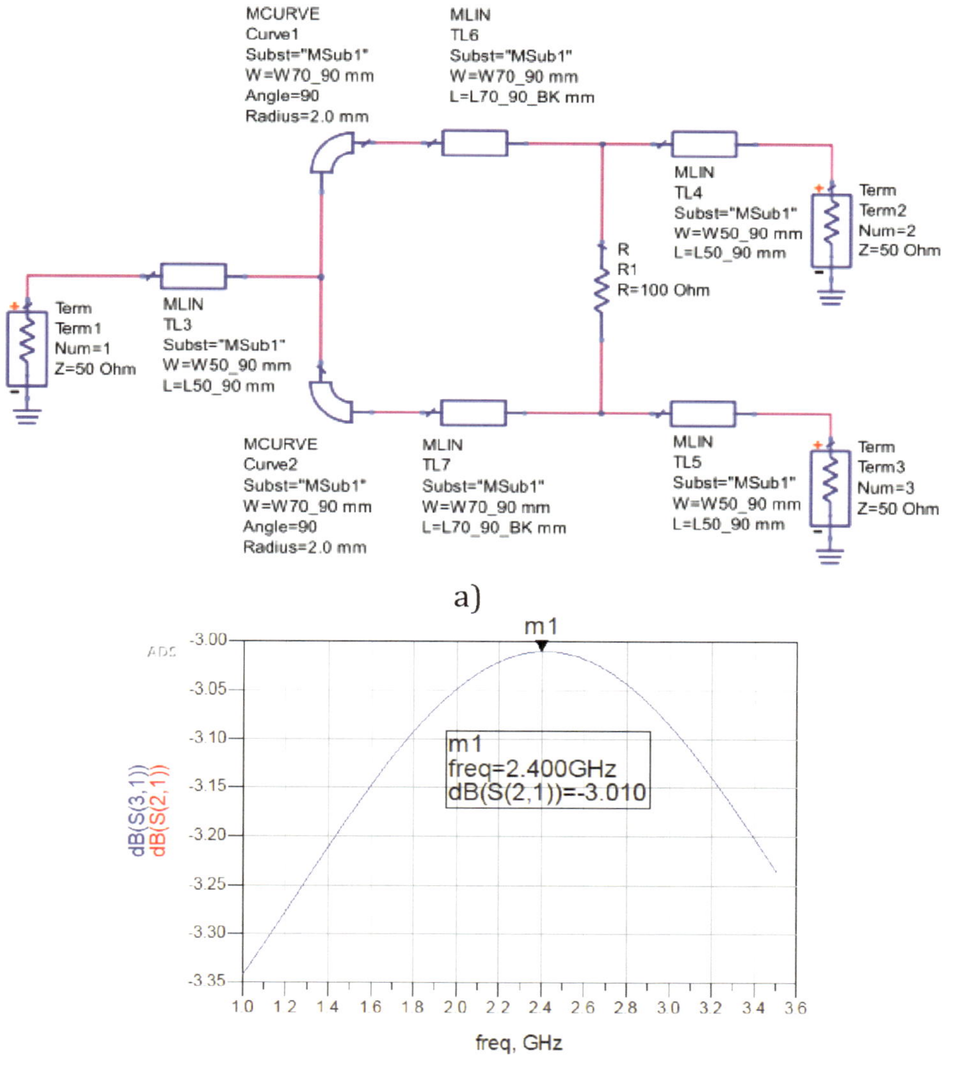

a)

b)

Figure 18.
The simulated 2-way WPD: (a) schematic circuit; (b) forward gains S_{21}, S_{31}.

components, changing the parameters of circuit. Thus, in simulation, it is necessary to insert discontinuity components such as Bend and Tee so that the simulated results are close to measured results. After that, dimension of transmission lines is slightly adjusted by using optimization tool in ADS software in order to get the desired results. The two-way WPD is shown in **Figure 19** and S_{21} = S_{31} = −3.23 dB.

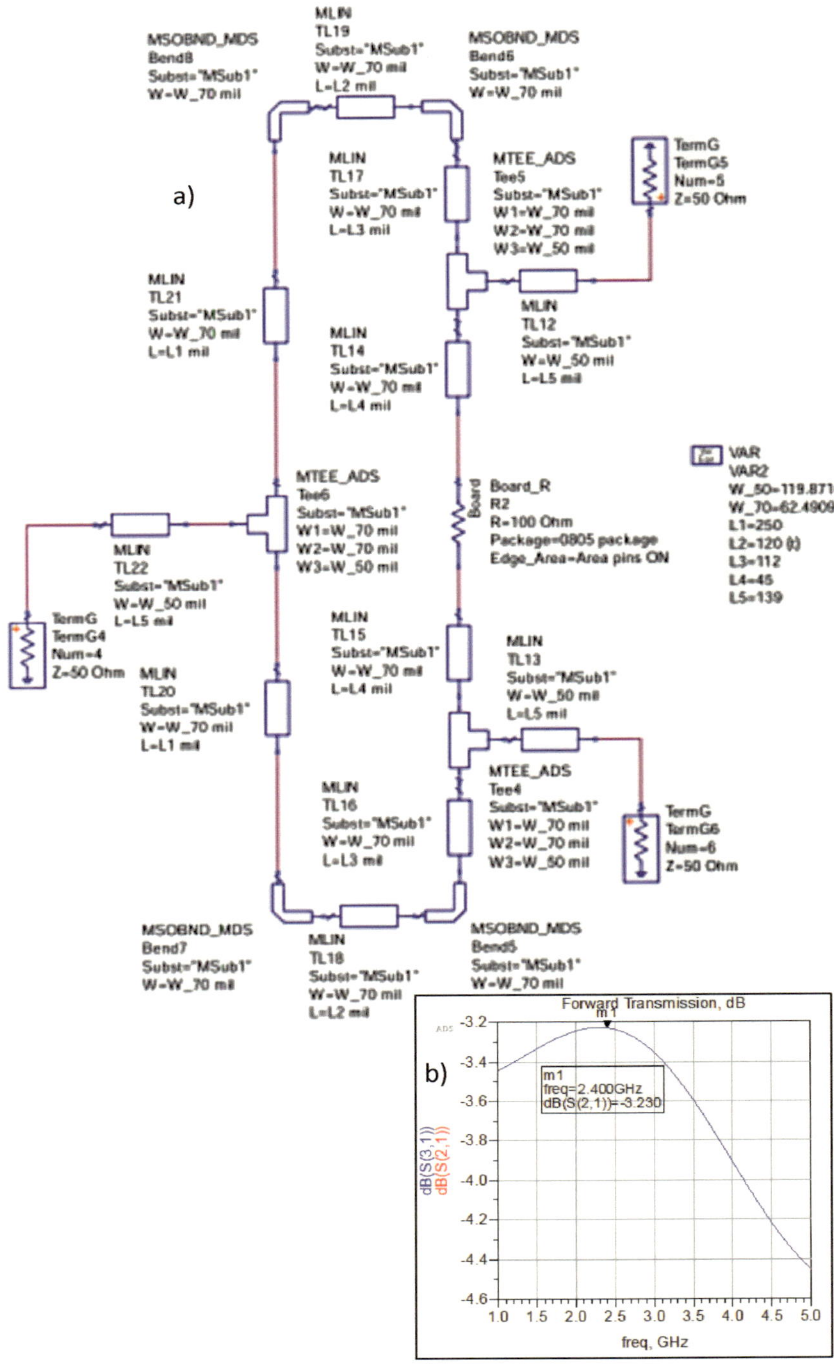

Figure 19.
The 2-way WPD: (a) schematic circuit; (b) forward gains S_{21}, S_{31}.

According to the cascade structure, the eight-way WPD is formed from the two-way WPDs and 50 Ω transmission lines. In addition, in microwave circuit, when transmission lines are placed closely to each other, the coupling between them occurs, which affects to characteristics of circuit. The coupling is an issue not easy to calculate and estimate and does not appear on schematic simulation, so the EM simulation and EM co-simulation are used to ensure the final simulated results (**Figure 20**).

Finally, the eight-way WPD is fabricated on FR4 substrate and then measured by PNA N5222A network analyzer. The forward gain for each port are shown in **Figure 21,** in which the thin red line, bold blue lines, and blue dash lines are forward gains of 8-way WPD in schematic simulation, EM co-simulation and fabrication, respectively. It can be seen that for schematic simulation, the forward gains in all output ports are identical and equal to −10.882 dB instead of −9 dB as in theory. The loss occurs during propagating on FR4 substrate with loss tangent coefficient tanδ = 0.025. For EM co-simulation, the forward gains at the output ports are not identical and change from −11.031 to −10.493 dB. As mentioned above, the difference comes from the coupling between transmission lines.

The forward gains of 8-way WPD made on FR4 material are relatively close to the simulation results in the schematic circuit and EM co-simulation at the designed frequency 2.45 GHz. The measurement results change around the simulation results with a difference less than 1 dB. The differences are due to the fabrication tolerance as well as the heterogeneity of the material in reality, causing the change of mismatch and loss compared with simulation. For the other frequency band, namely less than 2.4 GHz and greater than 2.5 GHz, there are large differences between simulation and reality. The reason is that the fabrication errors will make a great impact at undesired frequencies. However, this error does not affect the final result because the system is designed to operate at 2.45 GHz.

Additionally, in **Figure 22,** the thin red line, thick blue line and dash blue line represent the schematic simulated result, electromagnetic co-simulation, and the measured result, respectively. The isolation between output ports of the power divider is about 20 dB, similar to products on the market.

4.1.2 Reflection type phase shifter

For low loss variation, the important component – RTPS is designed on substrate Roger4003C with dielectric constant ε of 3.55, loss tangent tanδ of 0.0021 and substrate thickness h of 0.8 mm. The RTPS structure is simulated on Advanced Design Systems ADS software. ADS has built-in components such as microstrip lines, terminals, DC voltage, and stubs; however, the selected SMV1247 varactors are not supported, hence it is necessary to create an equivalent model of SMV1247 before designing and simulating the RTPS. The C-V characteristic curve of SMV1247 with the ADS model at 100 MHz is shown in **Figure 23b**. This curve is

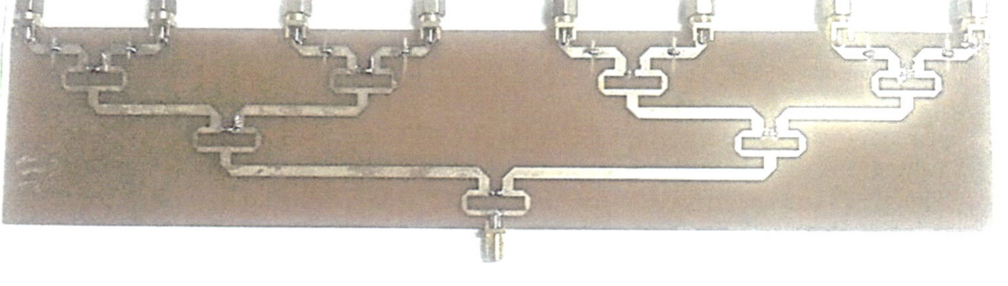

Figure 20.
Eight-way Wilkinson power divider.

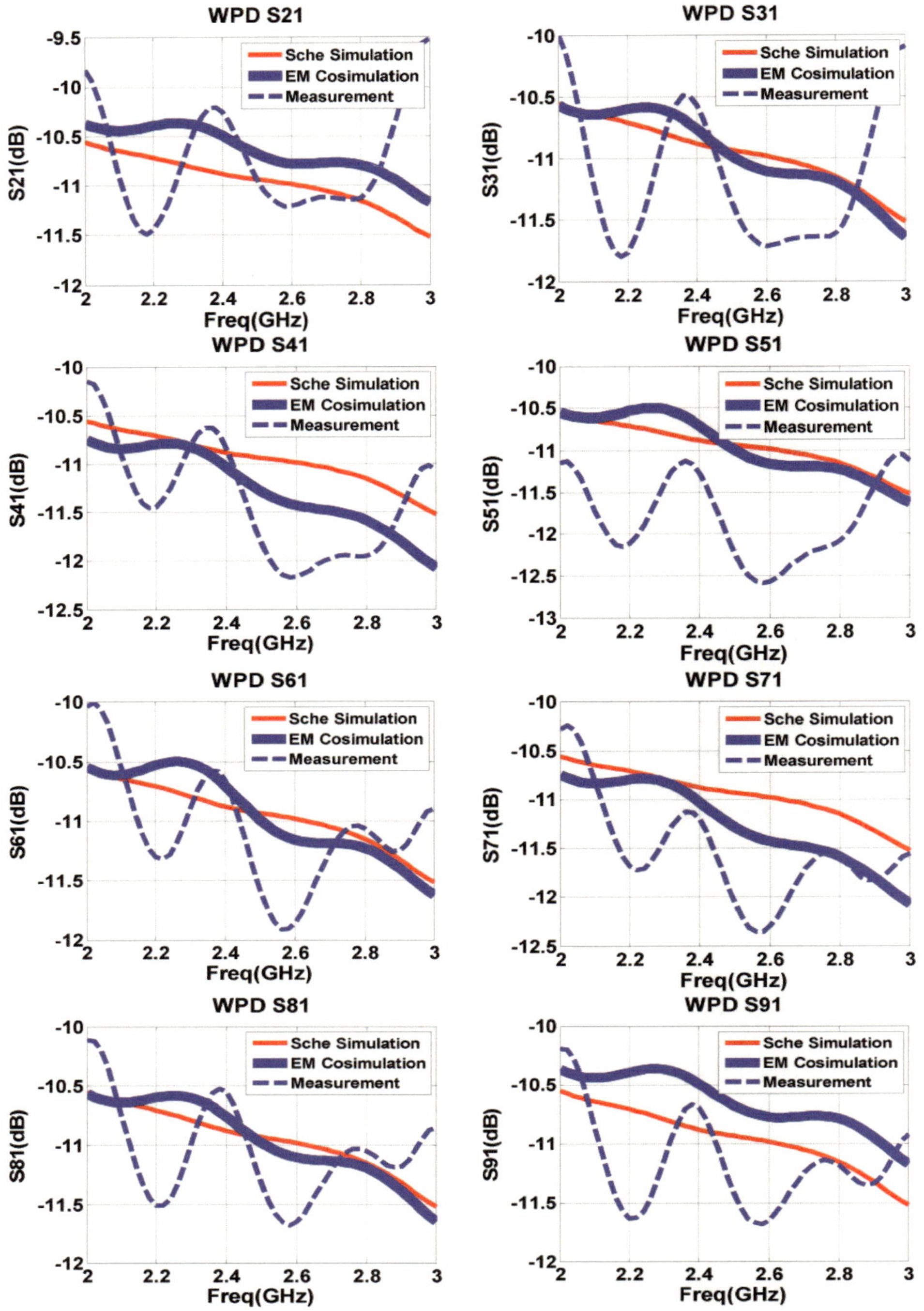

Figure 21.
Forward gain from port 2–9 of the WPD.

similar to the characteristic curve in the SMV1247 datasheet **Figure 23a**, so this SMV1247 model will be used on all ADS simulations. These characteristic curves were calculated from model in **Figure 24**.

In the technical documentation of SMV1247, Skyworks provides the SPICE model (**Figure 24a**). After importing this model into ADS software, the equivalent

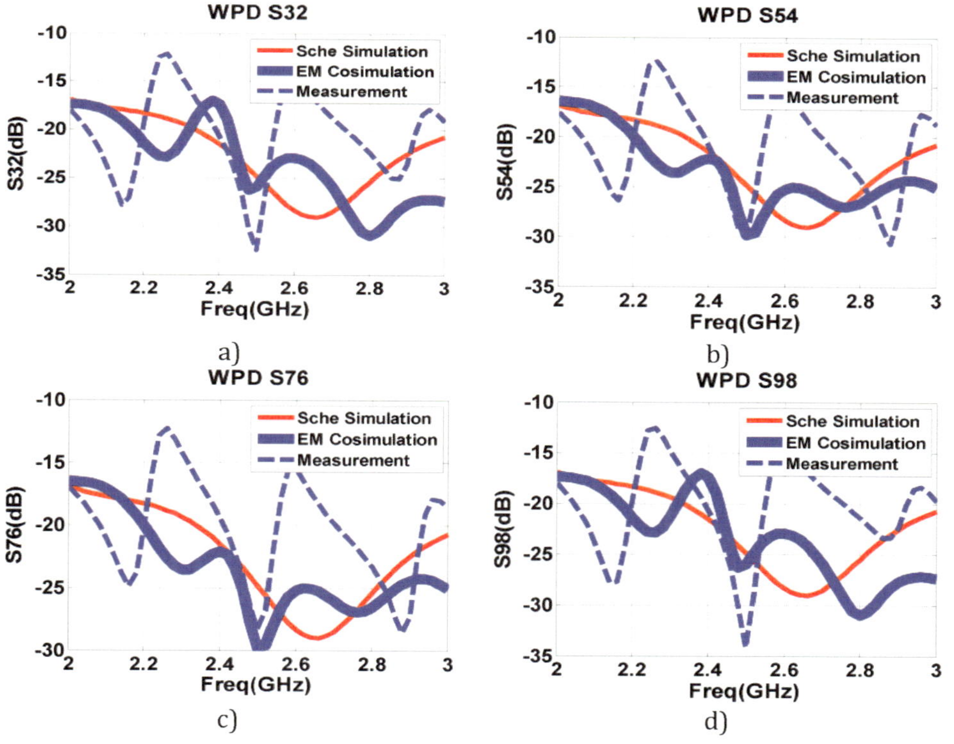

Figure 22.
Isolation between output ports.

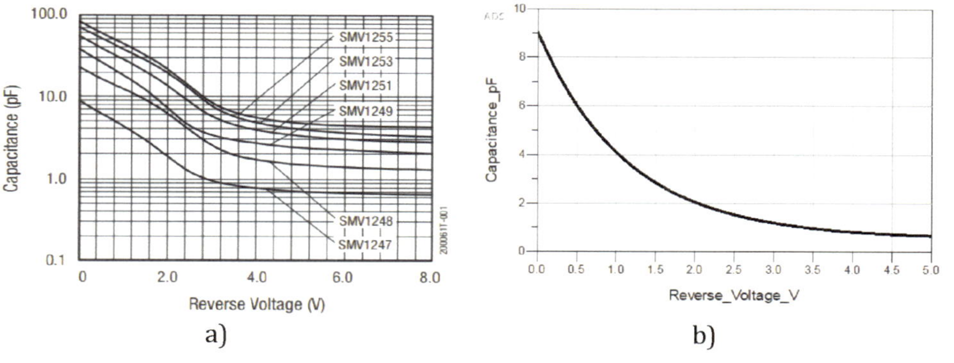

Figure 23.
C-V curve of SMV1247 on: (a) technical document [36]; (b) ADS.

model of SMV1247 is built with the L_s, C_p, and R_s values corresponding to SC-79 packages, L_s = 0.7nH, C_p = 0.54 pF, and R_s = 4.9 Ω (**Figure 24b**).

The RTPS is deployed both in schematic simulation and EM co-simulation to ensure the real operation (**Figure 25**). The microstrip discontinuity components such as bend, tee, and steps in width are inserted and their dimensions are adjusted in order to obtain the desired results. Then the RTPS is fabricated on Roger4003c (**Figure 26**) and measured by the Keysight PNA N5222A Network Analyzer.

The results of the phase shifter are shown in **Figure 27**, where the green dash line, blue dash dot line, and red circle line are schematic simulation, EM co-simulation, and measured results, respectively.

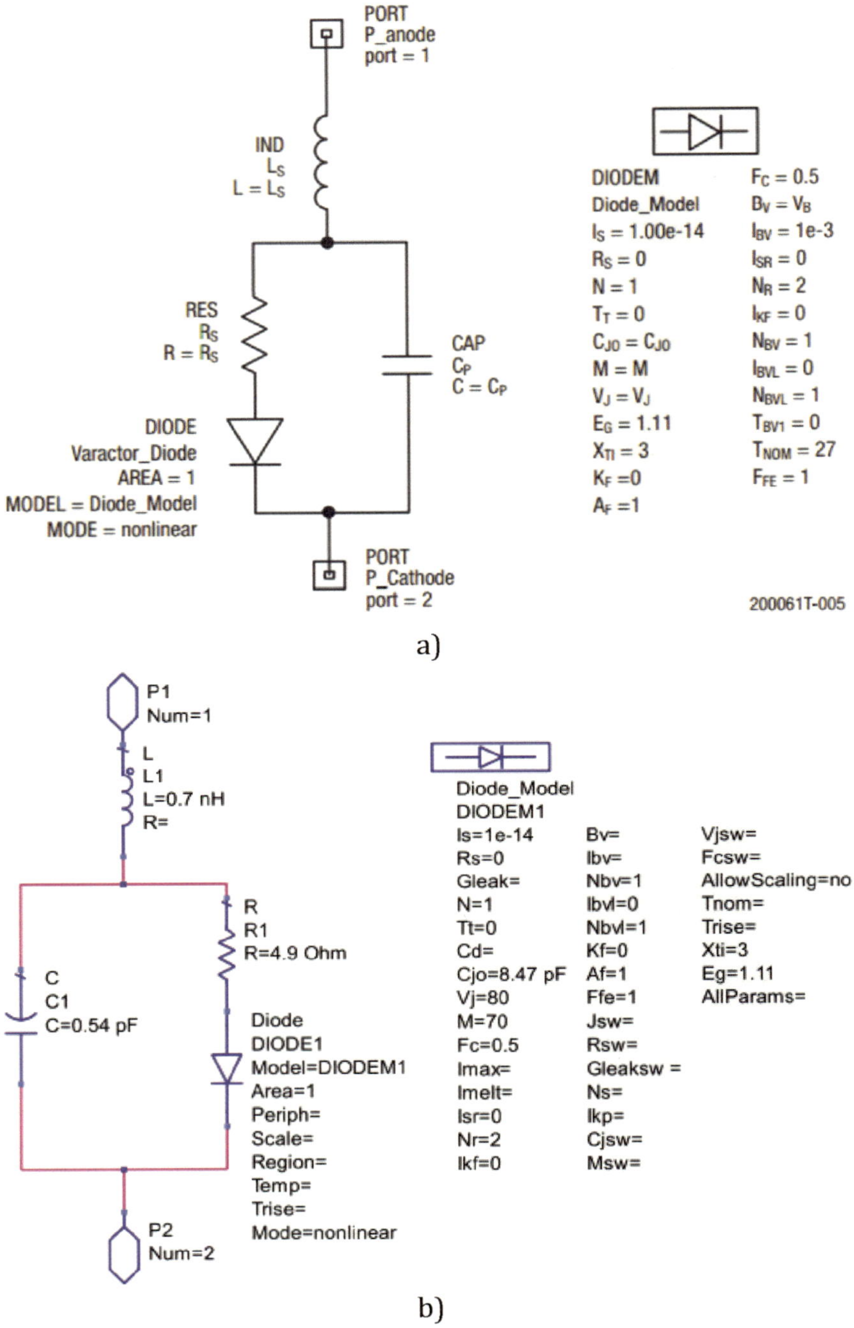

Figure 24.
Equivalent model of SMV1247: (a) on SPICE [36]; (b) on ADS.

Figure 27 shows the forward gains (S_{21}) in all three cases simulating the schematic, electromagnetic field simulation, and measurement results. I find that the energy transmitted through the phase shifter is little less than that of the schematic and electromagnetic simulation. The reason may be due to the error in fabrication process as well as contact between the varactors and the microstrip line, which is

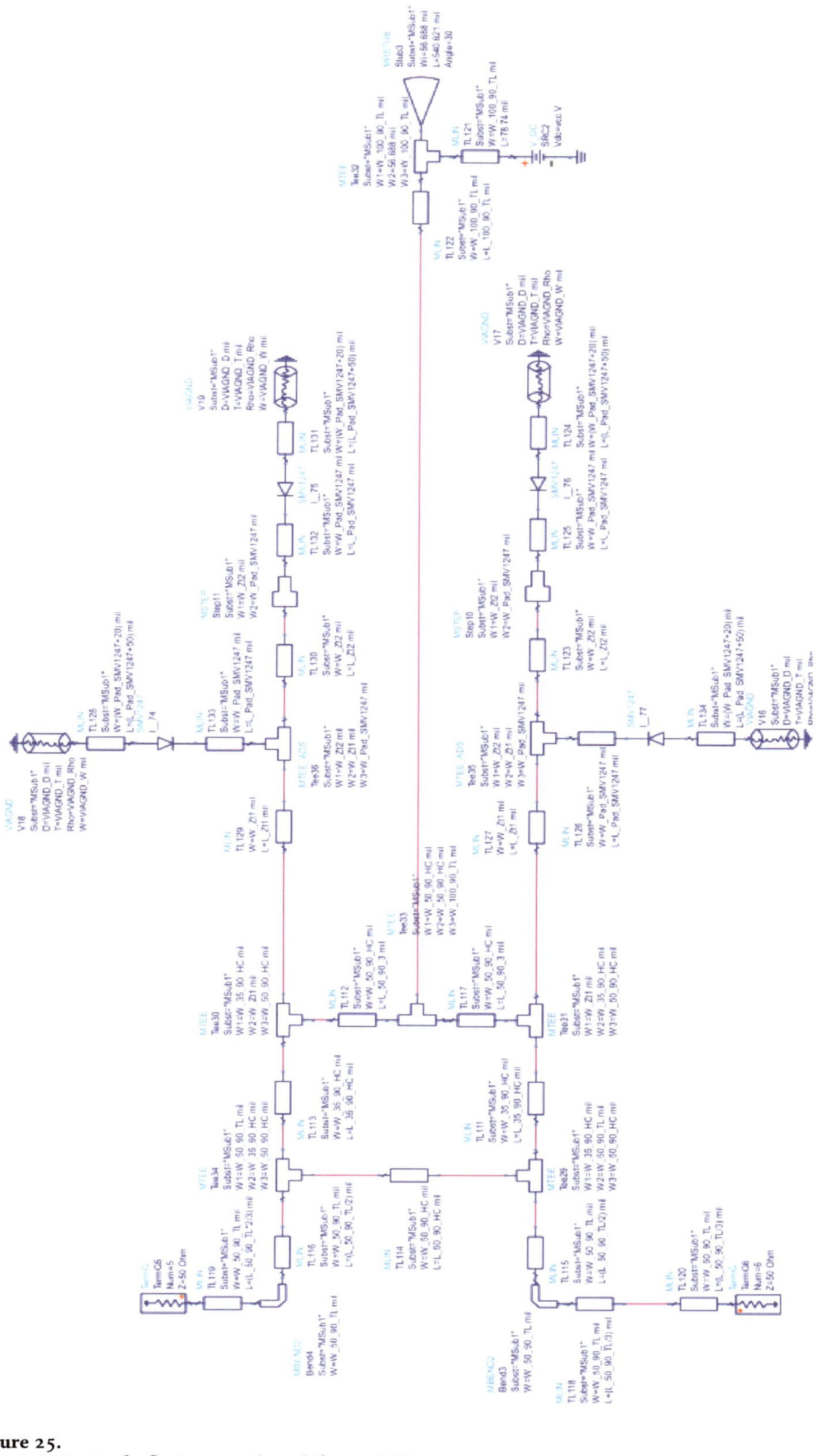

Figure 25.
Schematic circuit of reflection type phase shifter on ADS.

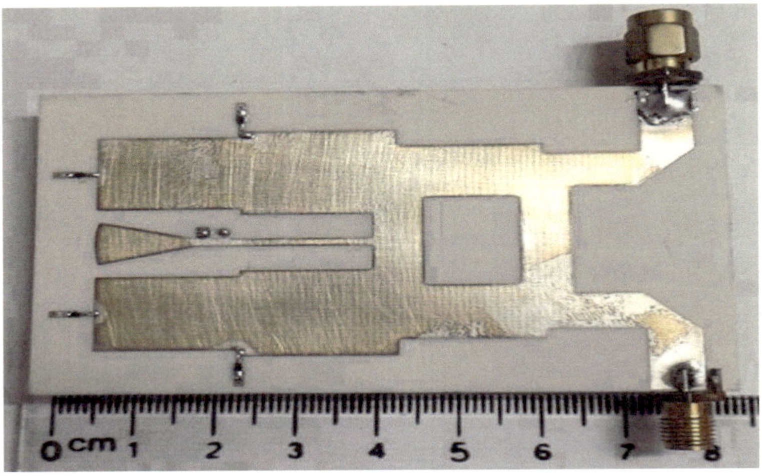

Figure 26.
Reflection type phase shifter fabricated on Roger4003c.

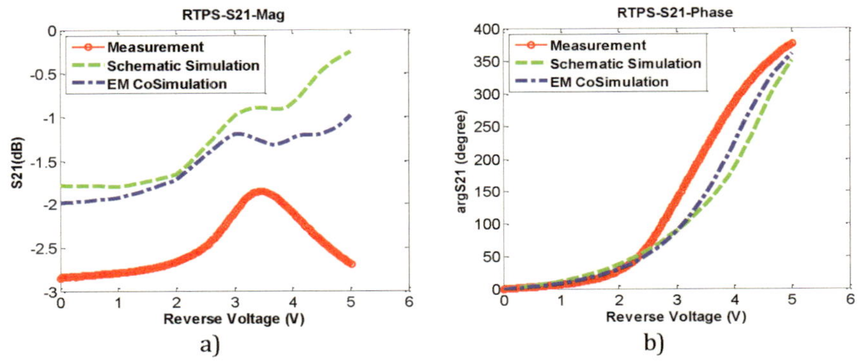

Figure 27.
Transmission coefficient S_{21} of RTPS: (a) magnitude; (b) phase.

not ideal and alter the resistance value in the equivalent circuit of the varactors and resulting in a loss of power that is much larger than simulation. The insertion loss is about 2.5 dB with nearly 0.5 dB of variation. The low insertion loss variation enables us to make a phased array antenna with uniform amplitude and spacing. **Figure 27b** shows phase shifts in three cases, the schematic, electromagnetic simulation, and measurement results. The phase shifters are capable of shifting about 370°. The phase shifter in reality although is not nearly as linear as the rest of the two cases, but with DAC12bits, the phase shifter can still be controlled with small resolution.

4.1.3 Microstrip patch antenna

The fundamental part – microstrip patch antenna is simulated on CST Microwave Studio software and then fabricated on FR4 material, as shown **Figure 28**. In order to determine the bandwidth of an antenna, two parameters, VSWR and return loss S_{11}, can be measured by the network analyzer PNA N5222A. At the operating frequency, the desired VSWR and S_{11} are less than 2 and −10 dB,

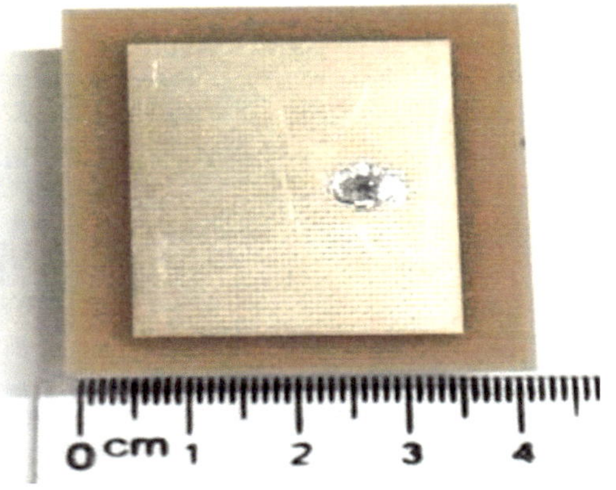

Figure 28.
Fabricated microstrip patch antenna.

respectively. The VSWR and S11 of patch antenna versus frequency are shown in **Figure 29**, where the thin red lines and dash blue lines are simulated and measured characteristics, respectively. It can be seen that in simulation, my microstrip patch antenna can operate at Wi-Fi band with resonant frequency from 2.4 to 2.484 GHz. However, due to the heterogeneity of the material in practice as well as errors in fabrication, my microstrip patch antenna in reality can only operate from 2.424 to 2.485 GHz. Although this antenna cannot be used in whole range of Wi-Fi band, it can still operate at several channels of Wi-Fi.

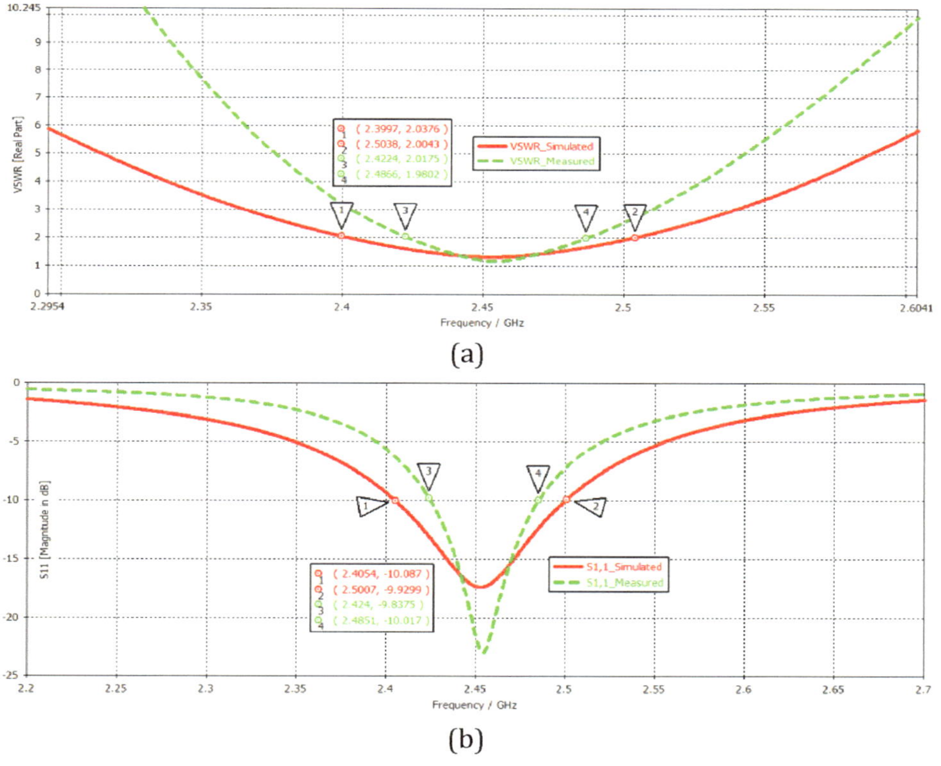

Figure 29.
Microstrip patch antenna parameters: (a) VSWR; (b) return loss.

4.2 Beamforming antenna measurement evaluation

4.2.1 Material and facility preparation

In general, the radiation pattern of an antenna is three-dimensional. Because it is impractical to measure a three-dimensional pattern, a number of two-dimensional patterns are measured. Patterns can be obtained by fixing one of angles (θ or ϕ) while varying the other. For a single antenna, the $\theta = \pi/2$ azimuthal pattern and $\phi = 0$ elevation pattern are usually chosen. For phased array antenna, only $\theta = \pi/2$ azimuthal pattern is measured to observe the direction of main beam, and the $\phi = 0$ elevation pattern does not provide useful information. Ideally, the measurement system would be placed in outdoor space in the far field region to cancel reflection waves. However, this ideal condition is not achievable; therefore indoor anechoic chambers have been developed for this antenna array measurement. The region inside anechoic chambers is covered with RF absorber. The reflection coefficient of anechoic chambers is about −40 dB [15, 37].

Structure of the measurement room is shown in **Figure 30**. A horn antenna, HF 906, is used as a transmitting antenna, and antenna under test (AUT) is received antenna. Thanks to the reciprocity property of antennas, the transmitting and receiving radiation patterns are identical. Therefore, the swap between two antennas is not necessary. The receiving and transmitting powers are measured by the PNA N5222A network analyzer in which the AUT is rotated through a rotating table. To automatically operate this system, a rotating table with an angle step of 5° is set up, and then the measured results are extracted and analyzed from the network analyzer.

Figure 31 shows the radiation pattern of my microstrip patch antenna. We can see that the half power beam width is larger than 90° and the measured radiation pattern is similar to simulated one. Therefore, in terms of radiation pattern, each antenna element meets the requirement for indoor positioning system (**Figure 32**).

4.2.2 Beamforming array measurement

All components of the array including power divider, phase shifter, antenna, and controller are assembled together to form a phased array antenna. The radiation

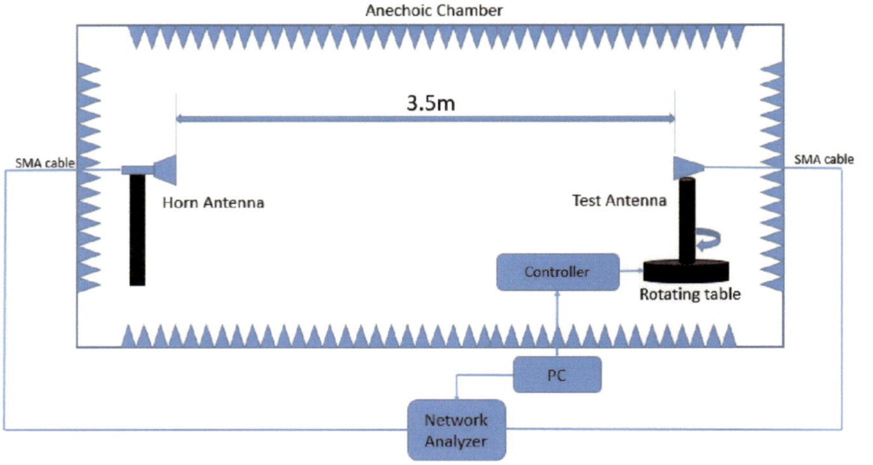

Figure 30.
Anechoic chamber configuration for measuring beamforming antenna array.

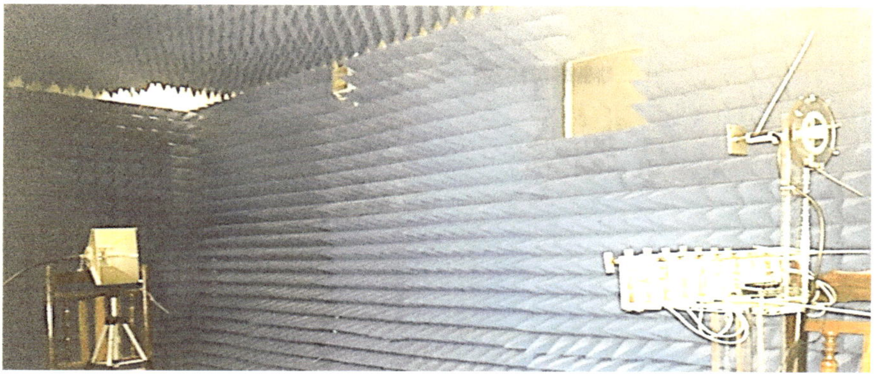

Figure 31.
Array antenna measurement in anechoic chamber.

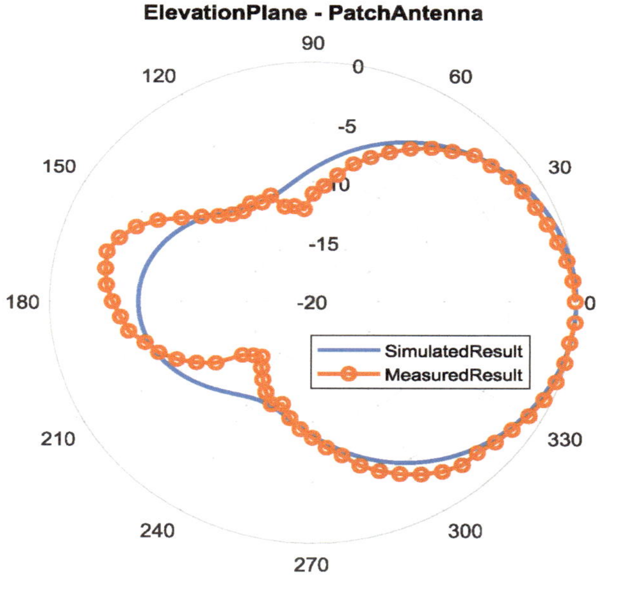

Figure 32.
Radiation pattern of microstrip patch antenna element.

patterns of phased array antenna are also measured in anechoic chamber at General Department of Technical Logistics of the Vietnamese Ministry of Public Security, similar to microstrip patch antenna at 2.45 GHz. The phased array antenna will be mounted in the rotating table to steer the main beam toward angles from −45 to 45° with step of 5°. The measured results are compared with simulated results in same coordinate to investigate the quality of the resulted array (**Figure 33**). The direction of main beam in measurement can track the simulated results. However, when phase shift is tuned, due to variation of electrical length change, and the amplitude variation on both WPD and RTPS causes the difference between direction of main beam in measurement and simulation. Additionally, fabrication tolerance also produces the beam angle error between the measured and calculated value. The measured beam scan angles are within ±5° tolerance (**Table 1**). In general, the measurement results are in good agreement with the simulated results.

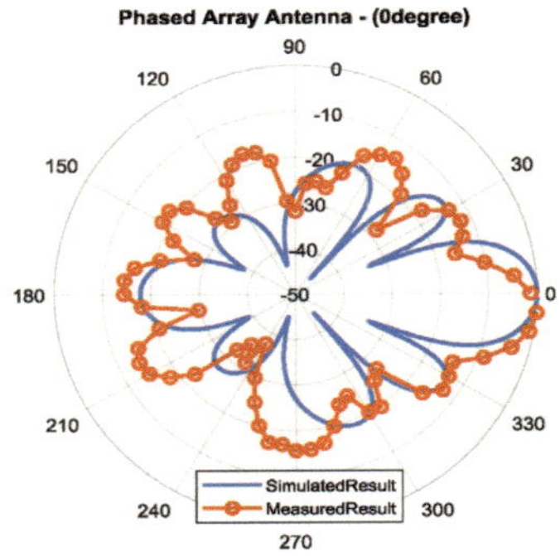

a) Radiation pattern at 0° angle

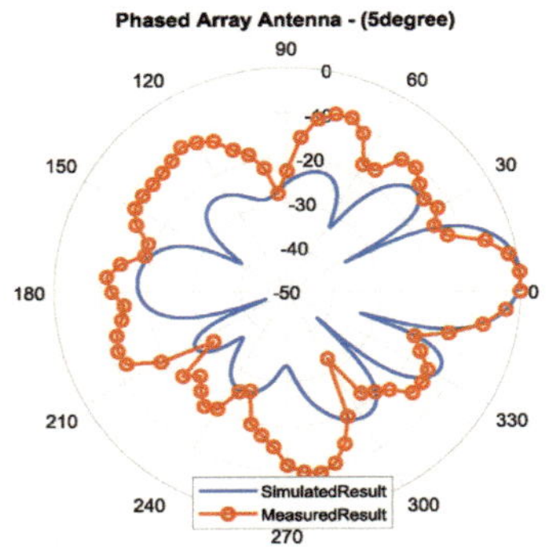

b) Radiation pattern at 5° angle

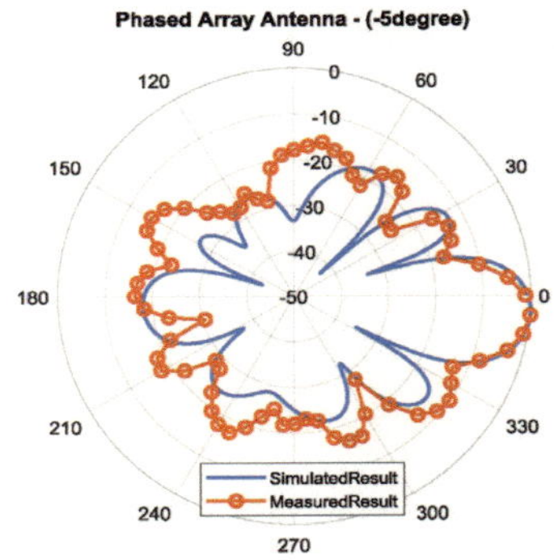

c) Radiation pattern at -5° angle

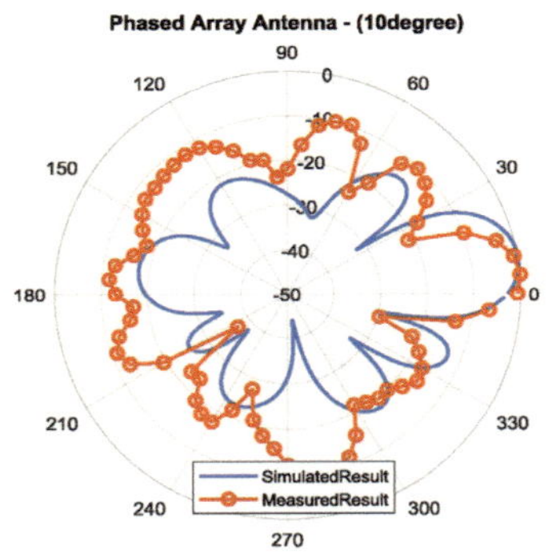

d) Radiation pattern at 10° angle

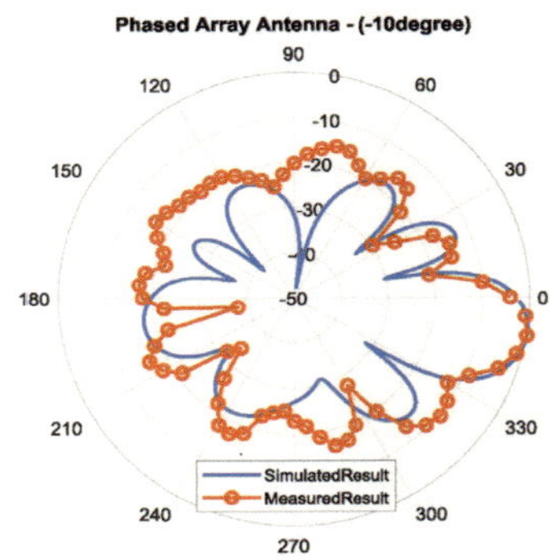

e) Radiation pattern at -10° angle

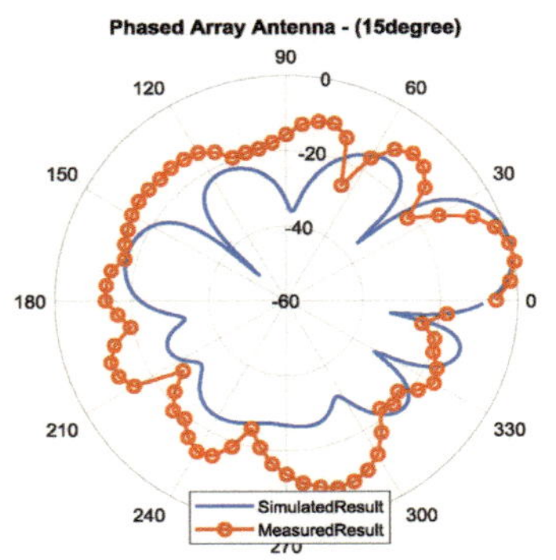

f) Radiation pattern at 15° angle

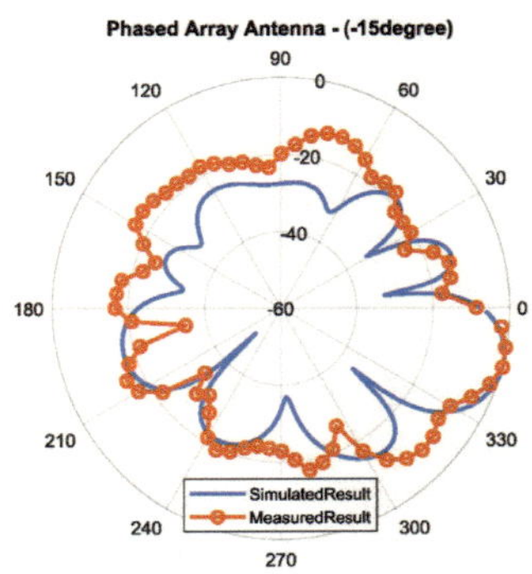

g) Radiation pattern at -15° angle

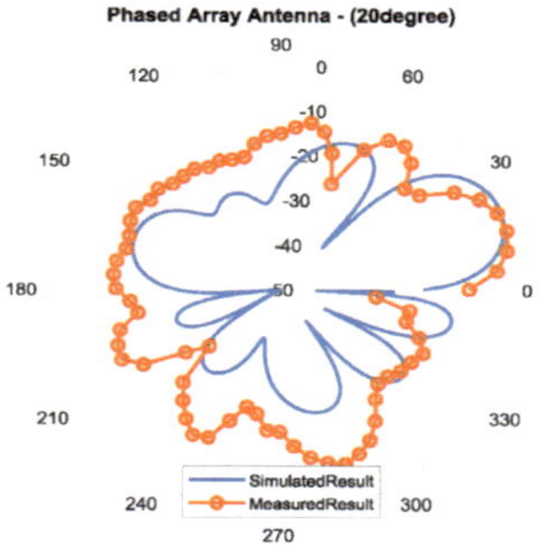

h) Radiation pattern at 20° angle

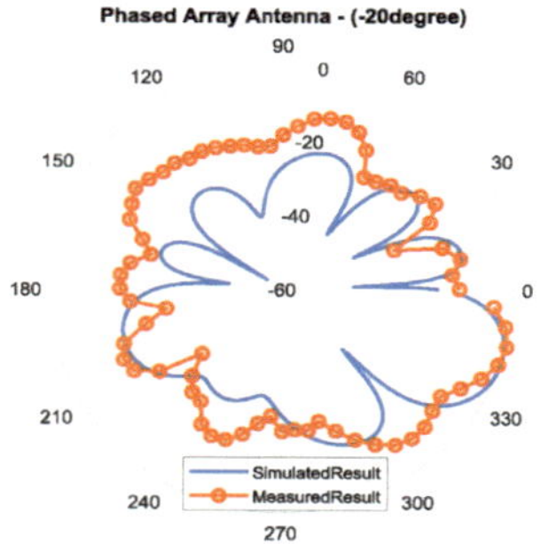

i) Radiation pattern at -20° angle

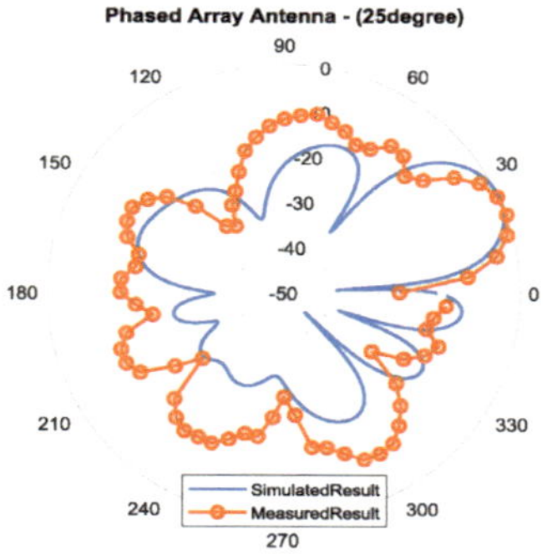

j) Radiation pattern at 25° angle

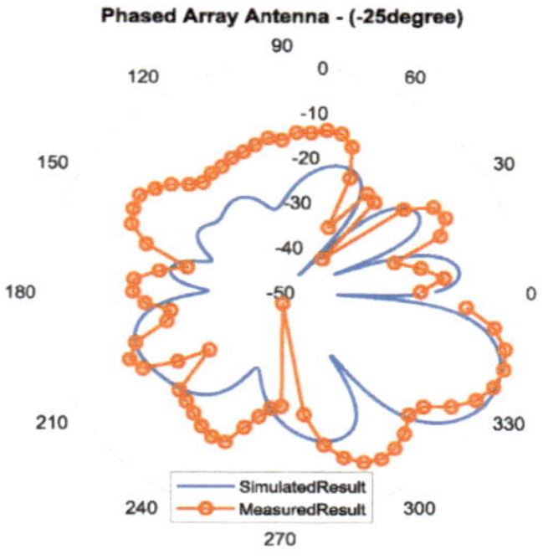

k) Radiation pattern at -25° angle

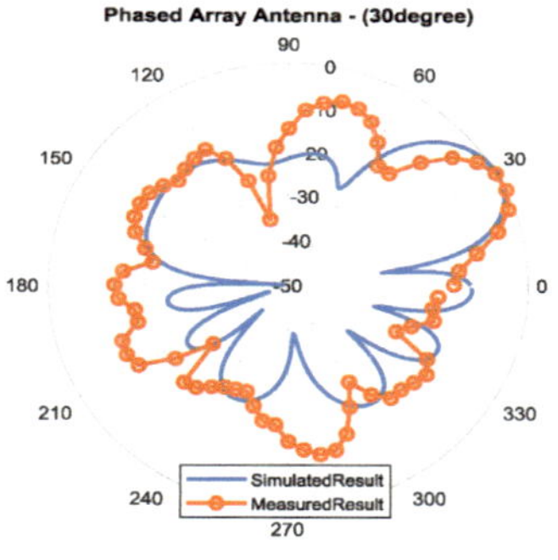

l) Radiation pattern at 30° angle

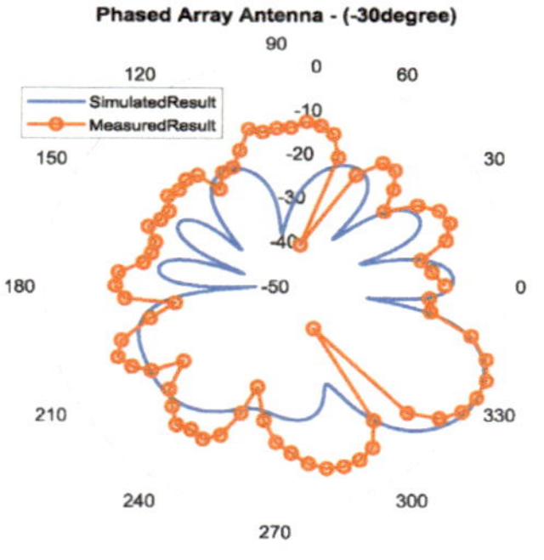

m) Radiation pattern at -30° angle

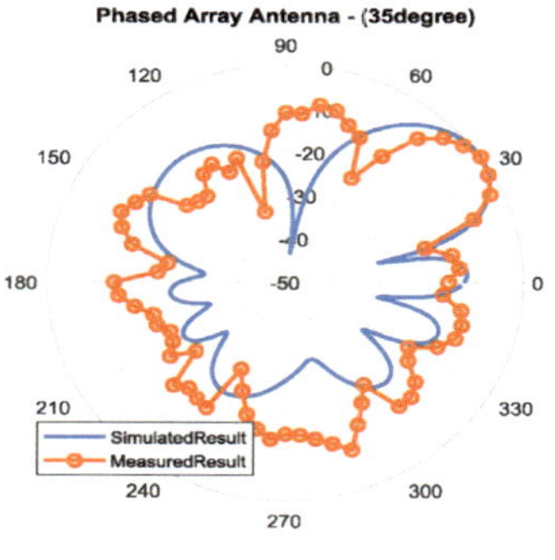

n) Radiation pattern at 35° angle

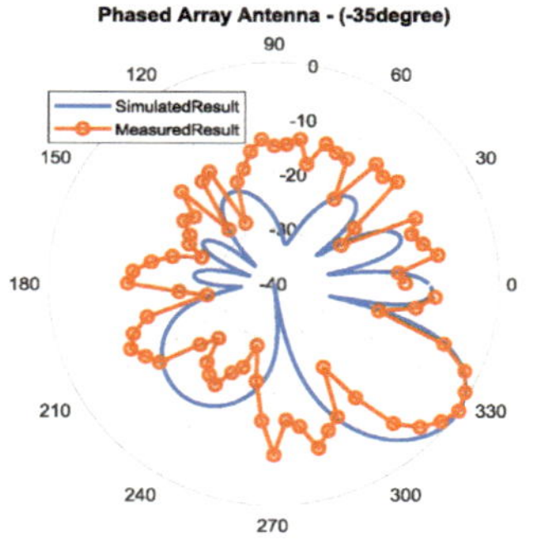

o) Radiation pattern at -35° angle

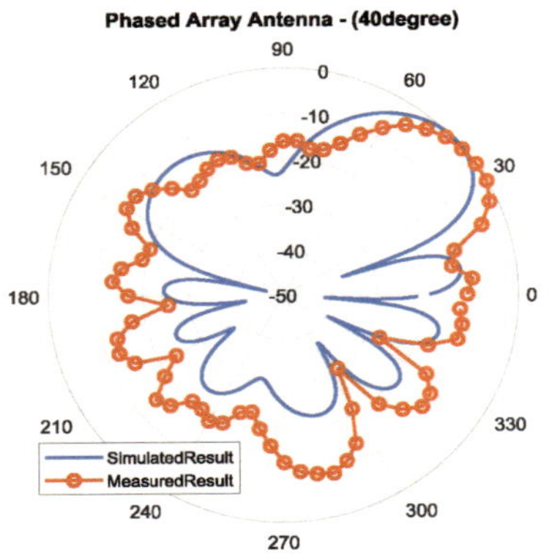

p) Radiation pattern at 40^0 angle

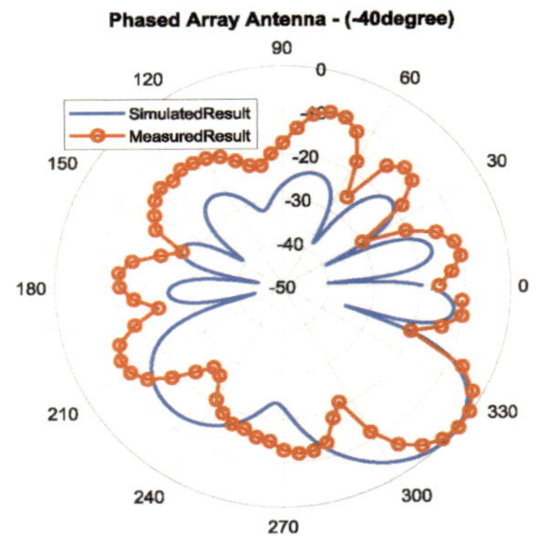

q) Radiation pattern at -40^0 angle

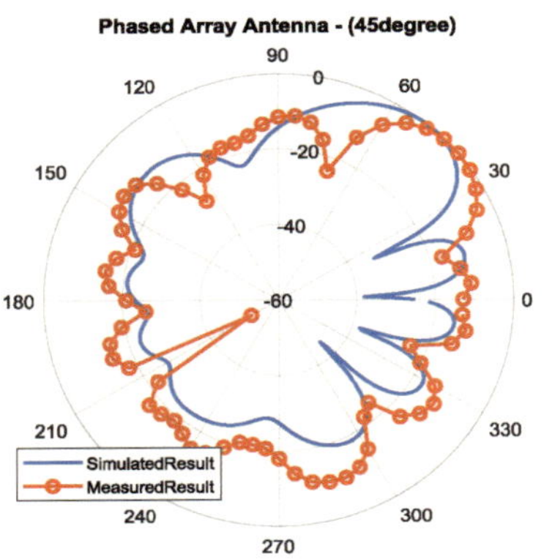

r) Radiation pattern at 45^0 angle

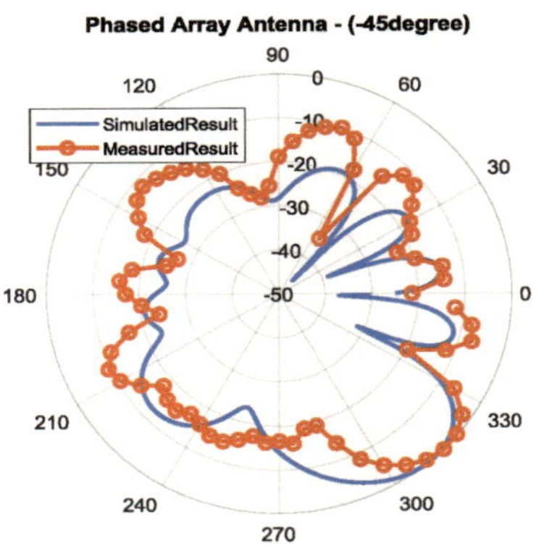

s) Radiation pattern at -45^0 angle

Figure 33.
Radiation pattern of phased array antenna at different angles.

Regarding side lobes, compared with simulated results, the gain of side lobes in reality is much higher. The reason is due to the difference of space in simulation and measurement. In simulation, the array antenna is computed in open space, and as a result, there is no affection of reflection on radiation pattern. In reality, although the antenna array is measured in an anechoic chamber, the reflection from wall and other objects in chamber still exist and affect to measured results. Compared with main lobe, side lobe level is less than about 10 dB in most case (**Figure 33**).

A comparison about performance among my antenna design with previous antenna designs for indoor localization is presented in **Table 2**. It can be seen that, in previous designs, based on switching predefined beams, the number of beam is similar to the number of antenna elements. This leads to the limitation of the number of lobes due to the size of an indoor antenna that cannot be too large. Hence, the scan step of these design is quite large, namely 60, 30, 90, and 18° corresponding to researches [10–13]. In this antenna array, by controlling the array factor of array antenna through phase shifter controlling of wave coming to antennas, the number of beam does not depend on the number of antenna element. Simultaneously, the full 360° continuous phase shifter enables me to arbitrarily adjust the phase shift, so the main beam can be steered continuously. The resolution of scanning angle step in steering can be achieved at 5°.

Main beam angle	Simulated side lobe level	Measured side lobe level
0°	−13 dB	−11 dB
5°	−12 dB	−10 dB
10°	−12 dB	−10 dB
15°	−15 dB	−9 dB
20°	−14 dB	−10 dB
25°	−11 dB	−10 dB
30°	−11 dB	−8 dB
35°	−11 dB	−10 dB
40°	−11 dB	−10 dB
45°	−13 dB	−12 dB
−5°	−12 dB	−10 dB
−10°	−12 dB	−10 dB
−15°	−15 dB	−9 dB
−20°	−14 dB	−9 dB
−25°	−11 dB	−9 dB
−30°	−11 dB	−9 dB
−35°	−11 dB	−10 dB
−40°	−11 dB	−10 dB
−45°	−13 dB	−8 dB

Table 1.
Comparison between simulated and measured main beam angle and side lobe level.

Antenna	Scanning range	No. of beam	Average step	No. of antenna
[10]	360°	6	60°	6
[11]	360°	12	30°	13
[12]	360°	4	90°	5
[13]	73°	4	18°	4
This work	**90°**	**19**	5°	**8**

Table 2.
Comparison with previously studied beamforming array antenna designs for indoor localization.

5. Conclusion

With the desire to improve the resolution of AoA-based indoor positioning system, this chapter has focused on the design of multi-port phased array antenna using the reflection type phase shifter. The studies demonstrate that the beamforming array antenna with two-way Wilkinson power divider that is able to limit the loss and increase the isolation between the output ports. In addition, the important part in the array, the reflection type phase shifter, can continuously control the phase shift in full 360° range with low insertion loss and small variation. The main beam of the array antenna can be steered in direction from −45 to 45°

with high resolution. This design can be commercially available for WLAN communication system in the frequency range from 2.424 to 2.484 GHz. This will facilitate an effective solution for indoor positioning system with high accurate localization. The future work will involve incorporating the phased array antenna for real positioning system for indoor object localization.

Acknowledgements

The author would like to thank Prof. Tan Phu Vuong and IMEP-LAHC Laboratory, Grenoble INP, France for their support in CST Microwave Studio and Keysight Advanced Design Systems (ADS) softwares. The author also thanks MSc. Nguyen Cong Thuan for his contribution to the work.

Conflict of interest

The author declares no conflict of interest.

Author details

Nguyen Thanh Huong[1,2]

1 School of Electrical Engineering, Hanoi University of Science and Technology, Hanoi, Vietnam

2 International Research Institute MICA, Hanoi University of Science and Technology, Hanoi, Vietnam

*Address all correspondence to: huong.nguyenthanh3@hust.edu.vn

References

[1] Lee C, Chang Y, Park G, Ryu J, Jeong SG, Park S, Park JW, et al. Indoor positioning system based on incident angles of infrared emitters. In: Proceeding of 30th Annual Conference of IEEE Industrial Electronics Society, Vol. 3, 2004. IECON; 02 November 2004. pp. 2218-2222

[2] Feldmann S, Kyamakya K, Zapater A, Lue Z. An indoor bluetooth-based positioning system: Concept, implementation and experimental evaluation. In: International Conference on Wireless Networks. Vol. 272; 23 June 2003

[3] Li H. Low-cost 3D bluetooth indoor positioning with least square. Wireless Personal Communications. 2014;**78**(2): 1331-1344. DOI: 10.1007/s11277-014-1820-1

[4] Woo S, Jeong S, Mok E, Xia L, Choi C, Pyeon M, et al. Application of WiFi-based indoor positioning system for labor tracking at construction sites: A case study in Guangzhou MTR. Automation in Construction. 2011; **20**(1):3-13. DOI: 10.1016/j.autcon.2010.07.009

[5] Mazuelas S, Bahillo A, Lorenzo RM, Fernandez P, Lago FA, Garcia E, et al. Robust indoor positioning provided by real-time RSSI values in unmodified WLAN networks. IEEE Journal of Selected Topics in Signal Processing. 2009;**3**(5):821-831. DOI: 10.1109/JSTSP.2009.2029191

[6] Alarifi A, Al-Salman A, Alsaleh M, Alnafessah A, Al-Hadhrami S, Al-Ammar MA, et al. Ultra wideband indoor positioning technologies: Analysis and recent advances. Sensors. 2016;**16**(5): 707. DOI: 10.3390/s16050707

[7] Medina C, Segura JC, De la Torre A. Ultrasound indoor positioning system based on a low-power wireless sensor network providing sub-centimeter accuracy. Sensors. 2013;**13**(3): 3501-3526. DOI: 10.3390/s130303501

[8] De Angelis G, Pasku V, De Angelis A, Dionigi M, Mongiardo M, Moschitta A, et al. An indoor AC magnetic positioning system. IEEE Transactions on Instrumentation and Measurement. 2014;**64**(5):1267-1275. DOI: 10.1109/TIM.2014.2381353

[9] Rishabh I, Kimber D, Adcock J. Indoor localization using controlled ambient sounds. In: 2012 International Conference on Indoor Positioning and Indoor Navigation (IPIN); 13 November 2012. pp. 1-10

[10] Giorgetti G, Cidronali A, Gupta SK, Manes G. Single-anchor indoor localization using a switched-beam antenna. IEEE Communications Letters. 2009;**13**(1):58-60. DOI: 10.1109/LCOMM.2009.081584

[11] Rzymowski M, Woznica P, Kulas L. Single-anchor indoor localization using ESPAR antenna. IEEE Antennas and Wireless Propagation Letters. 2015;**15**: 1183-1186. DOI: 10.1109/LAWP.2015.2498950

[12] Kamarudin MR, Nechayev YI, Hall PS. Onbody diversity and angle-of-arrival measurement using a pattern switching antenna. IEEE Transactions on Antennas and Propagation. 2009; **57**(4):964-971. DOI: 10.1109/TAP.2009.2014597

[13] Bui TD, Le MT, Nguyen QC. Electronically steerable antenna array for indoor positioning system. Journal of Electromagnetic Waves and Applications. 2019;**33**(7):838-852. DOI: 10.1080/09205071.2018.1555060

[14] Mailloux RJ. Phased Array Antenna Handbook 3e. USA: Artech House; 2017. ISBN: 978-1-6308-102

[15] Balanis CA. Antenna Theory: Analysis and Design 4e. NY: John Wiley & Sons; 2016. ISBN: 978-1-118-64206-1

[16] Miroslav DL, Milan JL, Borislav OL. Analysis of SDMA & Smart Antenna Techniques for Existing and New Mobile Communication Systems. November 2003. Available from: http://www.telfor.org.yu/telfor2001/radovi/4-1.pdf

[17] Tseng CH, Chen CJ, Chu TH. A low-cost 60-GHz switched-beam patch antenna array with Butler matrix network. IEEE Antennas and Wireless Propagation Letters. 2008;**7**:432-435. DOI: 10.1109/LAWP.2008.2001849

[18] Chang CC, Lee RH, Shih TY. Design of a beam switching/steering Butler matrix for phased array system. IEEE Transactions on Antennas and Propagation. 2009;**58**(2):367-374. DOI: 10.1109/TAP.2009.2037693

[19] Denidni TA, Libar TE. Wide band four-port Butler matrix for switched multibeam antenna arrays. In: 14th IEEE Proceedings on Personal, Indoor and Mobile Radio Communications. Vol. 3. Beijing, China: PIMRC; 2003. pp. 2461-2464. DOI: 10.1109/PIMRC.2003.1259161

[20] Butler J. Beam-forming matrix simplifies design of electronically scanned antenna. Electronic Design. 1961;**9**:170-173

[21] Rabinovich V, Alexandrov N. Antenna Arrays and Automotive Applications. Germany: Springer Science & Business Media; 2012. DOI: 10.1007/978-1-4614-1074-4

[22] Pozar DM. Microwave Engineering 3e. NY: John Wiley & Sons; 2009. ISBN: 978-0-4706-3155-3

[23] Nguyen TH, Nguyen CT. Beamsteering phased array antenna using a full 360° and programmable continuous phase shifter for indoor localization. In: 2018 IEEE Seventh International Conference on Communications and Electronics (ICCE). 2018. pp. 227-230

[24] Lambard T, Lafond O, Himdi M, Jeuland H, Bolioli S. A novel analog 360° phase shifter design in Ku and Ka bands. Microwave and Optical Technology Letters. 2010;**52**(8):1733-1736. DOI: 10.1002/mop.25307

[25] Bulja S, Mirshekar-Syahkal D. Analysis and design of a new reflection-type 360° phase shifter with combined switch and varactor. Microwave and Optical Technology Letters. 2010;**52**(3): 530-535. DOI: 10.1002/mop.24975

[26] Burdin F, Iskandar Z, Podevin F, Ferrari P. Design of compact reflection-type phase shifters with high figure-of-merit. IEEE Transactions on Microwave Theory and Techniques. 2015;**63**(6): 1883-1893. DOI: 10.1109/TMTT.2015.2428242

[27] Wadell BC. Transmission Line Design Handbook. USA: Artech House; 1991.ISBN: 978-0-8900-6436-8

[28] Gardner DW, Wickert MA. Microwave filter design using radial line stubs. In: IEEE Region 5 Conference, 1988: 'Spanning the Peaks of Electrotechnology'. 1988. pp. 68-72

[29] Atwater HA. The design of the radial line stub: A useful microstrip circuit element. Microwave Journal. 1985;**28**:149-156

[30] Weijun L, Xiaojuan C, Xiaoxin L, Xiaolin M, Xinyu L, Xiaoliang W. A radial stub test circuit for microwave power devices. Chinese Journal of Semiconductors. 2006;**27**(9):1557-1561

[31] Sorrentino R, Roselli L. A new simple and accurate formula for microstrip radial stub. IEEE Microwave

and Guided Wave Letters. 1992;**2**(12): 480-482. DOI: 10.1109/75.173401

[32] Afridi MA. Microstrip patch antenna– designing at 2.4 GHz frequency. Biological and Chemical Research. 2015;**2015**:128-132

[33] Casu G, Moraru C, Kovacs A. Design and implementation of microstrip patch antenna array. In: 2014 10th International Conference on Communications (COMM). 2014. pp. 1-4

[34] Hong JS, Lancaster MJ. Microstrip Filters for RF/Microwave Applications. NY: John Wiley & Sons; 2004. DOI: 10.1002/9780470937297

[35] Bahl IJ. Lumped Elements for RF and Microwave Circuits. USA: Artech House; 2003. ISBN: 978-1-5805-3661-5

[36] SMV1247 SERIES Hyperabrupt Junction Tuning Varactor. Available from: https://www.skyworksinc.com/en/Products/Diodes/SMV1247-Series

[37] Zhang Z. Antenna Design for Mobile Devices. John Wiley & Sons; 2017. ISBN: 978-1-1191-3232-5

Section 3

Novel RF Antennas Technologies

Chapter 7

Antennas for Space Applications: A Review

Volkan Akan and Erdem Yazgan

Abstract

It is well known that antennas are inevitable for wireless communication systems. After the launch of Sputnik-1 which was the first artificial satellite developed by USSR (Union of Soviet Socialist Republics), telecommunication technologies started to develop for space excessively. However, significance of the antennas as first or final RF-front end element has not been altered for the space communication systems. In this chapter, after introducing telecommunication and antenna technologies for space, which space environmental conditions are to be faced by these antennas are summarized. Then, frequency allocation that is a crucial design factor for antennas is explained and tabulated. And finally at the last part, different types of antennas used in different space missions are presented with their functional parameters and tasks.

Keywords: space, antenna, communication, satellite, spacecraft

1. Introduction

With the II World War, especially radar and military communication technologies gained thoughtful importance, and studies on antenna design, analysis and measurement caught a great momentum. Particularly, low volume antennas were needed for installation in air and naval combat vehicles. By the development of digital computers aftermath of the war, antenna design and analysis began to be carried to the computer environment to realize analytical and computational electromagnetic techniques. This era accelerated development process of antennas like other communication elements.

Nowadays, studies on microwave integrated circuits and components, which are frequently used in communication systems designed to operate at high frequencies, are going ahead at a dizzying pace. In addition, theoretical and experimental studies focusing on solid-state devices and planar transmission lines have greatly contributed to the development of microwave integrated circuits and components and expanded their use. In the light of these developments, interest to antennas, particularly for planar ones, has increased and thus, in order to reduce the cost and also volume occupied by microwave and RF systems, printed antennas are generally preferred to use them with high frequency circuits in the same plane.

As is well known, the space race of humankind started with the first artificial satellite *Sputnik-1* which was developed and placed into orbit by the USSR in 1957. A model of Sputnik-1 is shown in **Figure 1**. In the same year, *Sputnik-2* placed a living being into orbit of Earth for the first time. In the following year, the USA

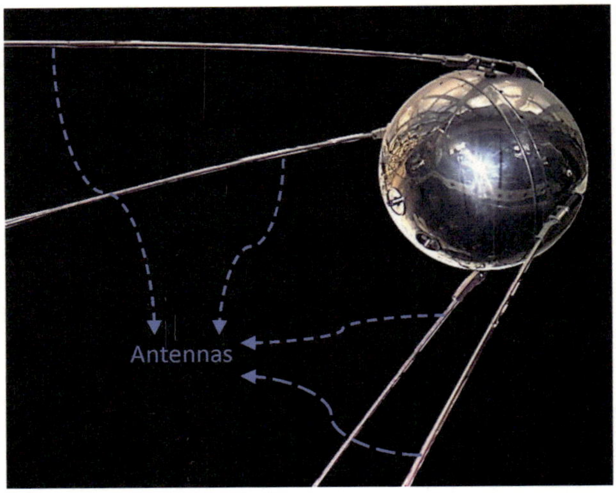

Figure 1.
The Sputnik-1 with monopole antennas (image courtesy of NASA/NSSDCA).

joined this race with *Explorer-1*. The Explorer-1 was placed into an orbit with a perigee of 360 km and apogee of 2535 km as mentioned in [1]. It had a cosmic ray sensor, an internal temperature sensor, a temperature sensor on the cone nose, and a microphone to detect the micro-meteoroid effect. Its most important task was to detect high-energy atomic particles with a cosmic ray sensor. The sensor therefore contained a Geiger-Müller tube. There were two transmitters at 108 and 108.03 MHz. It transmitted the relevant data to the station with those modules. As can be seen from **Figure 2**, the fiberglass slot antenna is positioned toward the nose of the vehicle. In addition, four flexible monopoles are placed in the middle of the vehicle. Its most important achievement was the discovery of the generations of Van Allen belt around the world. Afterward, this discovery was confirmed by Explorer-3 spacecraft [1].

By 2017, there were 4635 satellites orbiting the planet for different purposes based on the published Index of Objects Launched into Outer Space maintained

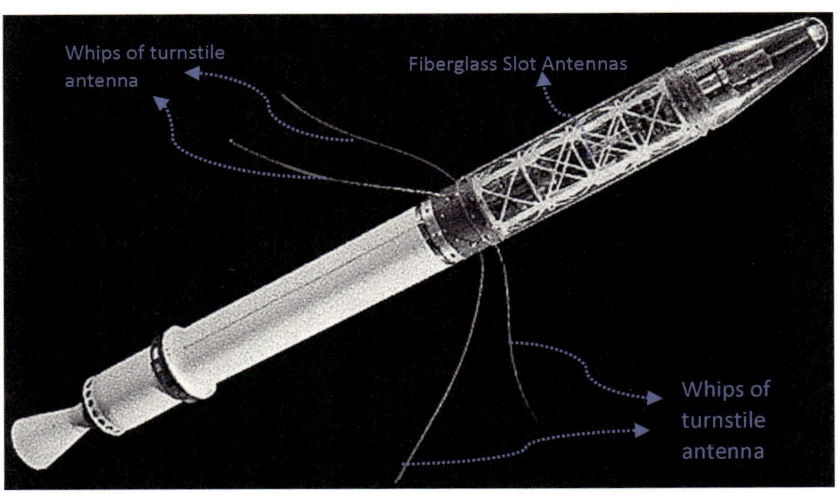

Figure 2.
The Explorer-1 with its turnstile antenna for VHF communication (image courtesy of NASA).

by the United Nations Office for Outer Space Affairs (*UNOOSA*). According to the report of 2019, while this chapter was being written, about 5000 satellites have been evolving around the world.

In recent years, especially antennas used in satellite communication systems are expected to have low volume, lightweight, low cost, high gain and directivity. Since the antennas used here are the last elements of the transmitting and receiving systems, they enable the connection of both sides over the space. They must be therefore suitable to the structure on which they are used, both electrically and physically. In addition, the gain and radiation pattern characteristics must be considered together with the general approaches used in the design of these antennas. The characteristics of the printed circuit antennas in meeting these criteria are more appropriate. Another important feature is that it is compatible with printed circuit technology and can be produced as a persistence of RF and high frequency circuit topology. Another advantage of printed circuit antennas is that they can be easily mounted on non-planar surfaces or manufactured using flexible printed circuit boards. In order to realize matching circuits, in very small areas inductive, resistive and capacitive surface mount device (SMD) components can be used with the printed circuit technology. Similarly, the frequency tuning of the antennas can be achieved electrically and mechanically in a variety of ways, which makes it particularly advantageous for the printed circuit antennas.

Depending on the orbit around the world in general, space vehicles and satellites can be divided into six main groups:

- Satellites with low earth orbiting (LEO) satellites
- Satellites with middle earth orbiting (MEO) satellites
- High elliptical orbiting (HEO) satellites
- Geostationary (GEO) satellites,
- Scientific research and exploring for solar system, deep space and others,
- Manned space flights-for now generally in LEO for example International Space Station (ISS).
- In addition, space launch vehicles can be added to this list.

There are also telemetry/telecommand (TM/TC) communication units in different frequency bands, global positioning systems and other telecommunication modules for transmitting and receiving the RF signals in launch vehicle, respectively. Consequently, suitable antennas should be designed and utilized according to planned mission for the space launch vehicles similar to the satellites.

One of the most needed satellites is LEO satellite. LEO satellites orbit between 160 and 1600 km from the Earth's surface. These satellites are usually small compared to communication GEO satellites, easy to launch and put into orbit. They can be used for different purposes. For ground monitoring purposes, satellite constellation can be placed into orbit and used for voice, fax and data communications. In addition, due to the limited surface area and volume available on the satellite, the antenna must be as small as possible in weight and volume. Finally, considering the limited power budget of the satellite, it is important that the antenna may have a passive and conical radiation pattern to direct the electromagnetic energy to low elevation angles.

Antennas used in LEO-type satellites can be divided into three types: payload data transmission (PDT) antennas for downloading high-density data to the ground station or inter satellite link (ISL) communication, payload antennas for special missions like mobile communication, GNSS services or remote sensing operations and TM/TC antennas to control the satellite and receive health parameters to monitor its functionality. The frequency ranges allocated for LEO satellites vary according to the characteristics of the payload on the satellite, but are determined by International Telecommunication Union (ITU).

After a LEO satellite is launched, it must be brought to desired position or orbit in order to fulfill the function of the satellite or be required to stabilize tumbling. This phase is called as Launch and Early Orbiting Phase (LEOP). In this case, hemispherical or omnidirectional antennas are very beneficial and used for the transmitting and receiving TM/TCs since they have wide coverage capability [2]. Antennas having directional or shaped conical radiation pattern are preferred in order to transmit telemetry and payload data to the ground station after it has been commissioned. Since line of sight (LOS) communication time interval is limited over one ground station, it is desirable to use this interval in the most efficient manner. This can be acquired by starting downlink and uplink communication at low elevation angles of the satellite. Because it will provide more time to download high-density payload data from the satellite. Therefore, at lower elevation angles of the satellite, a higher antenna gain is required, whereas in the case where the satellite is at higher elevation angle with respect to ground station, the antenna gain may be relatively lower so as to maintain link margin positively.

2. Antennas exposed to effects of space environment

In space not only functionality should be taken into consideration but also durability and reliability of antennas should be taken into account. Consequently, in design phase of antennas to be used in space applications, environmental conditions are decisive factors. Materials to be used on space antennas should meet requirements based on space qualifications and factors [3]. These factors can be listed under two main subjects: *effects due to the launching activity* and *space environment.*

2.1 Launch phase

During launch of spacecraft, acoustic vibrations, shocks, mechanical stress based on static loads, dynamic loads and sudden atmospheric pressure fall occur and those effects should be taken into account in the course of antenna design step. In addition, in commissioning phase pyrotechnical shocks are generated while deploying solar panels and payloads like deployable antennas. All of those may affect objects, for example antennas, detached to surface of spacecraft, adversely.

2.2 Space environment

After LEOP, antennas will be exposed to harsh space environment. Those can be listed as vacuum, high temperature changes regarding nonconductive thermal feature of vacuum typically between -150 and 150°C, outgassing or material sublimation which can create contamination for payloads especially on lens of cameras, ionizing or cosmic radiation (beta, gamma, and X-rays), solar radiation, atomic oxygen oxidation or erosion due to atmospheric effect of low earth orbiting.

2.3 Verification for launch and environmental effects

In order to verify that antennas can perform functionally in space environment and withstand launch effect mentioned above, some tests should be performed as addition to functional tests before mission started. These environmental verifications can be listed as:

- thermal qualification,
- sine vibration,
- random vibration or acoustic,
- quasi-static acceleration,
- stiffness measurement, and
- low outgassing compatibility [4].

To verify the modules, requirements and tests have been defined by NASA and ESA in their published standards. For space programs, the related requirements and tests are prepared based on those standards. Some important and general ones can be listed as:

- ECSS-E-ST-32-08C—materials
- ECSS-Q-ST-70-02—thermal vacuum outgassing test for the screening of space materials
- ECSS-Q-ST-70-71C Rev.1—materials, processes and their data selection
- ECSS-E-ST-10-03C—testing
- ECSS-Q-ST-70-04C—thermal testing for the evaluation of space materials, processes, mechanical parts and assemblies

which have been published by ESA and

- GSFC-STD-7000A—General Environmental Verification Standard (GEVS) for GSFC Flight Programs and Projects
- Outgassing Data for Selecting Spacecraft Materials
- NASA-STD-7002B—Payload Test Requirements
- NASA-STD-5001—Structural Design and Test Factors of Safety for Spaceflight Hardware
- NASA-STD-7001—Payload Vibroacoustic Test Criteria
- NASA-STD-7003—Pyroshock Test Criteria

which have been published by NASA.

2.4 Other effects

2.4.1 Multipaction and corona discharge

Multipaction is, basically, an event that can be reason of breakdown because of high power RF signal in a vacuum or near vacuum medium. It can reduce RF output power of device, cause noise in RF signal and even *corona discharge* because of ionization in presence of electromagnetic wave. Therefore, it can result a catastrophic failure of an antenna, RF component and even another payload module. There are two main factors for multipaction: high RF power and vacuum medium. Thus, related RF components including antennas should be either analyzed or tested for these phenomena. There is an analysis tool designed by ESA/ESTEC named as "ECSS Multipactor Tool". By using this tool one can calculate threshold and safety margin levels for pre-defined structures according to the operating frequency, impedance, RF power level, material finishing and minimum distance between metal tips or edges.

2.4.2 Passive intermodulation

As is known, in active RF devices there can occur intermodulation products of applied two or more tones at the output of the device. Similar phenomenon can be seen at antennas because of two main reasons: nonlinearity of material and nonlinearity of contact.

To avoid multipaction and passive intermodulation there are some published standards for design and verification phases. One of them is ECSS-E-20-01A Rev.1—*multipaction design and test*.

3. Frequency allocations for space missions

Radio frequency spectrum usage must be regulated to guarantee a cost-effective and high-capacity utilization for terrestrial and satellite communication systems all over the world. Although, the frequency intervals used for the spacecrafts vary according to the characteristics of the TM/TC modules and payload on the space vehicle, they are determined and coordinated by International Telecommunication Union (ITU). ITU, which is a United Nations structure, acts under the contract accepted by the executives of member states. For this aim, ITU holds a forum and owing to this forum it can do coordination of radiofrequency spectrum, define standards for communication protocols, device or equipment characteristics and interference levels between private sector and member states [5].

There are radio communication services defined for space applications to transmit or receive radio signals by the Radio Communication Regulations. Those are

- ASS: amateur satellite service
- BSS: broadcasting satellite service
- EES: Earth exploration satellite service
- FSS: fixed satellite service
- ISS: inter-satellite service

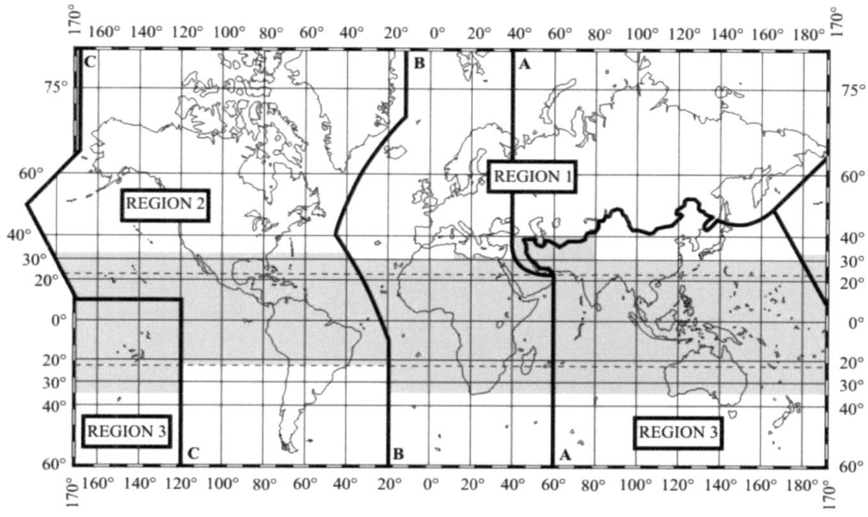

Figure 3.
ITU regions of the world for frequency allocation as given in [6] (image courtesy of ITU, source: ITU radio regulations (2016 edition)).

- MSS: mobile satellite service
- RSS: radio determination satellite service
- SOS: space operation service
- SRS: space research service

Based on the radio communication services mentioned above, the defined or allocated frequency bands can be private for one of them or shared between these services. Moreover, the world has been divided into three parts for the coordination as accepted and published in 1988. It is seen in **Figure 3**.

Region 1 consists of Europe, Africa, the Middle East and the land of former USSR; Region 2 is the Americas and finally region 3 is Asia Pacific but not including the Middle East and land of former USSR. In **Table 1**, some specific frequency bands used for space communication are presented. An interested reader to see full frequency spectrum allocations of terrestrial and space application can visit the web site created by IEEE Geoscience and Remote Sensing Society given in [7].

4. Antenna types used on spacecrafts

There are many different spacecrafts to conduct various missions for humankind. Consequently, they need different subsystems like communication subsystems, microwave imaging payloads, instrument landing systems which use radar technology, scientific and experimental research devices to explore deep-space and many others. This leads to a need for different type of antennas to fulfill defined missions successfully. Spacecrafts can be divided into four main groups: missile launchers, satellites, radio astronomy and deep space vehicles. Based on this categorization some antenna types used on those vehicles will be reviewed in the following sections.

Frequency band (MHz)	Primary allocation	European common allocation	Applications	Standard
144–146	Amateur Amateur satellite 5.216	Amateur Amateur satellite	Amateur Amateur satellite	EN 301 783
434.79–438	Amateur Radiolocation Earth exploration-satellite (active) 5.279A 5.138 5.271 5.276 5.277 5.280 5.282	Amateur Amateur satellite Radiolocation Earth exploration satellite (active) 5.279A 5.277 EU2 EU12	Active sensors (satellite) Amateur Amateur satellite	EN 301 783
1164–1215	Aeronautical radio navigation 5.328 Radio navigation satellite (space-to-Earth) (space-to-space) 5.328B 5.328A	Aeronautical radio navigation 5.328 Radio navigation-satellite (space-to-Earth) (space-to-space) 5.328B 5.328A	Aeronautical navigation GALILEO GLONASS GNSS repeater	EN 302 645
1215–1240	Earth exploration satellite (active) Radiolocation Radio navigation satellite (space-to-Earth) (space-to-space) 5.328B 5.329 5.329A Space research (active) 5.330 5.331 5.332	Earth exploration satellite (active) Radiolocation Radio navigation satellite (space-to-Earth) (space-to-space) 5.328b 5.329 5.329a Space research (active) 5.331 EU2 5.332	Active sensors (satellite) Defense systems GLONASS GNSS repeater GPS Radiolocation (civil)	EN 302 645
1559–1610	Aeronautical radio navigation Radio navigation satellite (space-to-Earth) (space-to-space) 5.208B 5.328B 5.329A 5.341 5.362B 5.362C	Aeronautical radio navigation Radio navigation satellite (space-to-Earth) (space-to-space) 5.208B 5.328B 5.329A 5.341 5.362B	GALILEO GLONASS GNSS pseudolites GNSS repeater GPS	EN 302 645
2025–2110	Earth exploration satellite (Earth-to-space) (space-to-space) Fixed Mobile 5.391 Space operation (Earth-to-space) (space-to-space) Space research (Earth-to-space) (space-to-space) 5.392 5.392	Earth exploration satellite (Earth-to-space) (space-to-space) Fixed Mobile 5.391 Space operation (Earth-to-space) (space-to-space) Space research (Earth-to-space) (space-to-space) 5.392 EU2 EU15 EU27	Defense systems Fixed PMSE Space research	EN 302 217 EN 302 064

Frequency band (MHz)	Primary allocation	European common allocation	Applications	Standard
2200–2290	Earth exploration satellite (space-to-Earth) (space-to-space) Fixed Mobile 5.391 Space operation (space-to-Earth) (space-to-space) Space research (space-to-Earth) (space-to-space) 5.392	Earth exploration satellite (space-to-Earth) (space-to-space) Fixed Mobile 5.391 Space operation (space-to-Earth) (space-to-space) Space research (space-to-Earth) (space-to-space) 5.392 EU15 EU27	Defense systems Fixed PMSE Radio astronomy Space research	EN 302 217 EN 302 064
8025–8400	Earth exploration satellite (space-to-Earth) Fixed Fixed satellite (Earth-to-space) Mobile 5.463 5.462A	Earth exploration satellite (space-to-Earth) Fixed Fixed satellite (Earth-to-space) Meteorological satellite (Earth-to-space) Mobile 5.463 5.462A EU2 EU27	Defense systems Earth exploration satellite Fixed Radio astronomy Radio determination applications	EN 302 217 EN 302 729

Table 1.
Some specific frequency bands for space communication defined by ITU.

4.1 Antennas for missile launchers

In order to acquire TM/TC communication, guidance, transmitting and receiving radar signals, sending video and image, communicating with satellite after departing, there are many antennas used on missile launchers. Up to now, a lot of different antenna types have been designed for this purpose. However, since the main objective of missile is military usage (ballistic missiles a good example), it is hard to find adequate info about subsystems on them in open literature. In the following subsections some examples are given.

4.1.1 Transmission line antennas

Particularly for TM/TC communication subsystems, missiles need Omni-directional antennas to communicate with ground stations. Since antennas are the final or first component of RF transmitter or receiver, respectively, they must be on outward or just underneath surface of missiles with RF transparent radome. Nevertheless, they must comply with aerodynamic structure of missile. Otherwise, it will increase air-drag during trip along the atmosphere. Therefore, if antennas will be used over the surface of a missile, they must be compatible with aerodynamic structure. A well-known type of antenna for this goal is *transmission line antenna* [8] which is also commonly used for other aerospace vehicles. It is known that radiation resistance of a transmission line is quite small. In order to increase the radiated power rather than power dissipated as heat, a transmission line can be terminated with reactive elements like capacitors, conducting bridges or open ends. Based on this technique, in [8] four types of them have been presented. These are *inverted L antenna, shunt-driven inverted L antenna with open end, shunt-driven inverted L antenna with capacitive end loading* and finally *m antenna.*

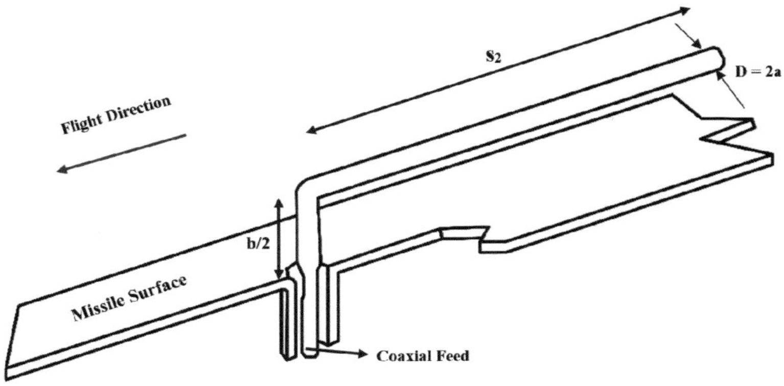

Figure 4.
Inverted L antenna for using on missiles [8].

Inverted L antenna is a simple wire antenna structure, which is folded on a point of a transmission line as illustrated in **Figure 4**. It should be emphasized that its open end is at the reverse direction to flight in order to afford aerodynamic conformity to the missile. It is fed by coaxial line. Distance between the ground surface of missile and the antenna wire is b/2, horizontal length of the antenna is s_2 and the diameter of wire that is used to form the antenna is 2a. Inverted L antenna has been analyzed based on the transmission line theory in [8]. Therefore, the diameter of wire 2a and height over surface b/2 must be smaller than operating wavelength. The derived relations for transmission line modeling of these antennas are valid for $2a < 0.01\lambda$ and $b \le 0.1\lambda$ as mentioned in [8].

Equivalent circuit model of inverted L antenna is shown in **Figure 5**.

Using transmission line theory radiation resistance of the antenna can be derived as

$$R^e = \frac{30\beta^2 b^2}{\sin^2 \beta s_2}\left\{1 - \frac{\sin 2\beta s_2}{2\beta s_2}\right\} \tag{1}$$

where β is phase constant for the transmission line [8].

The second type of transmission line antenna is shunt-driven inverted L antenna with open end as illustrated in **Figure 6**. In this case, there is a shorting line with a distance of s_1 from center of the feeding line. Again open end is opposite to the direction of flight.

Its equivalent circuit model is presented in **Figure 7**.

By using transmission line theory radiation resistance of the antenna can be derived as given in [8].

$$R^e = \frac{30\beta^2 b^2}{\cos^2 \beta s_2}\left\{\frac{1}{2}\left[\cos^2 \beta s + \sin^2 \beta s_1 + \cos^2 \beta s_2\right] - \cos\beta s \cos\beta s_2 \frac{\sin\beta s_1}{b s_1}\right\}, \tag{2}$$
$$\text{where } \cos\beta s \neq 0$$

Here, $s = s_1 + s_2$. If this antenna is modified by adding a tunable capacitor instead of open end it will be easy to tune input impedance of the antenna. It is shunt-driven inverted L antenna with capacitive end loading. Its illustration is shown in **Figure 8**.

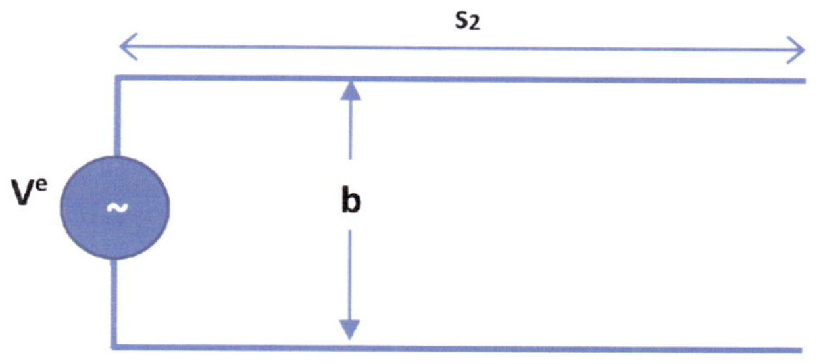

Figure 5.
Equivalent circuit model of inverted L antenna: end-driven open-end section of line [8].

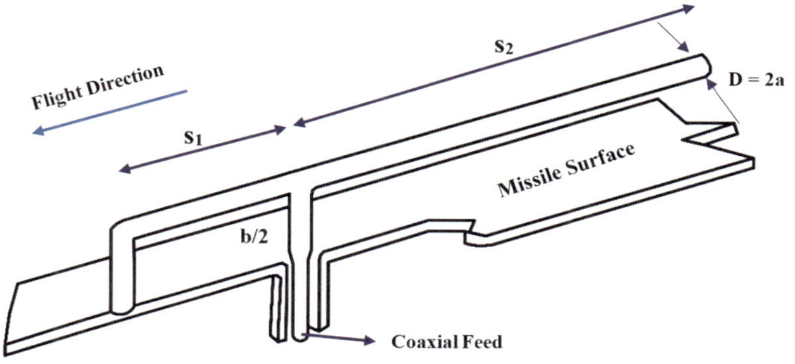

Figure 6.
Shunt-driven inverted L antenna with open end for using on missiles [8].

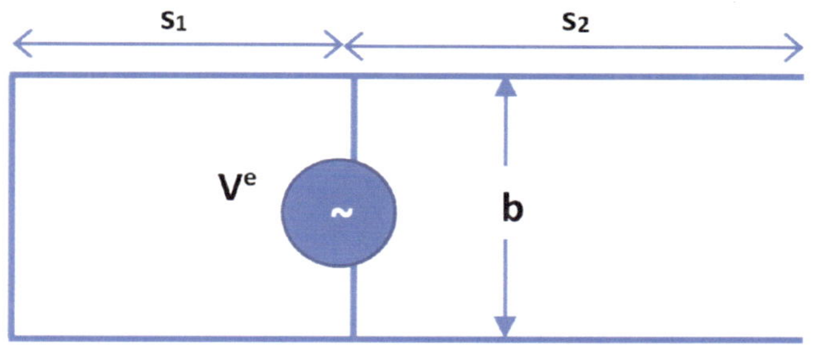

Figure 7.
Equivalent circuit model of shunt-driven inverted L antenna with open end: shunt-driven line with an open and a short-circuited termination [8].

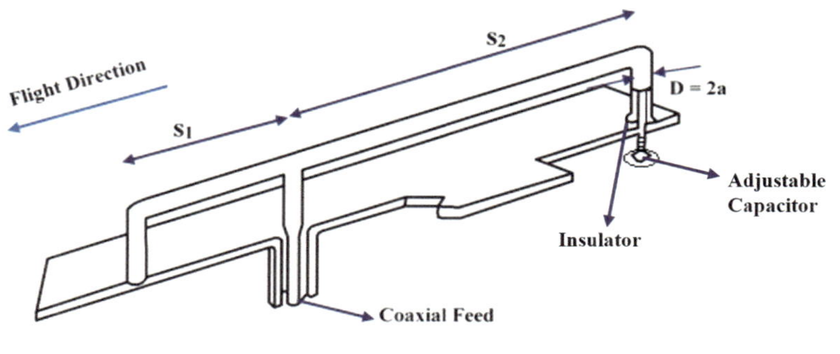

Figure 8.
Shunt-driven inverted L antenna with capacitive end loading for using on missiles [8].

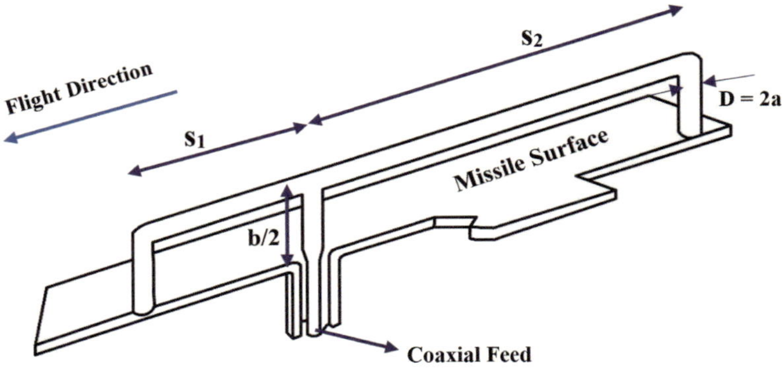

Figure 9.
m-Antenna with capacitive end loading for using on missiles [8].

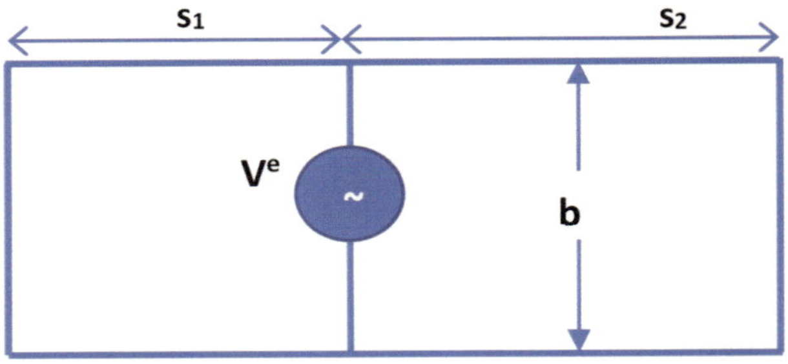

Figure 10.
Equivalent circuit model of m antenna: shunt-driven line with short-circuited terminations [8].

In order to derive the radiation resistance of reactive element ended inverted L antenna the relations given in [9] can be used. For the sake of brevity, derivation of radiation resistance has not been given within text.

The final type of transmission line antenna is m-antenna. Its illustration is given in **Figure 9**.

Its equivalent circuit model is presented in **Figure 10**.

Similarly, as used in previous antenna types, using transmission line theory radiation resistance of the antenna can be derived as [8].

$$R^e = \frac{30\beta^2 b^2}{\sin^2 \beta s_2}\left\{\frac{1}{2}\left[\sin^2 \beta s + \sin^2 \beta s_1 + \sin^2 \beta s_2\right] - \left(\frac{s_1^2 + s_1 s_2 + s_2^2}{\beta s_1 s_2}\right) sin\beta s\ sin\beta s_1 sin\beta s_2\right\} \quad (3)$$
$$(sin\beta s \neq 0)$$

and $s = s_1 + s_2$.

Up to date, many different versions of those basic antennas have been published in the literature. Particularly, their printed circuit versions are also available and used for different communication devices in cellular phones, tablets and portable computers as well.

4.1.2 Extremely thin and omnidirectional slot antenna array for launch vehicles

Another basic antenna used on missiles is conformal slot array structure. In order to get enhanced coverage for launch vehicles, array antennas are versatile

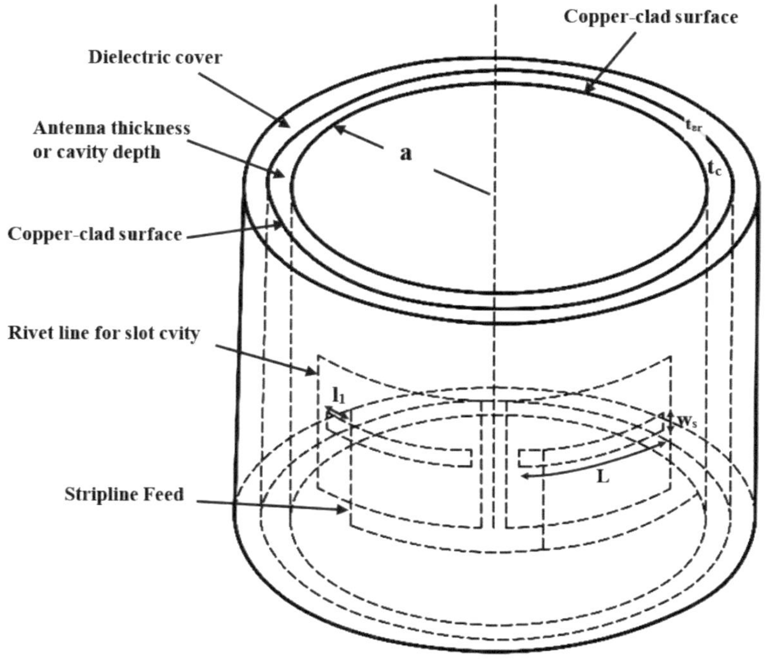

Figure 11.
Outline drawing of stripline-fed cavity-backed slot array antenna on space launch vehicle [10].

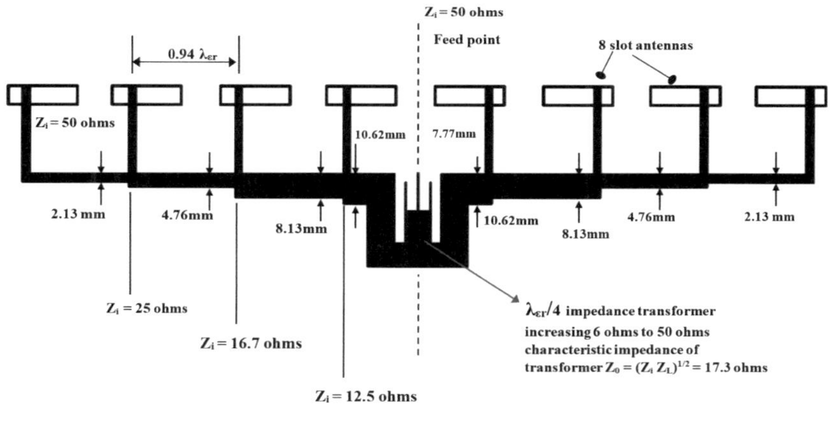

Figure 12.
Series stripline feed structure for the array of eight slots [10].

and effective. Almost omnidirectional pattern can be achieved using circumferential or conformal array antennas on the launch vehicles. There are various examples for this type in the general literature. One of them is presented in [10] Its outline illustration is shown in **Figure 11**. As shown in this figure eight slot elements were used and to decrease the number of branch lines the author used series feedline as shown in **Figure 12**. Input impedance of each slot antenna element is designed as 50 Ω. Moreover, in **Figure 12** matching of the main input port of the array is explained in detail. For matching 6 Ω input impedance of eight-element slot array to 50 Ω, λ/4 transformer is employed at the input. To suppress higher order modes and edge excitation of the slots, bolts are inserted circumferentially to form a rectangular transmission line. In Ref. [10], it is particularly

Description	Dimensions (mm)
Single slot antenna	
Slot size without coating	64.77 × 2.54
Minimum cavity size	50.8 × 83.82 × 2.39
Feeding point position from end of slot to obtain 50 Ω	15.24
L (slot length) where $t\varepsilon_r$ = 0.15	57.4
L (slot length) where $t\varepsilon_r$ = 0.10	58.17
Array of slots	
L (slot length) where $t\varepsilon_r$ = 0.10	55.88
Cavity size	50.8 × 83.82 × 2.310
Feeding point position from end of slot to obtain 50 Ωs	15.24

Table 2.
Physical design parameters of conformal stripline fed slot array antenna [10].

highlighted that spacing between each bolt and number of bolts are crucial to obtain continuous short circuit around slot aperture and also at transition area from RF connector to stripline feed.

Design parameters of this conformal array are presented in **Table 2**. Those physical design parameters were obtained for a substrate having a dielectric permittivity of 2.1 and a dielectric cover whose permittivity is 2.54. The dielectric cover was used to protect the array antenna from heat arising because of atmospheric friction. To obtain 50 Ω input impedance for each slot antenna, offset feeding technique was used.

Azimuth and elevation patterns of the array are presented in **Figures 13** and **14**, respectively. In these figures, omnidirectional radiation pattern characteristic is seen clearly around the launch vehicle.

In those figures red line represents the radiation pattern with superstrate and black dashed line represents radiation pattern without superstrate.

Today, different types of wrap around slot arrays are still being used on launch vehicles to form omnidirectional radiation pattern around the vehicle. The number of the array element can be increased depending on the circumferential electrical

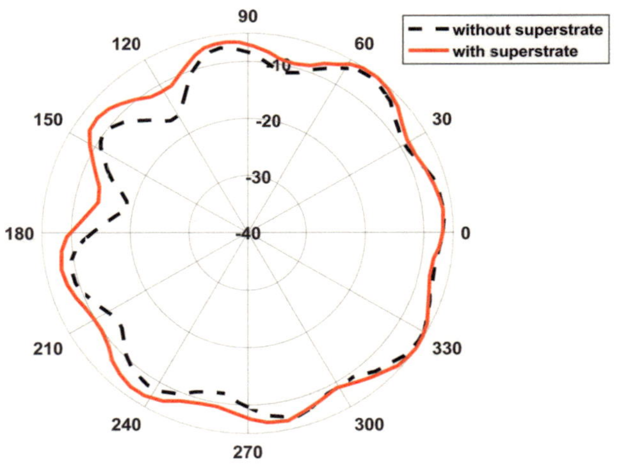

Figure 13.
Radiation pattern of the slot array in the azimuth plane with and without dielectric cover [10].

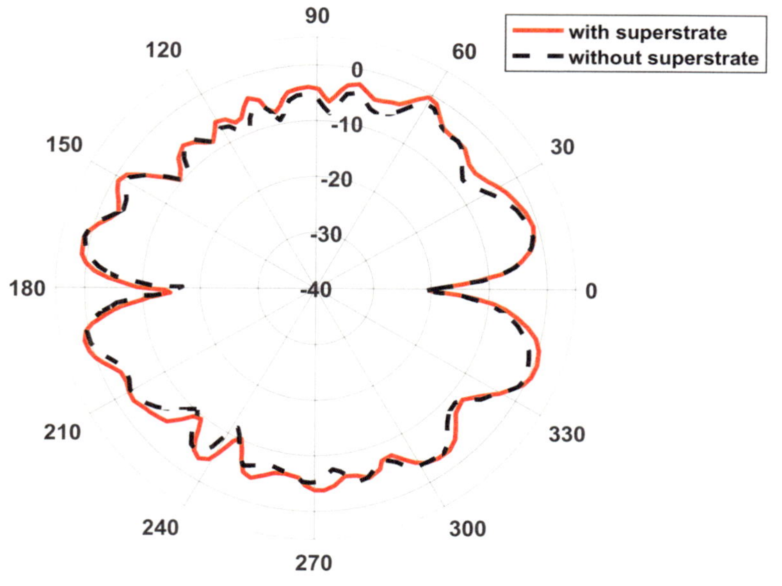

Figure 14.
Radiation pattern of the slot array in the elevation plane with and without dielectric cover [10].

length. Thanks to those antennas, TM/TC RF signals can be sent or received easily. An interested reader can find commercial products based on this type, since it has many advantages like ease of manufacturing and good conformity to aerodynamic structure of the vehicle.

4.2 Satellite antennas

Up to date, numerous antennas have been designed and employed for different space missions. As mentioned in previous parts, most of them are for satellite communication systems like TM/TC and PDT subsystems, broadcast payload etc. As given in the first section, satellites are usually categorized according to their orbits. Those orbits define and affect general characteristics of satellites to be designed and manufactured for power generation from their solar panels, communication period and slot with ground station, radiation endurance, parts to be used because of atmospheric effects like atomic oxygen and their payload specifications.

After frequency definition for subsystems, types of antennas to be used for communication, remote sensing instrument and scientific instruments are selected. For example, circularly polarized antennas are usually preferred for TM/TC antennas not to be affected from polarization mismatch, which can be caused by maneuvers during low earth orbiting phases and atmospheric effects like Faraday rotation [11]. Besides antennas used on small satellites should be as low profile as possible due to surface and volume restrictions. However, for PDT and remote sensing applications medium and high gain antennas are needed. To use high gain and therefore narrow beamwidth antennas efficiently, they should be steered whether directing whole satellite platform or using additional steering mechanism like electromechanical structures or electronically steerable phased array antenna systems.

As is well known there are different types of antennas used for satellites but since there is very limited place in this part, only some noteworthy ones will be taken into consideration and presented for the sake of brevity.

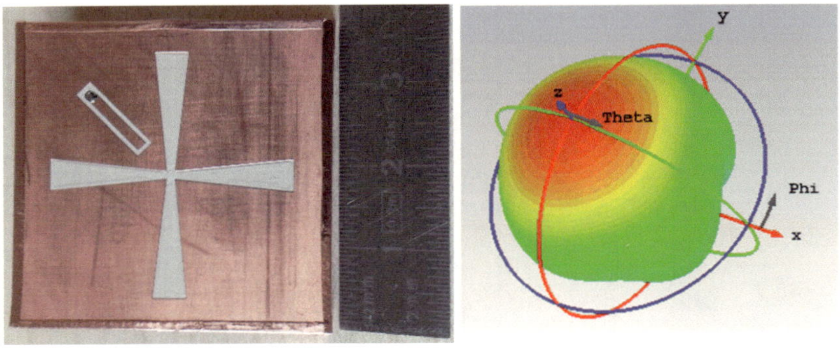

Figure 15.
Manufactured prototype of cavity-backed antenna with tapered crossed-slot aperture and its 3D radiation pattern [13].

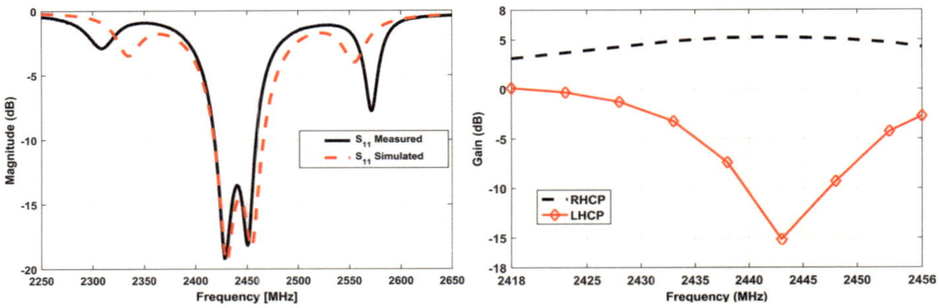

Figure 16.
Measured S11, co-pol and cross-pol gain results versus frequency for cavity-backed antenna with tapered-slot aperture [13].

4.2.1 Small antennas for tiny satellites

There is tremendous demand to accomplish space research at reasonable prices for universities and commercial entities therefore CubeSat is a practical and functional platform for this objective. Dimensions of a 1 U CubeSat are 100 mm x 100 m and it has aluminum T6061 structure with a total mass of up to 1 kg. However, 1 U can be easily enlarged to larger sizes like 2 U, 3 U, etc. Comparing to other satellite platforms, CubeSats have limited volume therefore submodules and antennas should fit into those tiny platforms.

There are good small antenna study examples developed under GAMALINK[1] [12] project for CubeSat antennas and one of them is given in [13]. In this study a miniaturized cavity-backed tapered cross-slot antenna has been presented. 38×38 mm^2 and 30×30 mm^2 footprints have been obtained on substrates having dielectric permittivity 6 and 9.2, respectively, at operating frequency about 2.44 GHz. Manufactured prototype of the antenna and its simulated 3D radiation pattern are shown in **Figure 15**.

Its maximum gain is at boresight and efficiency is small as expected because of miniaturization. However, its tiny dimensions make this antenna beneficial to save space on surfaces of small spacecrafts like CubeSats. In addition, prototype was manufactured to verify the simulation results shown in **Figure 16** as mentioned

---

[1] GAMALINK (Generic SDR-bAsed Multifunctional spAce LINK) is a project for European Union's FP7-Seventh Framework Programme for research, technological development and demonstration under grant agreement no 312830.

in [13]. S11, gain and cross pol gain are presented in this figure and the measured values support electromagnetic simulation results. These types of miniaturized antennas have low efficiency. Nonetheless, since they have small footprints they can be installed to the CubeSats easily especially for TM/TC communication and even for Inter Satellite Link (ISL). Therefore, for distinctive missions even 3 or 4 of them can be gathered to form an array.

However, an interested reader should review other studies where mesh and optically transparent or mesh type antennas are proposed as well as small antennas [14–22].

4.2.2 Deployable large antennas for tiny satellites

For some specific operations electrically large antennas can be needed on CubeSats. Those antennas are folded, stowed or packed in a CubeSat before and during launch process. After satellite platform is placed into orbit they are deployed to conduct their missions. For this aim, there are deployable antenna examples where cutting edge mechanical technologies are employed (**Figure 17**).

A stowed 0.5 m Ka-band mesh reflector antenna was installed into RaInCube platform to initiate usage of Ka-Band radar for meteorology on a low-cost and fast

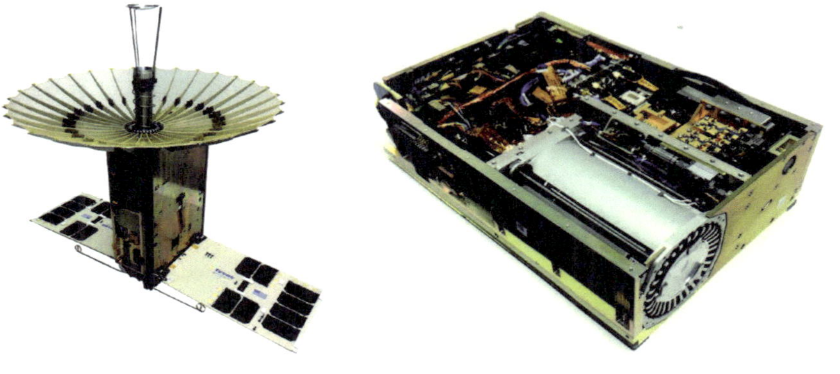

Figure 17.
Ka-band deployable parabolic dish antenna which has 30 ribs similar to an umbrella to be stowed and installed into RaInCube platform. Left deployed and right stowed into RaInCube (image courtesy of NASA/JPL-Caltech).

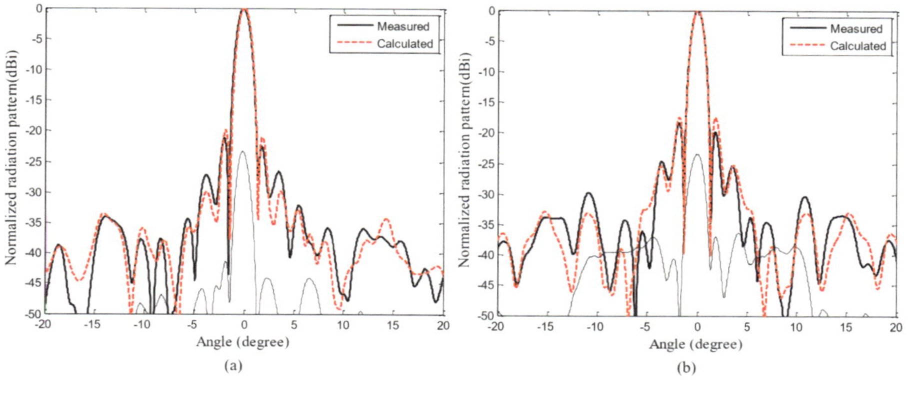

Figure 18.
Measured and calculated radiation pattern of the deployable mesh reflector antenna model. (a) ϕ = 0°. (b) ϕ = 90°. (Reprinted, with permission, from [25], © 2016 IEEE).

applicable 6 U CubeSat platform of NASA [23–25]. The measured gain and efficiency of this antenna are 42.6 dBi and 52%, respectively, at 35.75 GHz [23].

Radiation pattern of the reflector antenna is given in **Figure 18**. Its beamwidth is about 1.2° in E- and H-planes. RaInCube was launched into a near circular orbit where its altitude of ~400 km and with inclination of 51.6° on 21 May 2018 and the mesh reflector antenna deployment process achieved on 28 July 2018.

4.2.3 GEO satellite communication antennas

In the past, GEO satellites' main mission was only television broadcasting and voice data transmission. Therefore, there are many communication satellites as geosynchronous. In the last decade, they have started evolving and internet communication mission has begun to take place instead of TV broadcasting. The main reason for this is that the internet goes into all areas of life like business, education, entertainment, etc.

Since GEO satellites are about 36,000 km away from earth, they need high effective isotropic radiated power (EIRP) levels. So usually large aperture reflector antennas are employed. Based on ITU regulations generally these antennas shape their beams according to geographical regions. This is because to reuse frequencies allocated over regions as shown in **Figure 3**.

There are many different applications of reflector antennas for GEO communication satellites [26–33].

There is a good example to illustrate evolutionary change of GEO communication satellites. To provide high speed internet data communication JAXA started The "KIZUNA" – Wideband InterNetworking engineering test and Demonstration Satellite (WINDS) project. Its main mission was to enable super high-speed data communications of up to 1.2 Gbps. In this way, everybody can reach high-speed communications, no matter in which geographical region of Japan they live. Its illustration and its payload antennas are shown in **Figure 19**.

As shown in **Figure 19** there are three payload antenna structures on the satellite. Two of them are Ka-Band multibeam reflector antennas and the other one is Ka-Band Active Electronically-Controlled Scanning Array (AESA) antenna. The multibeam antennas have 2.4 m diameter and in type of 2 offset-feed Cassegrain. The AESA has 128 elements to establish transmit and receive beams. Multibeam reflector antennas are responsible for Japan with Asia and AESA is operational for Pacific and partly Asia.

KIZUNA was launched and put into Geosynchronous Orbit to acquire the highest-speed data communication of the world in 2008. Its planned operational life was 5 years and failed in February 2019 and started to drift. Therefore, it exceeded its planned operational life successfully.

4.2.4 Reflector antennas

A big number of scientist and communication antenna specialists are working on the increase of performance properties of reflector antennas for the widely usage in deep space communication, satellite communication stations, radio astronomy, current microwaves such as radio-links and radars. Parabolic reflector antennas are preferred to use as main reflector in communication systems due to its high gain and directivity properties. Also, these types of reflector antennas can give the opportunity for usage in multi-band and multi beam applications. For these reasons parabolic reflector antennas have attracted the intense interest of researchers for many years [34, 35]. A simple and known physical structure of a parabolic reflector antenna is in **Figure 20**.

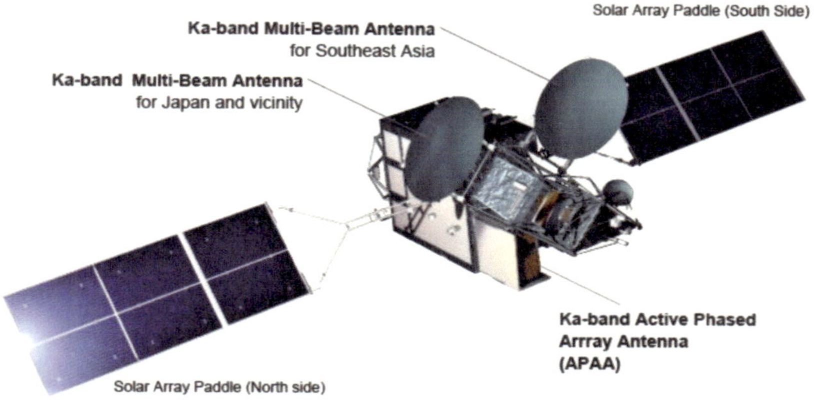

Figure 19.
KIZUNA" – WINDS high speed data communication satellite (image courtesy of JAXA).

Figure 20.
Parabolic reflector antennas at The NASA Deep Space Network Goldstone Ground Station Complex (image courtesy of NASA).

A parabolic reflector antenna consists of many important sections such as main reflector, feed, struts and control units, pedestal or support. Each of these should be carefully analyzed and designed.

Additionally, it is possible to use reflector antennas in various forms as:

- Receiver-transformer operation (single earth antenna at the end of down-up links on the same path) form
- Transmitter and receiver (two different earth antennas at the ends of uplink-satellite-downlink paths) form
- Transmitter, satellite control unit and receiver (three different earth antennas at the ends of uplink-satellite-control unit paths and control units–satellite-receiver paths) form
- A number of earth reflector antennas depending on coverage areas of satellites.

Parabolic reflector can be fed as in axisymmetric, asymmetric and off-focus fed forms as seen in **Figure 21**. Symmetric feeding causes aperture blockage effects of feed and struts. To avoid this blockage, asymmetric and off-focus fed forms are preferred. For multiple beam generation array type feedings have been used. The approximate maximum gain of a parabolic reflector is given as below:

$$G_{\max} = \frac{4\pi A_{eff}}{\lambda^2} \tag{4}$$

Here, A_{eff} is effective aperture area including the effect of losses due to spillover, aperture taper, Joule type heating and distortions on the surface of reflector.

Reflector antennas have different shapes such as parabolic, hyperbolic, elliptic, circular and line profiles. Although the shapes are quite different, for mathematical analysis they can be converted to each other by defining a parameter called eccentricity. Using eccentricity parameter, the whole family can be analyzed in a unified form [36]. Various shapes and their eccentricity values are illustrated in **Figure 22**.

Incident and reflected field directions for a feed located at the focus of parabolic and hyperbolic antennas are seen in **Figure 23**.

In order to increase gains of parabolic reflector antennas, dual antenna structures can be used. Cassegrain (main reflector parabolic antenna and sub-reflector hyperbolic reflector) and Gregorian (main reflector parabolic antenna and sub-reflector ellipsoid reflector) are two most common types of dual antennas. As an example, a Cassegrain type reflector antenna is presented in **Figure 24**.

A dual-antenna can be fed in symmetrical, asymmetrical and off-focus forms as given in **Figure 25**.

There are several methods for the analysis of reflector antennas. Physical optics (PO) comes first among the known methods [37]. As it is known, it gives the correct results for the main lobe and first side lobe patterns [38].

PO integral has main difficulty in application due to the complexity of the physical structure. The diffraction effects are also considered in order to obtain the full and correct radiation patterns particularly to determine side and back lobes. Physical theory of diffraction (PTD), equivalent edge currents (EEC) [39], aperture field method (AFM) and Jacobi-Bessel series method have been applied to find the radiation patterns for front space of reflector antenna [40]. The uniform theory of diffraction (UTD) [41–43]

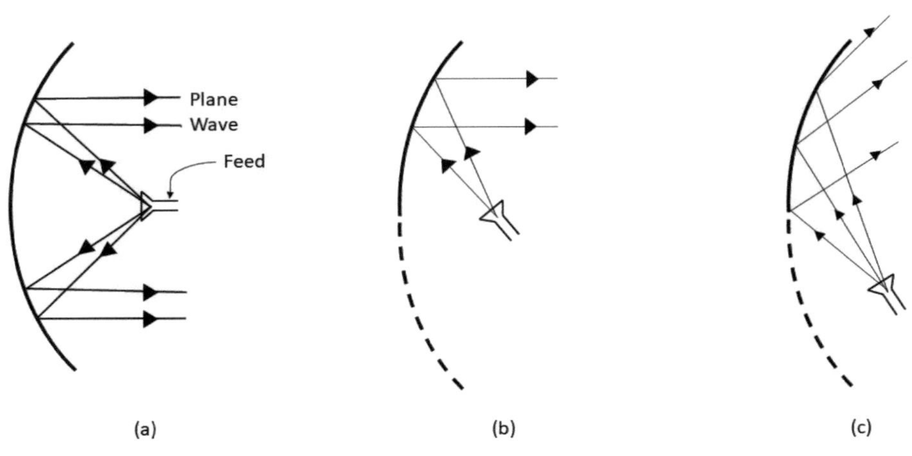

Figure 21.
Feeding types of a parabolic reflector antenna, (a) symmetric, (b) asymmetric, and (c) off-focus fed.

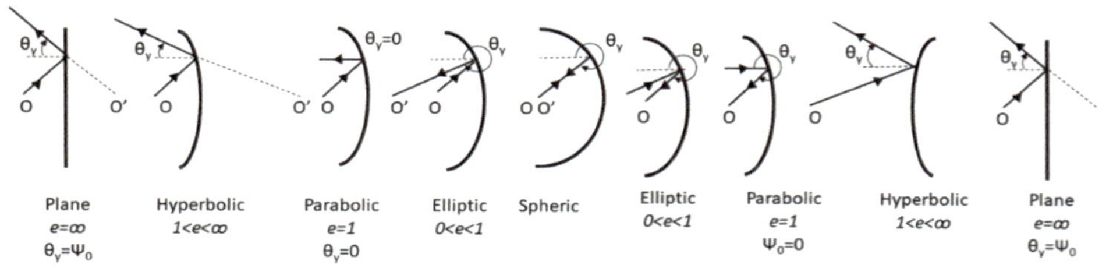

Figure 22.
Variation of reflector shapes with eccentricity.

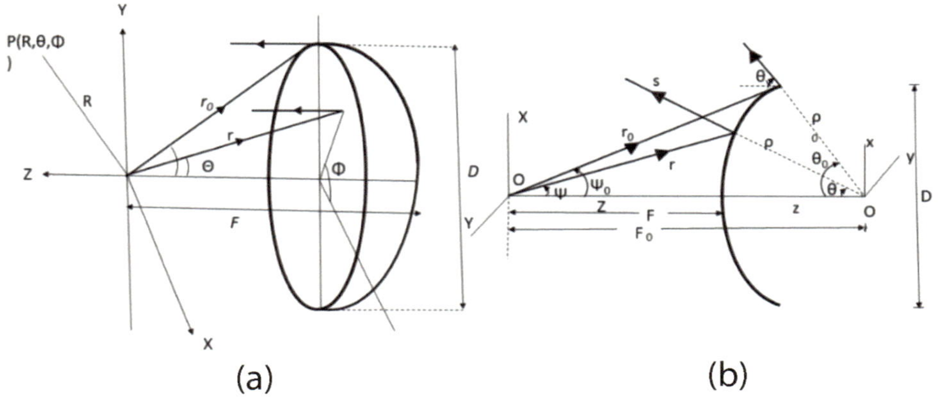

Figure 23.
Incident and reflected field directions of a reflector antenna (a) parabolic and (b) hyperbolic.

Figure 24.
DSS-43 dish at the Canberra Deep Space Communication Complex (CDSCC): a Cassegrain antenna whose diameter is 70 m (image courtesy of NASA).

and the uniform asymptotic theory (UAT) [44] based on the geometrical theory of diffraction (GTD) [45, 46] can be used in the analysis of radiation patterns of reflector antennas for various directions. These methods applied to the analysis of reflector antennas by many researchers [44, 47, 48] as seen in **Figure 26**.

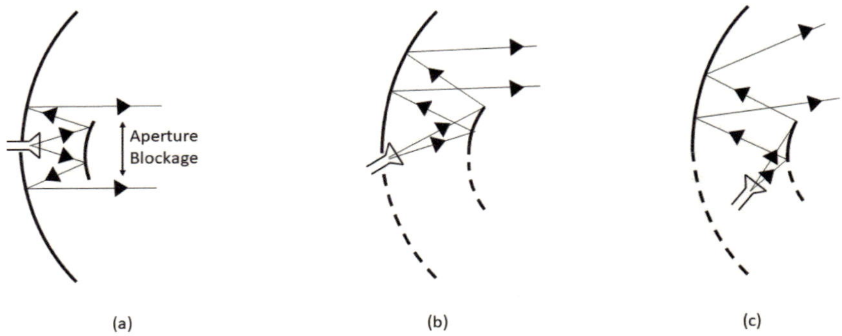

Figure 25.
Feeding types of dual antennas, (a) symmetrical, (b) asymmetrical, and (c) off-focus.

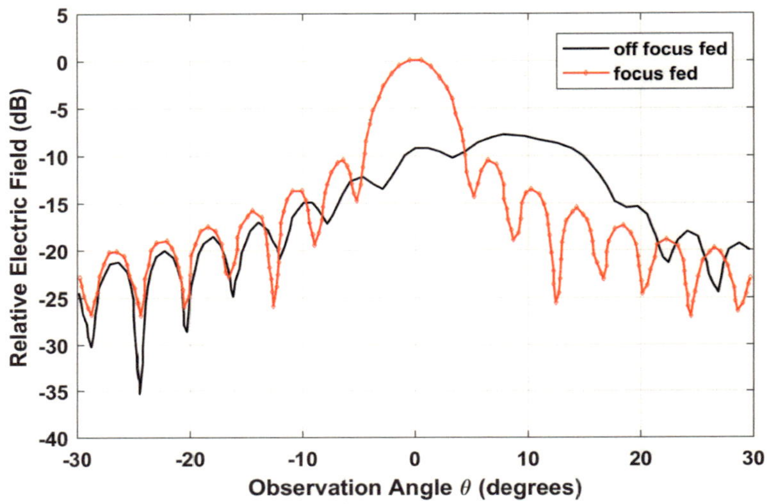

Figure 26.
Radiation pattern of a parabolic reflector antenna, $D = 15\lambda$,, $\alpha = \pi/3$, for focus fed [39] and off-focus fed with $fx = fy = fz = \lambda$. [40].

The use of array type feeding has been preferred for easy beam scanning in single and dual-reflectors by some scientists [29]. Also in recent years, some new techniques have been developed in the analysis of off-focus fed and array feeding in single and dual reflectors for preventing aperture blockage of feeds, struts and subreflectors. These techniques are called as equivalent feed [49, 50] and equivalent parabola techniques [51]. By using the combination of these two techniques, it is possible to minimize side and back lobes [52, 53] (**Figure 27**).

As well known, it is important to reduce fields due to such lobes. Additional advantages of these techniques are optimization of the location of off-focus feeders and array elements. Thus beam scanning and corrections for catching the best receive/transmit signal are achieved. All these give the ability of increasing the performance of the space communication systems from the aspects of reflector antenna systems.

4.3 Antennas for deep space vehicles

For exploring other planets, comets, moons etc., space vehicles carrying scientific instrumentation are designed and launched. To compensate overmuch free

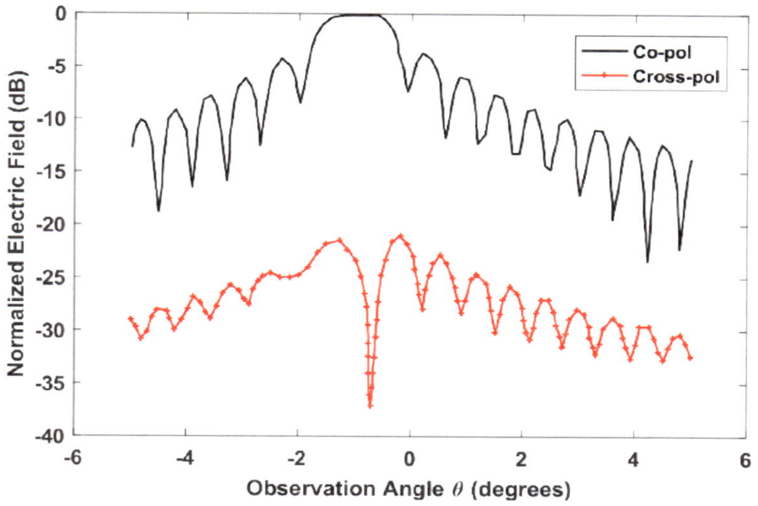

Figure 27.
Radiation pattern of a defocused fed Cassegrain antenna with D = 100λ, do = 90λ, F = 62.5λ, β = 150, e = 1.996, n = 8 and fx = fy = fz = 3λ by using equivalent feed and equivalent paraboloid techniques [40, 53].

space loss in communication budget, high gain antennas are needed. Therefore, challenging design and manufacturing technologies are employed for those antennas. Moreover, they have to comply with hard space qualification standards to operate in harsh space environment.

High gain antenna (HGA) on Mars rover **Curiosity** *of Mars Science Laboratory (MSL)* can be given as a pertinent example. HGA was developed by EADS CASA Espacio for NASA/JPL-Caltech [54]. This is circularly polarized microstrip patch array antenna consisting of 48 elements on a gimbal system to send and receive data between Mars and Earth at X-band.

In **Figure 28**, Engineering Qualification Model (EQM) of HGA is seen. The gimbal system has two degree of freedom for pointing. This RHCP antenna's gain values for Rx and Tx frequencies are listed in **Tables 3** and **4**, respectively, at different elevation angles.

This antenna is still being used on *Curiosity* at X-Band for uplink and downlink data from Mars surface to Earth. Its image taken by itself on Mars is shown in **Figure 29**.

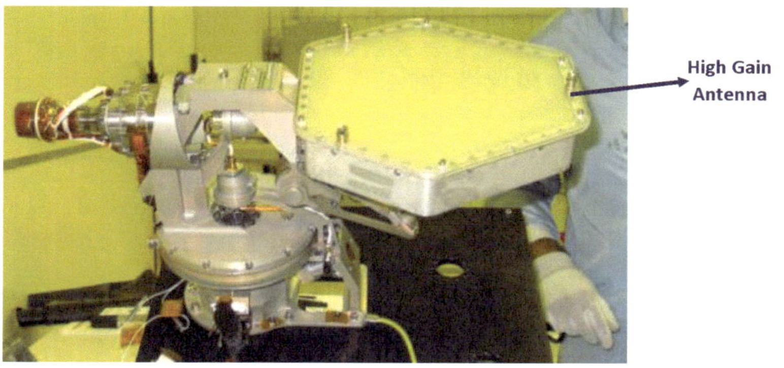

Figure 28.
EQM of HGA used for MSL (image taken from [54]; ©EurAAP; used with permission).

	At Θ = 2°		At Θ = 5°	
	Measured gain (dBi)	**Required gain (dBi)**	**Measured gain (dBi)**	**Required gain (dBi)**
7145 MHz	21.4	18	18.9	16
7170 MHz	21.5		18.8	
7195 MHz	21.5		18.8	

Table 3.
Rx gain values of HGA [54].

	At Θ = 2°		At Θ = 3.8°	
	Measured gain (dBi)	**Required gain (dBi)**	**Measured gain (dBi)**	**Required gain (dBi)**
8390 MHz	22.9	21.5	20.7	19.5
8420 MHz	22.9		21.1	
8455 MHz	22.7		20.7	

Table 4.
Tx gain values of HGA [54].

Figure 29.
HGA shown on Curiosity Mars rover (image courtesy of NASA/JPL-Caltech).

MSL was launched on 26 November 2011 and landed on 5 August 2012 onto Mars surface successfully. It was still working while this book chapter was being written.

Another challenging antenna design and application for deep space mission is *Mars Cube One* (MarCO) project of NASA/JPL-Caltech. NASA launched a Mars lander whose name is *Interior Exploration using Seismic Investigations, Geodesy and Heat Transport* (InSight) to Mars on 5 May 2018. There were two accompanying CubeSats: MarCO-A and MarCO-B to relay data to Earth from InSight on Mars [55]. As shown in **Figure 30** there is a deployable X-Band reflectarray antenna on the back surface of solar arrays.

The main task of this antenna with X-Band transponder is to support the communication of NASA's Mars Reconnaissance Orbiter (MRO) for downlink of the telemetries during InSight Rover's entry, descent and landing phases. This reflectarray antenna has 29.2dBic gain at X-Band as mentioned in [55].

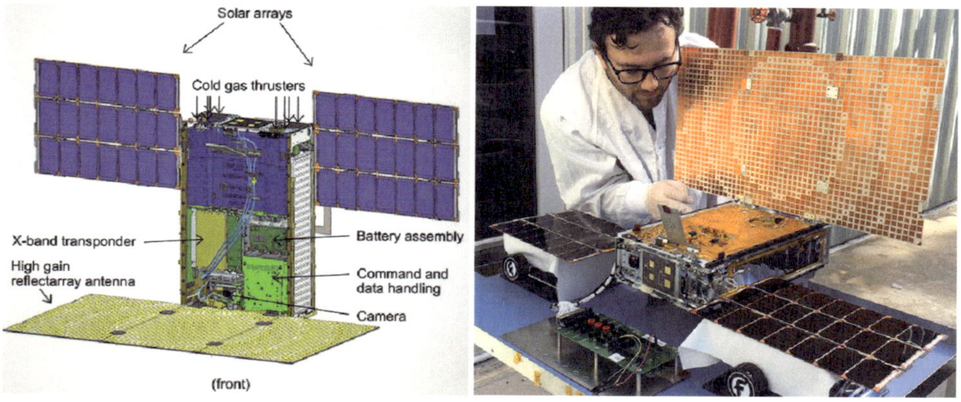

Figure 30.
Front view of MarCO CubeSat (image courtesy of NASA/JPL-Caltech).

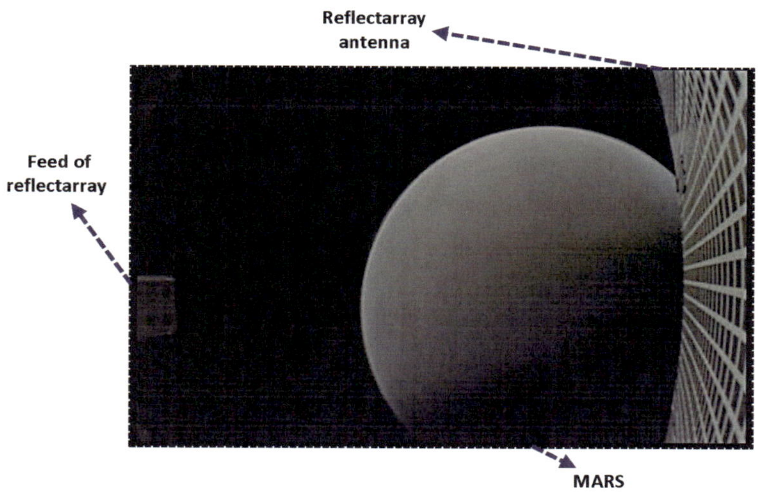

Figure 31.
Deployed reflectarray antenna with Mars scene. This photo was taken by onboard camera with fisheye lens of MarCO-B (image courtesy of NASA/JPL-Caltech).

An impressive image was taken by onboard camera with fisheye lens of MarCO-B after reflectarray antenna and its feed were deployed as given in **Figure 31**. In this figure, reflectarray elements and microstrip patch array feed of it are seen obviously. MarCO sent related telemetries from InSight successfully until its contact with the ground station was lost in February 2019.

Since reflectarray antennas have low stowage volume, manufacturing easiness using printed circuit board technology and lightweight mass, they became attractive in space industry. TUBITAK[2] Space Technologies Research Institute started a project named as YADAS in 2015 to develop X-Band reflectarray antenna to be used on LEO satellites [56]. Through this project many reflectarray prototypes in different element arrangements were designed, manufactured and measured. In **Figures 32** and **33**, simulation and measurement of the reflectarray antenna where the elements have been arranged in rectangular form at X-Band are shown.

[2] TUBITAK: The Scientific and Technological Research Council of Turkey.

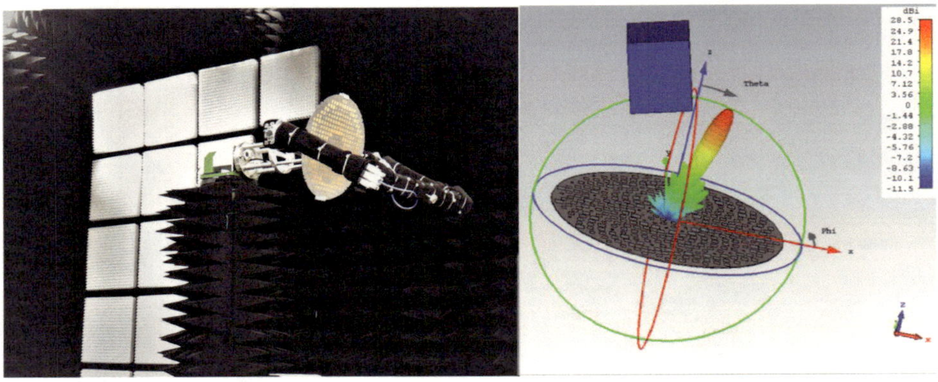

Figure 32.
Simulated and measured reflectarrays within YADAS project at X-band [56] (image courtesy of TUBITAK Space Technologies Research Institute).

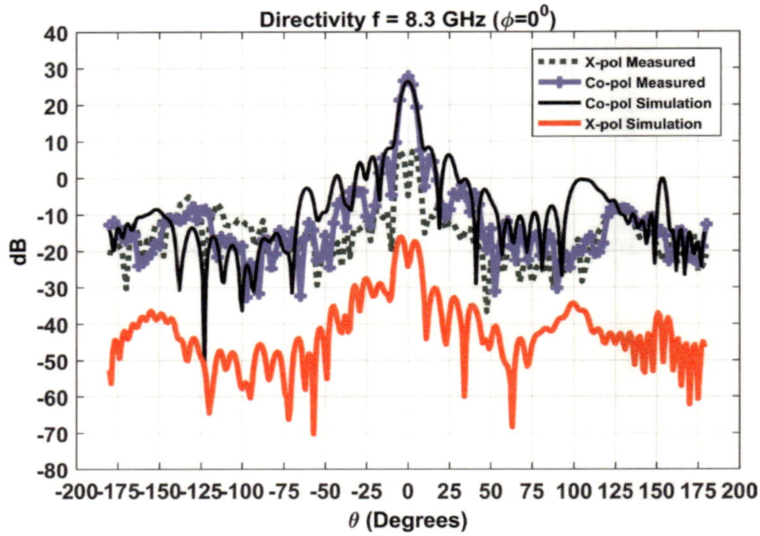

Figure 33.
Directivity comparison of simulation and measurement result for reflectarray antenna where the elements have been arranged in rectangular form (YADAS) [56] (image courtesy of TUBITAK Space Technologies Research Institute).

The developed antennas reached Technology Readiness Level (TRL) 4. Some studies are going on to bring beam-shaping capabilities to the designed reflectarrays.

5. Conclusions

Since Sputnik-1 was launched into orbit, the space technologies and antennas for spacecraft and satellite communications have been making progress rapidly. Thanks to this progress humankind have been learning about planets, comets, stars and moons, briefly near, deep and even interstellar space. Moreover, exploiting this technology, people can communicate with each other easily even if they are not in same country or continent. This is also an important cause of globalization.

In this chapter, antennas have been discussed as an RF-front end element of communication subsystems on spacecrafts and satellites. As can be seen from subsections of the chapter, there are many different antenna types used to provide communication services of space vehicles. Because of the limited space and for the sake of brevity only remarkable ones have been given as examples. An interested reader can find many other antenna examples used for space applications in open literature. Design parameters and their types are generally defined according to their frequency bands, transmit RF power, mass and volume requirements, mission type and environmental conditions. Therefore, like other equipment and subsystems to be used on the mission, they should be tested based on the published standards. Many of parts of these standards have been created from past mission experiences, experiments and trials. Two well-known space agencies, NASA and ESA have been publishing detailed standard documents and they are continuously updating them based on the "lessons learned" and experimental improvements.

Today countries are trying to enhance budget and qualified man power devoted to the space all over the world. Because they know that by this means they can develop many different technologies in fields of aerospace, defense, and mobile communication. Nevertheless, space industry is a long-term investment and it needs passion, sustainability, and also patience, because in this technological area the final product emerges after many detailed research and development stages.

Acknowledgements

The authors would like to express special thanks to TUBITAK Space Technologies Research Institute (TUBITAK UZAY), National Aeronautics and Space Administration (NASA), Japan Aerospace Exploration Agency (JAXA), The Institute of Electrical and Electronics Engineers (IEEE) and the European Association on Antennas and Propagation (EurAAP) for permissions of some specific photographs and figures to use in this chapter. The authors also thank to Dr. Aslı ER AKAN for editing and redrawing figures within the text.

Abbreviations

AESA	active electronically-controlled scanning array
AFM	aperture field method
ASS	amateur satellite service
bps	bit per second
BSS	broadcasting satellite service
CDSCC	Canberra Deep Space Communication Complex
EEC	equivalent edge currents
EES	Earth exploration satellite service
ECSS	The European Cooperation for Space Standardization
EIRP	effective isotropic radiated power
EQM	engineering qualification model
ESA	European Space Agency
ESTEC	European Space Research and Technology Centre
EurAAP	European Association on Antennas and Propagation
FSS	fixed satellite service
GEVS	general environmental verification standard
GEO	geostationary orbit
GNSS	global navigation satellite system

GTD	geometrical theory of diffraction
HGA	high gain antenna
HEO	high elliptical orbiting
IEEE	The Institute of Electrical and Electronics Engineers
InSight	Seismic Investigations, Geodesy and Heat Transport
ISL	inter satellite link
ISS	International Space Station
ITU	International Telecommunication Union
JAXA	Japan Aerospace Exploration Agency
JPL-Caltech	Jet Propulsion Laboratory-California Institute of Technology
LEO	low earth orbiting
LEOP	launch and early orbiting phase
LOS	line of sight
MarCO	Mars Cube One
MEO	middle earth orbiting
MRO	Mars Reconnaissance Orbiter
MSS	mobile satellite service
NASA	National Aeronautics and Space Administration
NSSDCA	NASA Space Science Data Coordinated Archive
PMSE	programme making and special events
PDT	payload data transfer
PO	physical optics
PTD	physical theory of diffraction
RF	radio frequency
RSS	radio determination satellite service
SMD	surface mount device
SOS	space operation service
SRS	space research service
TED	Türkiye Eğitim Derneği
TM/TC	telemetry/telecommand
TRL	technology readiness level
TUBITAK	The Scientific and Technological Research Council of Turkey
TV	television
UAT	uniform asymptotic theory
UNOOSA	United Nations Office for Outer Space Affairs
USSR	Union of Soviet Socialist Republics
UTD	uniform theory of diffraction
WINDS	Wideband Inter Networking engineering test and Demonstration Satellite

Author details

Volkan Akan[1]* and Erdem Yazgan[2]

1 Microwave and Antenna Systems Group, TUBITAK Space Technologies Research Institute, Ankara, Turkey

2 Department of Electrical and Electronics Engineering, TED University, Ankara, Turkey

*Address all correspondence to: volkan.akan@tubitak.gov.tr

References

[1] Imbriale WA, editor. Spaceborne Antennas for Planetary Exploration. New Jersey, Wiley: Hoboken; 2006. 592 pp. DOI: 10.1002/0470052783

[2] Öncü E, Akan V. Single stub tuning application via semi-rigid coaxial cables for VHF and UHF monopole antennas on RASAT. In: 10th Mediterranean Microwave Symposium (MMS'10); 25-27 August 2010; Guzelyurt, Cyprus. 2010

[3] Ozturk F, Akan V, Topalli K, et al. Complex permittivity measurements of dielectrics for space antenna radome and substrates in x-band. In: International Applied-Computational-Electromagnetics-Society Symposium – (ACES); 26-30 March 2017; Firenze, Italy. 2017

[4] NASA Outgassing Data for Selecting Spacecraft Materials [Internet]. 2018. Available from: https://outgassing.nasa.gov [Accessed: April 20, 2020]

[5] ITU Booklet on National Frequency Management. Geneva; 1988

[6] International Telecommunication Union (ITU). Radio Regulations. Articles. 2016;**1**

[7] Frequency Allocations. [Internet]. Available from: http://www.grss-ieee.org/frequency_allocations.html [Accessed: April 20, 2020]

[8] King R, Harrison CW, Denton DH. Transmission-line missile antennas. IRE Transactions on Antennas and Propagation. 1960;**8**:88-90. DOI: 10.1109/TAP.1960.1144802

[9] King RWP, Harrison CW. Antennas and Waves: A Modern Approach. Massachusetts, USA: The MIT Press; 1969. 778 pp. ISBN-10: 0262110334

[10] Campbell TG. An Extremely Thin, Omnidirectional, Microwave Antenna Array for Spacecraft Applications. Hampton, VA, United States: NASA Langley Research Center; 1969

[11] Akan V, Yazgan E. Analysis and design of circularly polarized and frequency tunable microstrip antenna having conical radiation pattern characteristic. In: 32nd ESA Antenna Workshop on Antennas for Space Applications, European Space & Technology Centre (ESA/ESTEC); 05-08 October 2010; Noordwijk, Netherlands. 2010

[12] Akan V, Dudak C. Antenna subsystem of GAMALINK platform. In: 6th European CubeSat Symposium; 14-16 October 2014; Estavayer-le-Lac, Switzerland. 2014

[13] Akan V. Electrically small printed antenna for applications on CubeSat and nano-satellite platforms. Microwave and Optical Technology Letters. 2015;**57**:891-896. DOI: 10.1002/mop.28989

[14] Mizuno TJ, Roque JD, Murakami BT, et al. Antennas for distributed nanosatellite networks. In: IEEE/ACES International Conference on Wireless Communications and Applied Computational Electromagnetics; April 2005; Honolulu, HI, USA. 2005

[15] Murakami BT, Ohta AT, Tamamoto MA, et al. Self-steering antenna arrays for distributed picosatellite networks. In: 17th Annual/USU Conference on Small Satellites; Utah, USA. 2003. pp. 1-5

[16] Simons RN, Lee RQ. Feasibility study of optically transparent microstrip patch antenna. In: IEEE Antennas and Propagation Society International Symposium and Digest; 13-18 July 1997; Montreal, Quebec, Canada. 1997

[17] Turpin TW, Baktur R. Meshed patch antennas integrated on solar cells. IEEE

Antennas and Wireless Propagation Letters. 2009;**8**:693-696. DOI: 10.1109/LAWP.2009.2025522

[18] Volakis JL, Chen C-C, Fujimoto K. Small Antennas: Miniaturization Techniques and Applications. New York: McGraw-Hill; 2010. 448 pp. ISBN-10: 0071625534

[19] Chen SY, Chou HT, Chiu YL. A size reduced microstrip antenna for the applications of GPS signal reception. In: IEEE Antennas and Propagation Society International Symposium; 9-15 June 2007; Honolulu, HI, USA. 2007

[20] Leisten O, Vardaxoglou JC, McEvoy SP, Wingfield RA. Miniaturized dielectrically loaded quadrifilar antenna for GPS. Electronics Letters. 2001;**37**:1321-1322. DOI: 10.1049/el:20010906

[21] Sievenpiper DF, Dawson DC, Jacob MM, et al. Experimental validation of performance limits and design guidelines for small antennas. IEEE Transactions on Antennas and Propagation. 2012;**60**:8-9. DOI: 10.1109/TAP.2011.2167938

[22] Akan V, Köse S, Kuzu L. Simulation and realization of a miniaturized tunable microstrip patch antenna. In: 31st International Review of Progress in Applied Computational Electromagnetics; 22-26 March 2015; Williamsburg, USA. 2015

[23] Chahat N et al. CubeSat deployable Ka-band mesh reflector antenna development for earth science missions. IEEE Transactions on Antennas and Propagation. 2016;**64**:2083-2093. DOI: 10.1109/TAP.2016.2546306

[24] Chahat N. A mighty antenna from a tiny CubeSat grows. IEEE Spectrum. 2018;**55**:33-37. DOI: 10.1109/MSPEC.2018.8278134

[25] Chahat N et al. Ka-band high-gain mesh deployable reflector antenna enabling the first radar in a CubeSat: RainCube. In: 2016 10th European Conference on Antennas and Propagation (EuCAP); 10-15 April 2016; Davos, Switzerland. 2016

[26] Wan J, Lu S, Wang X, Ai Y. A steerable spot beam reflector antenna for geostationary satellites. IEEE Antennas and Wireless Propagation Letters. 2016;**15**:89-92. DOI: 10.1109/LAWP.2015.2430894

[27] Jensen L, Sekora R, Schröder N. Advanced reflector antennas for geostationary spacecraft. IETE Technical Review. 1999;**16**:19-25. DOI: 10.1080/02564602.1999.11416798

[28] Greda LA, Winterstein A, Dreher A, et al. A satellite multiple-beam antenna for high-rate data relays. Progress in Electromagnetics Research. 2014;**149**:133-145. DOI: 10.2528/PIER14072502

[29] Saka B, Yazgan E. Pattern optimization of a reflector antenna with planar-array feeds and cluster feeds. IEEE Transactions on Antennas and Propagation. 1997;**45**:93-97. DOI: 10.1109/8.554245

[30] Alexovich J et al. The Hughes geo mobile satellite system. In: Proceedings of the Fifth International Mobile Satellite Conference; June 1997; Pasadena, CA. 1997

[31] Ekelman EP, Lee BS. An array-fed, dual-reflector antenna system (of offset confocal paraboloids) for satellite antenna applications. In: IEEE Symp. Antennas Propag.;, 26-30 June 1989; San Jose, CA, USA. 1989

[32] Thomson M. The Astromesh deployable reflector. In: IEEE Symp. Antennas and Propag.; 11-16 July 1999; Orlando, FL, USA. 1999

[33] Chandler C, Hoey L, Hixon D, et al. Ka-band communications satellite

antenna technology. In: 20th International AIAA Communication Satellite Systems Conference; 12-15 May 2002; Montreal, Quebec, Canada. 2002

[34] Balanis CA. Antenna Theory. 3rd ed. New York, NY, USA: Wiley; 2005, 1136 pp. ISBN-13: 978-0471667827

[35] Rahmat-Samii Y, Haupt RL. Reflector antenna developments; a perspective on the past, present, and future. IEEE Antennas and Propagation Magazine. 2015;**57**:85-95. DOI: 10.1109/MAP.2015.2414534

[36] Yazgan E. A simple formulation of UTD for some reflector antennas. International Journal of Electronics. 1989;**66**:283-288. DOI: 10.1080/00207218908925385

[37] Love AW. Reflector Antennas. Piscataway, NJ: IEEE Press; 1978, 428 pp. ISBN-13: 978-0471046066

[38] Scott C. Modern Methods of Reflector Antenna Analysis and Design. Boston: Artech House; 1990, 144 pp. ISBN-13: 978-0890064191

[39] James GL, Kerdemelidis V. Reflector antenna radiation pattern analysis by equivalent edge currents. IEEE Transactions on Antennas and Propagation. 1973;**21**:19-24. DOI: 10.1109/TAP.1973.1140409

[40] Wood PJ. Reflector Antenna Analysis and Design. London: Peter Peregrinus Ltd.; 1980, 221 pp. ISBN-13: 978-0906048214

[41] Yazgan E, Şafak M. Comparison of UTD and UAT in axially symmetric reflectors. IEEE Transactions on Antennas and Propagation. 1987;**35**:113-116. DOI: 10.1109/TAP.1987.1143965

[42] Yazgan E, Göksel H. The effect of surface loading on the radiation pattern of a spherical reflector. IEEE Transactions on Electromagnetic Compability. 1988;**30**:561-563. DOI: 10.1109/15.8770

[43] Yazgan E. The radiation pattern of a surface loaded ellipsoidal reflector. IEEE Transactions on Antennas and Propagation. 1987;**35**:437-439. DOI: 10.1109/TAP.1987.1144102

[44] Lee SW. Comparison of uniform asymptotic theory and Ufimtsev's theory in electromagnetic edge diffraction. IEEE Transactions on Antennas and Propagation. 1977;**25**:162-170. DOI: 10.1109/TAP.1977.1141559

[45] Keller JB. Geometrical theory of diffraction. Journal of the Optical Society of America. 1977;**52**:116-130. DOI: 10.1364/JOSA.52.000116

[46] Kouyoumjian RG, Pathak PH. A uniform geometrical theory of diffraction for an edge in a perfectly conducting surface. Proceedings of the IEEE. 1974;**62**:1448-1461. DOI: 10.1109/PROC.1974.9651

[47] Rudge AW, Adatia NA. Offset-parabolic-reflector antennas: A review. 1978;**66**:1592-1618. DOI: 10.1109/PROC.1978.11170

[48] Şansal GT, Yazgan E, Afacan E. Determination of the radiation pattern of an array fed paraboloidal reflector antenna by using equivalent feed in 3-D. In: NATO Symposium on Smart Antennas RTA; 7-10 April 2003; Chester, United Kingdom. 2003

[49] Rusch WVT. An equivalent focused reflector feed in place of any generalized defocused or extended feed. Electronics Letters. 1978;**14**:5-6. DOI: 10.1049/el:19780004

[50] Yazgan E, Şansal GT. Three-dimensional equivalent focused feed for an offset paraboloidal reflector antenna. Electromagnetics. 2004;**24**:597-605. DOI: 10.1080/02726340490513329

[51] Rusch WVT, Prata A, Rahmat-Samii Y, Shore R. Derivation and application of the equivalent paraboloid for classical offset Cassegrain and Gregorian antennas. IEEE Transactions on Antennas and Propagation. 1990;**38**:1141-1149. DOI: 10.1109/8.56949

[52] Turgut G, Yazgan E. Analysis of the asymmetric dual reflector antenna systems by using equivalent feed and antenna concepts. In: ELECO 2017; 30 Nov.-2 Dec. 2017; Bursa, Turkey. 2017

[53] Turgut G, Yazgan E. Application of equivalent paraboloid antenna and equivalent feed methods for off-focus fed dual asymmetric reflectors. Journal of Faculty of Engineering and Architecture of Gazi University. 2019;**34**:1929-1938. DOI: 10.17341/gazimmfd.571637 (in Turkish)

[54] Olea A, Montesano A, Montesano C, Arenas S. X-band high gain antenna qualified for Mars atmosphere. In: Proceedings of the Fourth European Conference on Antennas and Propagation; 12-16 April 2010; Barcelona, Spain. 2010

[55] Hodges RE, Chahat N, Hoppe DJ, Vacchione JD. A deployable high-gain antenna bound for Mars: Developing a new folded-panel reflectarray for the first CubeSat mission to Mars. IEEE Antennas and Propagation Magazine. 2017;**59**:39-49. DOI: 10.1109/MAP.2017.2655561

[56] Akan V et al. Technical Report for Research Project: YADAS, 7I150800-MASG-RPR-2017007-01. Ankara, Turkey: TUBITAK Space Technologies Research Institute; June 2017 (in Turkish)

Chapter 8

A Novel Class of Super-Elliptical Vivaldi Antennas for Ultra-Wideband Applications

Abraham Loutridis, Simay Kazıcı, Oleg V. Stukach, Arman B. Mirmanov and Diego Caratelli

Abstract

A novel class of complex-shaped antipodal Vivaldi antennas with enhanced impedance matching characteristics and an equivalent fractional bandwidth (FBW) of 166.8% is proposed. The reported antenna geometry is designed using the super-ellipse equation and implemented using an inexpensive FR4 laminate having a size of 170.7 mm × 134 mm × 0.2 mm. The presented low-cost, easy-to-fabricate radiating structure yields typical total efficiency, realized gain, and front-to-back ratio of 68%, 8.2 dBi, and 21.1 dB, respectively, across the operational frequency range. A parameter study of key geometrical features of the antenna is detailed in order to provide useful design guidelines while getting a better insight into the relevant physical behavior. Finally, a prototype is realized and characterized. The numerical results collected by full-wave simulation of the antenna structure are found to be in good agreement with the experimental measurements taken on the physical demonstrator.

Keywords: antipodal Vivaldi antennas, ultra-wideband, super-ellipse formula

1. Introduction

High data rates (up to 50 Mbps within a range of 10 m), low power consumption, and minimal interference make ultra-wideband (UWB) technology an attractive solution for short-range wireless communication [1]. Although the spectrum allocated for UWB technology overlaps existing frequency bands, the very low power levels (regulated by the Federal Communications Commission) prevent interference.

Among various broadband radiating structures such as printed dipoles [2], monopoles [3], bow-ties [4–6], patch antennas [7, 8], dielectric resonator, and lens antennas [9–11], tapered slot antennas (TSAs) exhibit the most favorable performance in terms of bandwidth and radiation pattern characteristics. For example, Vivaldi antennas, a special subclass of TSAs, offer large bandwidth and flat gain over the frequency range; in addition, they present compact and robust system integration properties in a wide range of fields, including wireless communication, microwave imaging, millimeter-wave applications, ground penetrating radar, remote sensing [12], and biomedical screening and diagnosis [13–18].

Vivaldi antennas were first proposed by Gibson in 1979 [19]. Subsequently, Gazit developed antipodal Vivaldi antennas (AVAs) comprising two metallic plates placed on the top and bottom sides of a low-permittivity substrate, forming an antipodal slotline transition [20]. This antenna configuration is characterized by a simpler feeding mechanism and delivers a wider operational bandwidth than Gibson's proposal.

Several AVA configurations have since been proposed in the literature [21–25]. These solutions have improved the radio-frequency (RF) characteristics of the antennas, but they have also unduly complicated the antenna geometry, thus increasing the manufacturing costs and even performance deviations due to tolerances and inaccuracies during fabrication.

As is well known from theory, the tapering of the aperture strongly influences the antenna radiation characteristics [26–29]. In this paper, a new type of AVA is proposed, namely the super-elliptical antipodal Vivaldi antenna (SAVA). Compared with conventional AVAs, the SAVA has more favorable RF characteristics, such as extended bandwidth, larger front-to-back ratio (FBR), and higher gain flatness. Herein, the proposed antenna is designed, and its circuital characteristics and radiation properties optimized for the 1.14–12.6 GHz frequency band.

2. Antenna design and simulation

2.1 Factors influencing antenna performance

The radiation mechanism of a Vivaldi antenna (**Figure 1**) is such that it works as a resonator at low frequencies and as a traveling-wave radiator at high frequencies. The lower cut-off frequency depends strongly on the width of the antenna (i.e., the maximum separation between the two arms), whereas the lower cut-off wavelength is around $\lambda/2$ for a given low-frequency band.

The thickness of the substrate strongly influences antenna performance. As the thickness increases, undesired modes become increasingly excited. These modes change the phase of the traveling waves along the two flares to create pattern distortion, resulting in high cross-polarization. Therefore, thin substrates composed of low-dielectric materials are preferred. Furthermore, the electrical length of the radiating antipodal flares determines the phase difference of the traveling-wave currents. Hence, this factor affects the broadband performance of the antenna and determines the higher and lower cut-off frequencies as well as antenna gain [26–29].

2.2 Antenna design and fabrication

The proposed Vivaldi antenna was modeled using CST Microwave Studio (version 2019) as follows. The antenna was printed on a thin 171 mm ($0.6\lambda_0$ at 1.14 GHz) $\times$ 134 mm ($0.5\lambda_0$ at 1.14 GHz) FR4 substrate (ϵ_r = 4.15, tanδ = 0.02, thickness = 0.2 mm) with a double-sided copper metallization thickness of 0.035 mm. The antenna is fed by a 50-Ω 0.36-mm-long microstrip on the front side of the substrate and connected directly to a 50-Ω SMA connector. A metallic thin radiator, placed on the front side of the PCB layer, is connected to the microstrip line using a highly smooth transition. The antipodal metallic radiator and the grounding plane of the microstrip feeding line are placed on the rear side of the PCB layer.

2.3 Super-ellipse formula in antenna design

In 1818, the French mathematician Gabriel Lamé introduced the following super-ellipse formula:

$$\left|\frac{x}{a}\right|^m + \left|\frac{y}{b}\right|^n = 1 \tag{1}$$

In Eq. (1), a and b denote the semi-axes of the super-ellipse, whereas the positive real-valued parameters m and n control the convexity of the curve. In this study, to improve RF characteristics, we use Eq. (1) to describe the radiating flares and feeding structure of the antenna (**Figure 1**). The resulting antennas are referred to as super-elliptical antipodal Vivaldi antennas (SAVAs). A parametric version of Eq. (1) is given hereafter:

$$\begin{cases} x(t) = a|\cos t|^m \\ y(t) = b|\sin t|^n \end{cases} \tag{2}$$

The default values of m and n are set to 2. Analytical curves are created on the basis of Eq. (2) and implemented in the antenna model built in CST Microwave Studio. The super-elliptical analytical curves shape the antenna contour to the special geometry shown in **Figure 1**. The radiating structure in **Figure 1** consists of quarter-ellipses with major semi-axes b1, a2, a3, and a3_1 and minor semi-axes a1, b2, and b3.

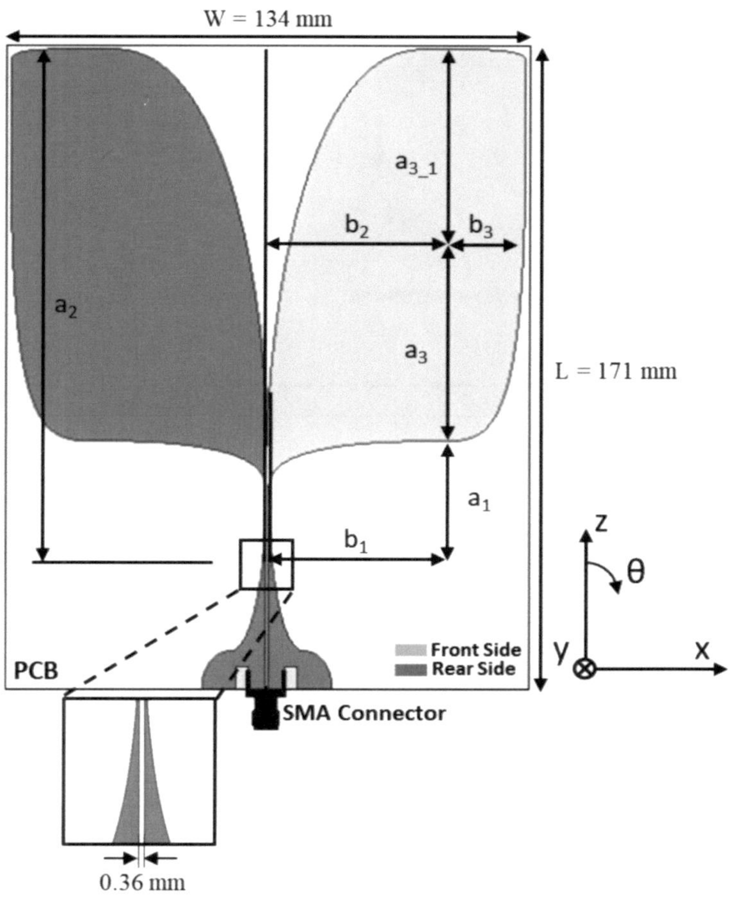

Figure 1.
Antenna geometry of a Vivaldi antenna and its coordinate system.

Similarly, the antenna feeding structure is designed using Eq. (1). Both the ground plane and the feeding microstrip line consist of quarter-ellipses. Note that the various sections of the antenna in **Figure 2** are characterized by different ellipticity parameters m and n, depending on the parametric representation (2).

2.4 Parametric investigation

This section discusses the parametric investigation of the variation of the voltage standing wave ratio (VSWR) with respect to six key parameters—namely n1, n2, n3, m1, m2, and m3—of the SAVA antenna shown in **Figure 2**. In the parametric simulations that follow, all parameters apart from the parameter of interest were set to 2.

Figure 3 shows the effect of parameter n1, which was varied from 0.5 to 3. The simulated results show that the impedance-matching properties of the antenna are strongly dependent on n1, with strong variations evident in the 3–12 GHz frequency

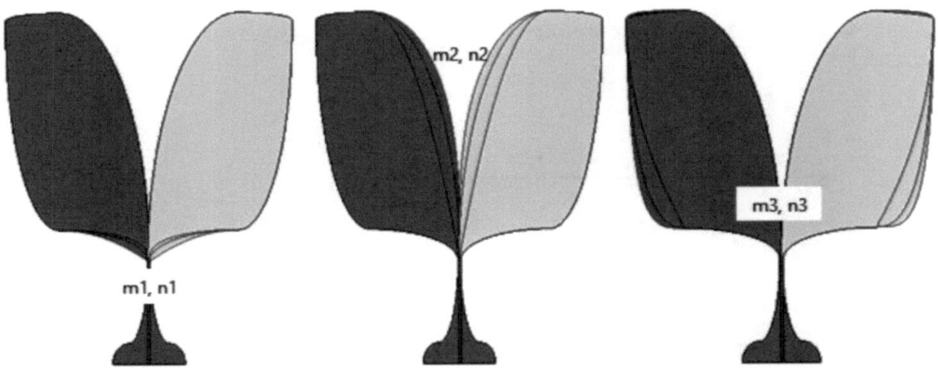

Figure 2.
Application of super-ellipse formula to antenna geometry.

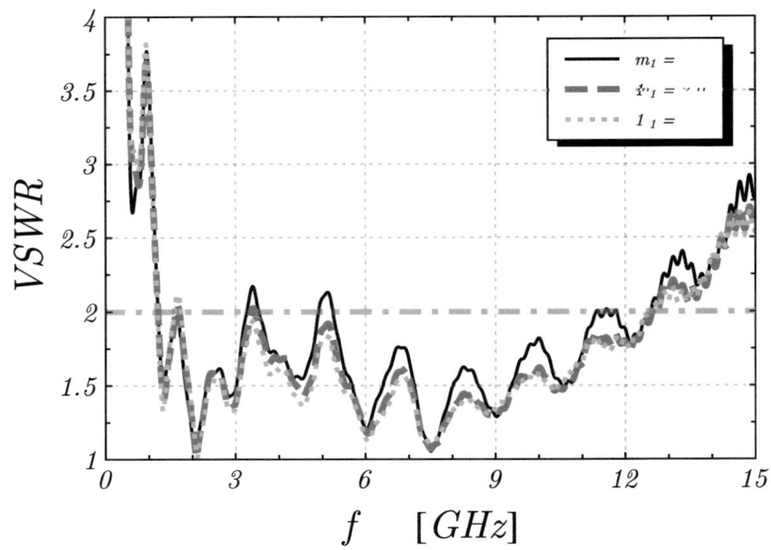

Figure 3.
Simulated VSWR dependence on n1.

range. The optimal value of n1 was identified to be 3, with this enabling a VSWR smaller than 2:1 over across a frequency range of nearly three octaves (an octave is a doubling of frequency).

Figure 4 depicts the simulated VSWR dependence on parameter m1, which was varied from 0.5 to 4. One can notice that the impedance-matching characteristics of the antenna tend to improve, especially in the upper band of the operational frequency range, as m1 becomes larger. Optimal performance is achieved for m1 = 2.

Similarly, **Figures 5** and **6** show the effect of n2 and m2 on VSWR, which were varied from 1 to 3 and from 0.5 to 2, respectively. From the simulated data, it is

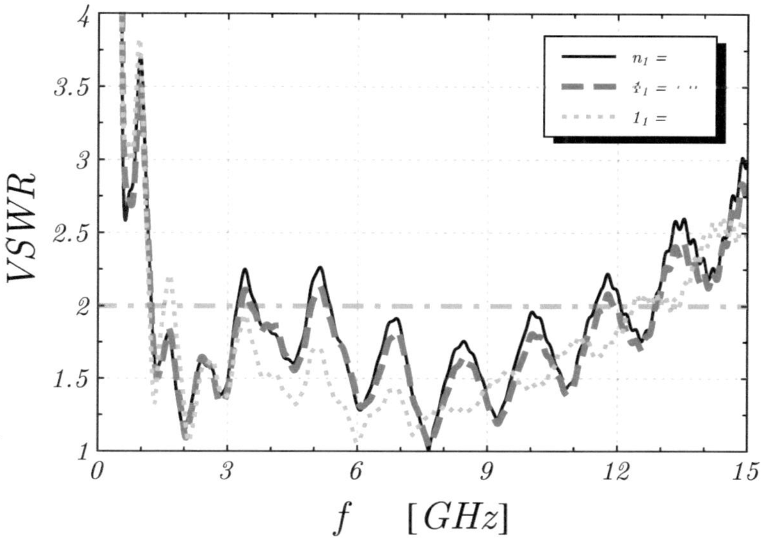

Figure 4.
Simulated VSWR dependence on m1.

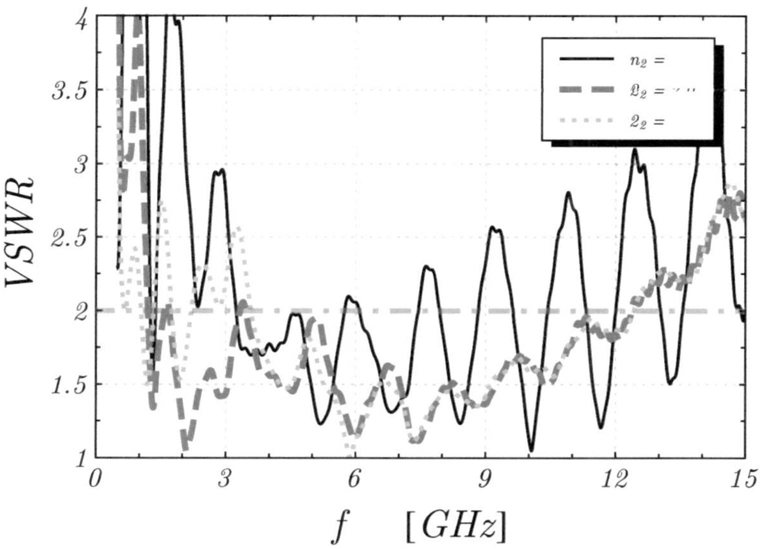

Figure 5.
Simulated VSWR dependence on n2.

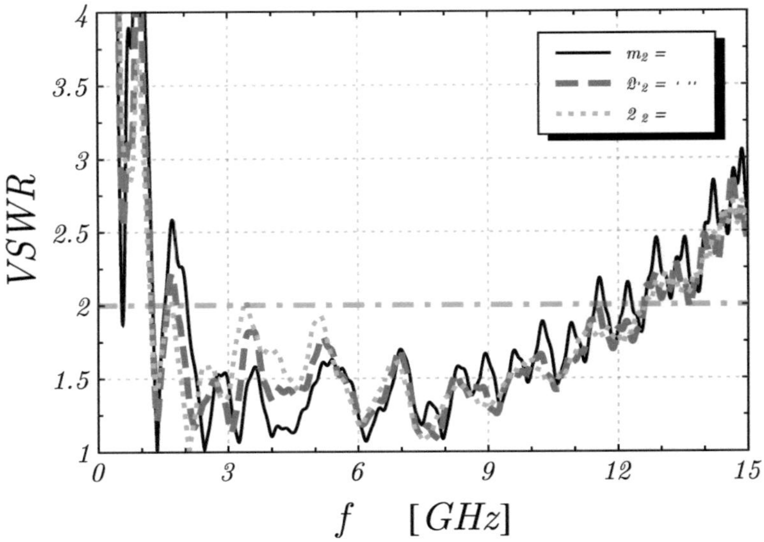

Figure 6.
Simulated VSWR dependence on m2.

apparent that improved performance in terms of reduced VSWR is achieved, wherein both n2 and m2 are selected to be 2.

Figures 7 and **8** similarly illustrate the effect of n3 and m3, which were varied from 0.5 to 3 and from 0.5 to 4, respectively. The simulations clarify that the VSWR antenna response does not depend heavily on either n3 or m3. The optimal values of n3 and m3 were identified to be 3 and 4, respectively.

Finally, to optimize the shape of the two antenna flares, the aforementioned optimized values of n1, n2, n3, m1, m2, and m3 were used as inputs in a particle swarm optimization (PSO) procedure.

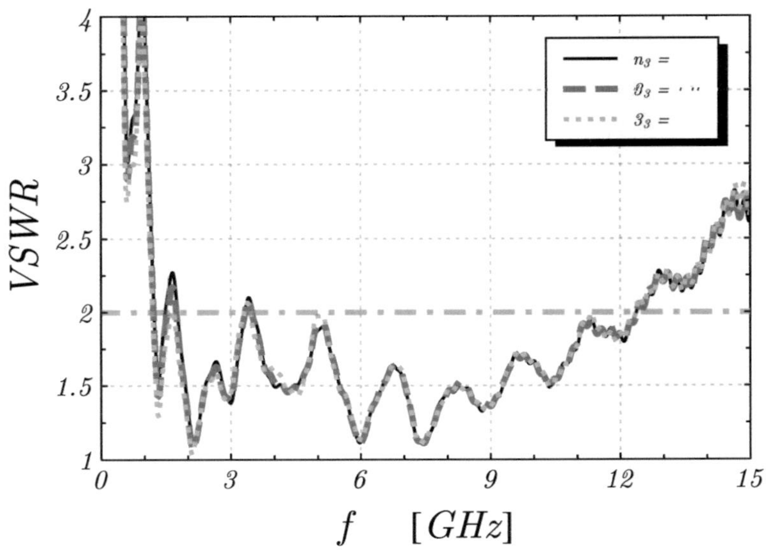

Figure 7.
Simulated VSWR dependence on n3.

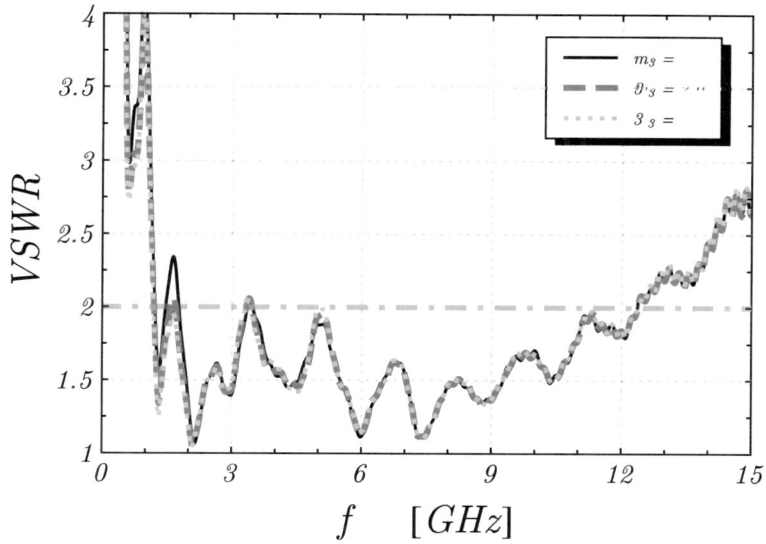

Figure 8.
Simulated VSWR dependence on m3.

With the end goal of improving antenna performance, the two main objectives of the PSO procedure were as follows:

1. achieve broader bandwidth (BW) $[f_H - f_L]$ in absolute terms (f_H = higher cut-off frequency at a 2:1 VSWR level, f_L = lower cut-off frequency at a 2:1 VSWR level) and

2. decrease the lower cut-off frequency f_L.

The PSO-optimized values for the six parameters were as follows: n1 = 3.15, m1 = 2.25, n2 = 2.05, m2 = 2.35, n3 = 3.451, and m3 = 4.3.

3. Antenna prototype characterization

The SAVA optimized in Section II was prototyped and characterized comparatively against conventional antipodal Vivaldi antenna (CAVA) configurations.

3.1 VSWR, efficiency, realized gain, and FBR

Figure 9 presents a comparison of the VSWR performance of the SAVA and CAVA configurations. Clearly, the SAVA features substantially more favorable impedance-matching properties and a lower cut-off frequency, without compromising on compactness. The SAVA had a measured impedance bandwidth (for VSWR < 2:1) of 11.46 GHz (1.14–12.6 GHz) with a fractional bandwidth of 166.8%. Please note that the reported performance figures are in excellent agreement (within 5% difference) with the numerical results obtained by full-wave simulation of the radiating structure. By contrast, the CAVA had a significantly smaller bandwidth of 9.46 GHz (1.21–10.67 GHz), with a fractional bandwidth of 159.2%.

The radiation properties of both the antenna configurations were characterized using a near-field scanner (MVG StarLab) in the 0.6–18 GHz frequency range. The

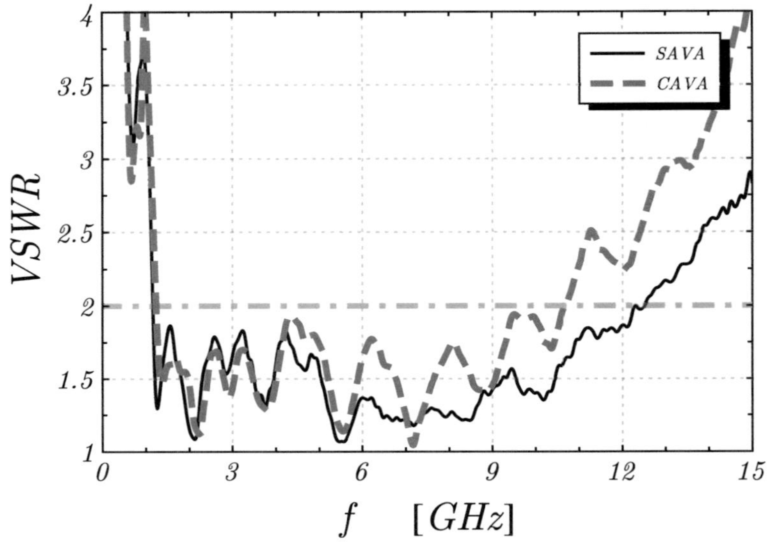

Figure 9.
Measured VSWR performance of the SAVA and CAVA antennas.

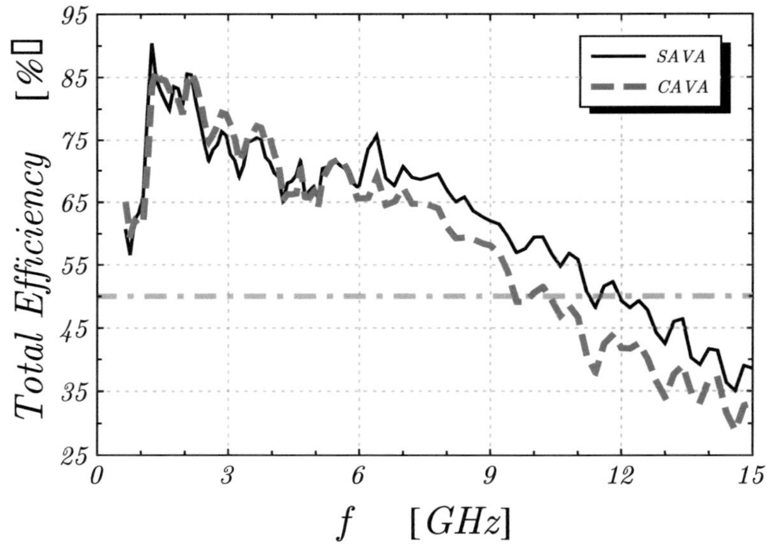

Figure 10.
Measured total efficiency of the SAVA and CAVA antennas.

average efficiency, realized gain, and FBR across the operational frequency band were 68.5%, 8.2 dBi, and 21.1 dB, respectively, for the SAVA, whereas the corresponding values were 62%, 7.9 dBi, and 19.5 dB, respectively, for the CAVA (**Figures 10–12**).

Figures 13 and **14** show the measured radiation patterns of the SAVA along two elevation planes (ϕ = 0° and 90°) at working frequencies of 2, 6, and 10 GHz. At low frequencies, the proposed antenna exhibits good directional radiation patterns, with smooth contours in both elevation planes. As the operating frequency increases, strong ripples appear in the peripheries of the radiation patterns due to the excitation of higher-order modes.

The SAVA antenna has end-fire characteristic with the main lobe in the axial direction (here, in z-direction) of the tapered slot.

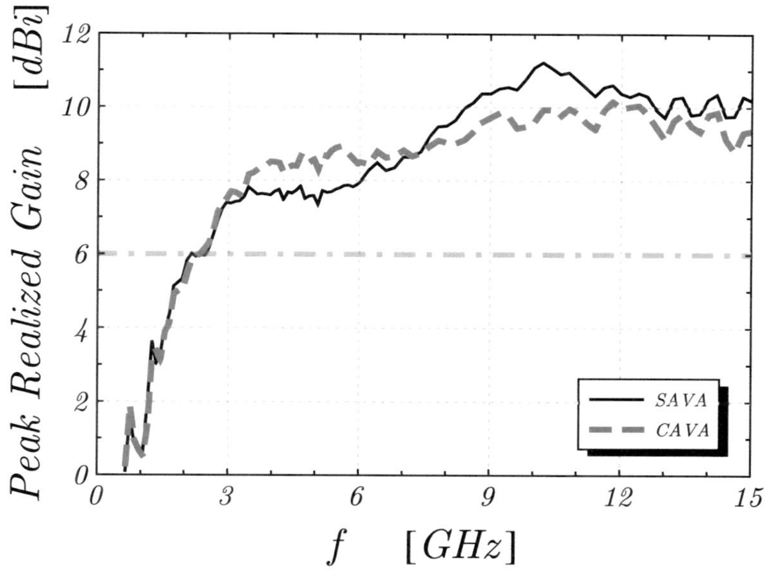

Figure 11.
Measured peak realized gain of the SAVA and CAVA antennas.

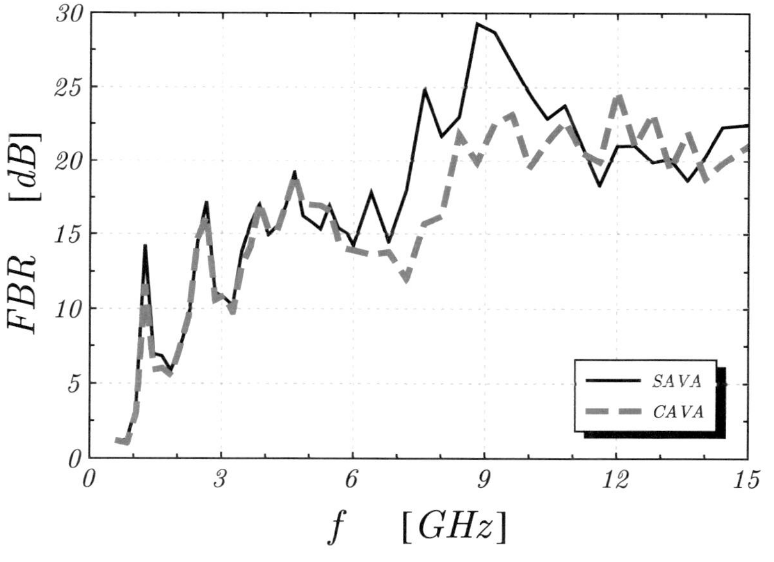

Figure 12.
Measured FBR of the SAVA and CAVA antennas.

3.2 Fidelity factor

The fidelity factor [30–31] quantifies the degree to which a radiated electric field (E-field) waveform of a transmitting antenna resembles the input pulse. As the amplitude of the E-field waveform is expected to differ from that of the input pulse, the E-field waveform and the input pulse are normalized to compare only the shape:

$$\hat{i}(t) = \frac{i(t)}{\left[\int_{-\infty}^{+\infty} |i(t)|^2 dt\right]^{\frac{1}{2}}} \tag{3}$$

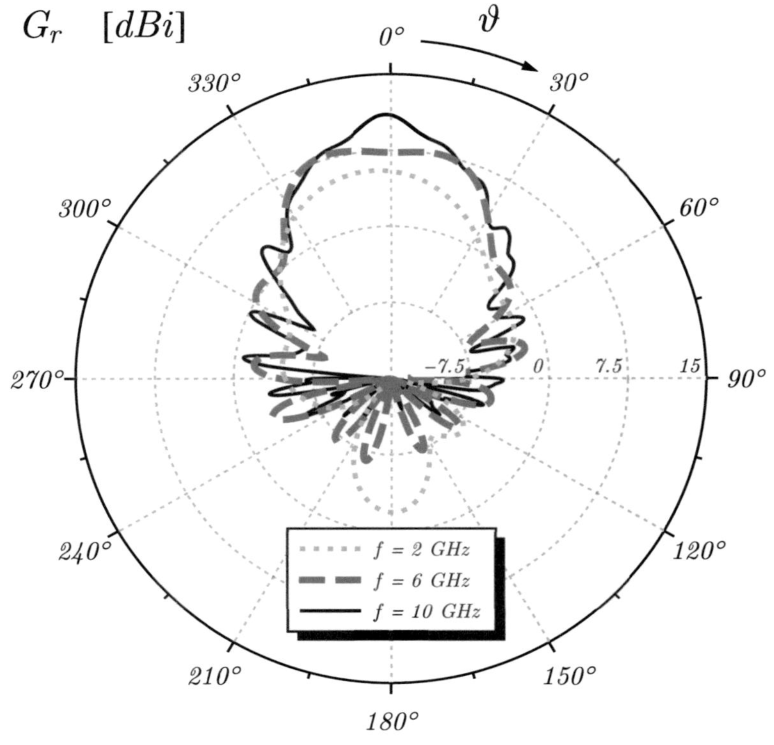

Figure 13.
Measured radiation patterns of the SAVA antenna along the cut plane ϕ = 0° at the working frequency of 2, 6, and 10 GHz.

$$\hat{e}(t) = \frac{e(t)}{\left[\int_{-\infty}^{+\infty} |e(t)|^2 dt\right]^{\frac{1}{2}}} \tag{4}$$

The degree of correlation between Eqs. (3) and (4) is quantified as the fidelity factor:

$$\begin{aligned} FF &= \max_{\tau} \int_{-\infty}^{+\infty} \hat{i}(t)\hat{e}(t-\tau)dt \\ &= \max_{\tau} \left\{ \frac{\int i(t)e(t-\tau)dt}{\sqrt{\int i^2(t)dt}\sqrt{\int e^2(t)dt}} \right\} \end{aligned} \tag{5}$$

The fidelity factor varies from 0 to 1, with 1 indicating that the E-field waveform is identical to the input pulse and that no distortion occurs during transmission.

To calculate the fidelity factor in this work, a Gaussian-modulated pulse in the desired frequency band (3.1–10.6 GHz) was used as the input signal, with ideal filed probes placed at the far field of the transmitting antenna. **Figure 15** illustrates the fidelity factor distribution along the H-plane of the SAVA and CAVA antennas. Clearly, the proposed SAVA configuration outperforms the CAVA configuration in both planes.

3.3 Group delay

Group delay, defined as the derivative of phase response ($\angle H(\omega)$) versus frequency [32], is used to characterize two port system (e.g., filters, amplifiers, and mixers).

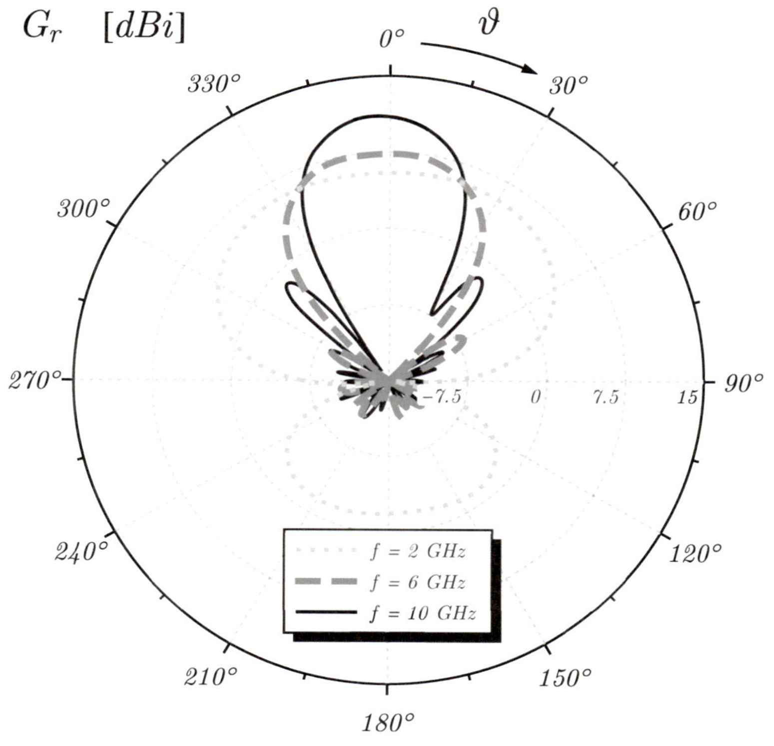

Figure 14.
Measured radiation patterns of the SAVA antenna along the cut plane $\phi = 90°$ at the working frequency of 2, 6, and 10 GHz.

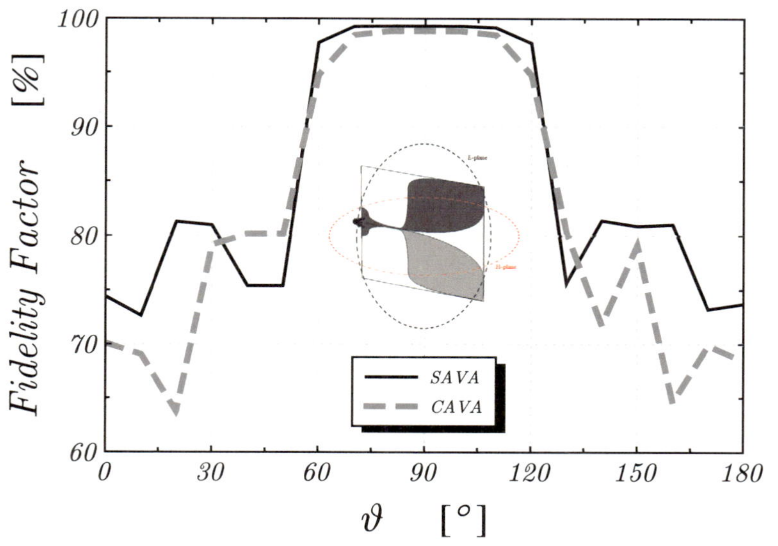

Figure 15.
Fidelity factor characteristics of the SAVA and CAVA antennas along the H-plane.

$$\tau = -\frac{d[\angle H(\omega)]}{d\omega} = -\frac{1}{360^{\circ}}\frac{d[\angle H(f)]}{df} \tag{6}$$

This factor has previously been applied to UWB antenna systems [32]. It measures the total phase distortion of the antenna system, using which the dispersion of

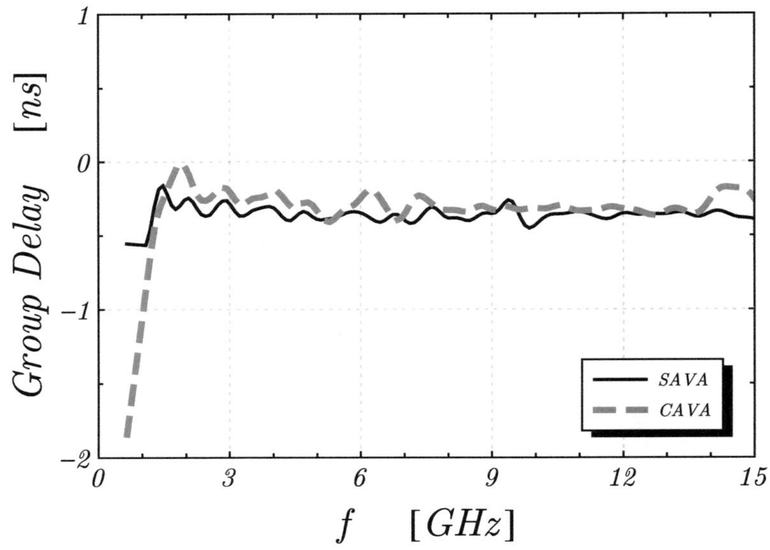

Figure 16.
Measured group delay of the SAVA and CAVA antennas.

the transmitted signal can be inferred. For a given frequency, the average group delay represents the time required for a signal to travel from one antenna terminal to the other.

The group delay of both the SAVA and CAVA configurations was measured in a free space environment. A network analyzer was used to measure the S21 group delay between two identical antennas (in each configuration) placed 2.5 m apart (i.e., far-field conditions) and aligned with the E-plane (**Figure 16**).

Clearly, the SAVA shows a smoother and more linear behavior, whereas the CAVA exhibits strong group delay variation, indicating phase nonlinearity along most of the frequency band.

4. Conclusion

This work has proposed a novel class of SAVAs. The geometry of the considered radiating structures is optimized by means of the super-ellipse formula, thus achieving UWB operation. Compared with current conventional Vivaldi antenna configurations, the proposed topology yields compelling benefits in terms of better impedance-matching properties, larger efficiency, gain, and front-to-back radiation ratio across a broad frequency range, without compromising on antenna compactness. The high directional radiation characteristics of SAVAs make them suitable for several applications, such as UWB communications and remote sensing, microwave imaging, ground penetrating radar, and biomedical screening.

Author details

Abraham Loutridis[1], Simay Kazıcı[2], Oleg V. Stukach[3], Arman B. Mirmanov[4] and Diego Caratelli[2,5*]

1 Alps Alpine Europe, Munich, Germany

2 The Antenna Company, Eindhoven, The Netherlands

3 Higher School of Economics (HSE), National Research University, Moscow, Russia

4 Seifullin Kazakh Agro Technical University, Astana, Kazakhstan

5 Eindhoven University of Technology, Eindhoven, The Netherlands

*Address all correspondence to: diego.caratelli@antennacompany.com

References

[1] Foerster J, Green E, Somayazulu S, Leeper D. Ultra-wideband technology for short-or medium-range wireless communications. Intel Technology Journal. 2001;**5**(2):1-11

[2] Paraforou V, Tran D, Caratelli D. A novel supershaped slot-loaded printed dipole antenna with broadside radiation for dual-band WLAN applications. In: Proceedings of the European Conference on Antennas and Propagation; The Hague, The Netherlands; 6–11 April 2014. pp. 3460-3463

[3] Caratelli D, Mescia L, Bia P. Design and full-wave characterization of supershaped printed monopole antennas. In: Proceedings of the IEEE AP-S/URSI Symposium; San Diego, California, U.S.A.; 9–14 July 2017. pp. 1769-1770

[4] Caratelli D, Yarovoy A, Ligthart LP. Full-wave analysis of cavity-backed resistively-loaded bow-tie antennas for GPR applications. In: Proceedings of European Radar Conference; 2008. pp. 204-207

[5] Caratelli D, Yarovoy A. Design and full-wave analysis of cavity-backed resistively loaded circular-end bow-tie antennas for GPR applications – Part I. Applied Computational Electromagnetics Society Journal. 2010; **25**:809-817

[6] Caratelli D, Yarovoy A. Design and full-wave analysis of cavity-backed resistively loaded circular-end bow-tie antennas for GPR applications – Part II. Applied Computational Electromagnetics Society Journal. 2010;**25**:818-829

[7] Paraforou V, Tran D, Caratelli D. A dual-band supershaped annular slotted patch antenna for WLAN systems. In: Proceedings of European Conference on Antennas and Propagation; The Hague, The Netherlands; 6–11 April 2014. pp. 2762-2764

[8] Samaras KA, Maximidis RT, Koutinos A, Caratelli D, Sahalos JN, Kyriacou GA. Characteristic mode analysis of drop-like supershaped patch antenna. In: Proceedings of International Conference on Modern Circuits and Systems Technologies; Thessaloniki, Greece; 7–9 May 2018

[9] Simeoni M, Cicchetti R, Yarovoy A, Caratelli D. Plastic-based supershaped dielectric resonator antennas for wide-band applications. IEEE Transactions on Antennas and Propagation. 2011;**59**(12): 4820-4825

[10] Bia P, Caratelli D, Mescia L, Gielis J. Analysis and synthesis of supershaped dielectric lens antennas. In: IET Microwaves, Antennas and Propagation. IET. 2015;**9**:1497-1504. DOI: 10.1049/iet-map.2015.0091

[11] Mescia L, Bia P, Caratelli D, Chiapperino MA, Stukach O, Gielis J. Electromagnetic mathematical modeling of 3D supershaped dielectric lens antennas. Hindawi Mathematical Problems in Engineering. 2016;**2016**: 8130160, 1-8130110. DOI: 10.1155/2016/8130160

[12] Lay-Ekuakille A, Vergallo P, Giannoccaro NI, Massaro A, Caratelli D. Prediction and validation of outcomes from air monitoring sensors and networks of sensors (Invited). In: Proceedings of 5th International Conference on Sensing Technology; Palmerston North, New Zealand; 28 November-1 December 2011. pp. 78-83

[13] Biswas B, Ghatak R, Poddar D. A fern fractal leaf inspired wideband antipodal Vivaldi antenna for microwave imaging system. IEEE Transactions on Antennas and Propagation. 2017;**65**(11):6126-6129

[14] Zhu S, Liu H, Chen Z, Wen P. A compact gain-enhanced Vivaldi antenna array with suppressed mutual coupling for 5G mmWave application. IEEE Antennas and Wireless Propagation Letters. 2018;**17**(5):776-779

[15] Baldi M, Cerri G, Chiaraluce F, Eusebi L, Russo P. Non-invasive UWB sensing of astronauts' breathing activity. Sensors. 2015;**15**(1):565-591

[16] Pumallica-Paro MA, Arizaca-Cusicuna JL, Clemente-Arenas M. Optimizing cutoff frequency in an antipodal Vivaldi antenna for GPR applications through a novel balun. In: IEEE XXV International Conference on Electronics, Electrical Engineering and Computing (INTERCON). IEEE; 2018

[17] Alzabidi MA, Aldhaeebi MA, Elshafiey I. Optimization of UWB Vivaldi antenna for tumor detection. In: 1st International Conference on Artificial Intelligence, Modelling and Simulation; IEEE; 2013

[18] Yang K, Loutridis A, Bao X, McEvoy P, Ammann MJ. A coplanar Vivaldi antenna with integrated filter for Ka-band. In: Loughborough Antennas & Propagation Conference (LAPC); UK; 1–4 November 2016

[19] Gibson P. The Vivaldi aerial. In: 9th European Microwave Conference; IEEE; 1979

[20] Gazit E. Improved design of the Vivaldi antenna. In: IEEE Proceedings H (Microwaves, Antennas and Propagation); IET; 1988

[21] Puskely J, Lacik J, Raida Z, Arthaber H. High-gain dielectric-loaded Vivaldi antenna for Ka-band applications. IEEE Antennas and Wireless Propagation Letters. 2016;**15**: 2004-2007

[22] Wang H, He S, Ding Z, Cao J, Yang Y. A miniaturized Vivaldi antenna with high gain for ultra-wideband applications. In: Sixth Asia-Pacific Conference on Antennas and Propagation; IEEE; 2017

[23] Belen MA, Evranos İO, Güneş F. Gain enhancement of antipodal Vivaldi antenna, In: 26th Signal Processing and Communications Applications Conference (SIU); IEEE; 2-5 May 2018

[24] Herzi R, Bouslama M, Osman L, Gharsallah A. Design of a compact antipodal Vivaldi antenna with band-rejected characteristic. In: Mediterranean Microwave Symposium (MMS); IEEE; 28-30 Nov. 2017

[25] Fernandez-Martinez P, Martin-Anton S, Segovia-Vargas D. Design of a wideband Vivaldi antenna for 5G base stations. In: IEEE International Symposium on Antennas and Propagation and USNC-URSI Radio Science Meeting; IEEE; 7-12 July 2019

[26] Nassar IT, Weller TM. A novel method for improving antipodal vivaldi antenna performance. IEEE Transactions on Antennas and Propagation. 2015;**63**(7):3321-3324

[27] De Oliveira AM, Perotoni MB, Kofuji ST, Justo JF. A palm tree antipodal Vivaldi antenna with exponential slot edge for improved radiation pattern. IEEE Antennas and Wireless Propagation Letters. 2015;**14**: 1334-1337

[28] Amiri M, Tofigh F, Ghafoorzadeh-Yazdi A, Abolhasan M. Exponential antipodal Vivaldi antenna with exponential dielectric lens. IEEE Antennas and Wireless Propagation Letters. 2017;**16**:1792-1795

[29] Karmakar A, Saha A, Bhattacharjee A, Bhawal A. In: Design of a fractal based wideband antipodal Vivaldi antenna with improved radiation characteristics for biomedical applications. In: Fourth International

Conference on Research in Computational Intelligence and Communication Networks (ICRCICN); IEEE; 22-23 Nov. 2018

[30] Kwon D-H. Effect of antenna gain and group delay variations on pulse-preserving capabilities of ultrawideband antennas. IEEE Transactions on Antennas and Propagation. 2006;**54**(8): 2208-2215

[31] Liu J, Eisele KP, Hay SG, Zhong S. Effects of printed UWB antenna miniaturization on pulse fidelity and pattern stability. IEEE Transactions on Antennas and Propagation. 2014;**62**(8): 3903-3910

[32] Zhu X, Li Y, Yong S, Zhuang Z. A novel definition and measurement method of group delay and its application. IEEE Transactions on Instrumentation and Measurement. 2009;**58**(1):229-233

Chapter 9

Additive Manufacturing for Antenna Applications

Gregory Mitchell and David Turowski

Abstract

This chapter describes the results of additive manufacturing (AM) for a multi-band antenna that effectively replaces two with a single footprint. The antenna achieves distinct modes of operation by achieving flexibility between horizontal and vertical polarizations on transmit and receive at both S-band and X-band frequencies. Low dielectric constants of commercial AM materials limit current AM antenna designs. Research into the composition of high-dielectric feedstocks for AM opens the design space for 3D printed hybrid material antennas. We compare the performance of an AM antenna to the same prototype using traditional methods and materials.

Keywords: additive manufacturing, dual-band antenna, dual polarization, 3D printing

1. Introduction

Additive manufacturing (AM), also known as 3D printing, allows engineers to rethink the traditional antenna design space. AM facilitates complex designs that require properties not achievable by current manufacturing methods. The 3D and hybrid-material approaches needed to achieve these designs makes AM critical to the future of radio frequency (RF) systems.

New research is spearheading development of RF AM technology to facilitate development of scalable AM antennas with built-in frequency and polarization agility. AM is a disruptive technology that facilitates complex designs requiring properties not achievable by current manufacturing methods. Strides in AM show robust structural and mechanical parts, but industry has yet to develop and fully characterize a large suite of RF materials compatible with AM methods. Low dielectric constants of commercial feedstocks limit current AM antenna designs. Recent research into the composition of high-dielectric feedstocks and for AM opens the design space for 3D printed hybrid material antennas [1–3]. This research includes loading low-loss polymer matrices such as acrylonitrile butadiene styrene (ABS) with varying volumes of high-dielectric ceramic nanoparticles. However, increasing the volume percentage of high-dielectric ceramics creates brittle filaments, which break when spooled and bind when printing, making dielectric constants greater than 8 unviable for fused deposition modeling (FDM) printers. However, initial research into incorporating the filament extrusion step directly at the print head shows promise by eliminating the need to print from a pre-fabricated spool of filament effectively sidestepping the ceramic volume loading limitation

IntechOpen

previously mentioned. This technique could open the doors for even higher dielectric AM filaments because designers will not need to spool or flex the highly loaded polymer in order to print with it.

Currently, AM produces robust structural and mechanical parts, but designers have yet to fully develop and characterize electromagnetic properties of AM feedstocks for printing antennas and other RF devices. Recent research into the composition of high-permittivity feedstocks for AM opens the design space for hybrid material antennas, and necessitates an emphasis on the measurement of electromagnetic properties for printed dielectric substrates.

Novel high dielectrics and conductive inks for AM enable complex and integrated antenna designs in all three dimensions. This leads to our integration of our S−/X-band antenna into a single aperture allowing for simultaneous multiband capabilities. The 3D and hybrid material approaches needed to achieve this scalable, agile, and multimode antenna demonstrate the necessity of AM for future of RF systems. Additionally, AM enables on demand supply closer to the point of need. This reduces the logistical burden, cost, and upgradeability of RF system maintenance in the field. These benefits result rapid development of new technologies, and increases agility to address new RF needs as they emerge in near real time.

This chapter focuses on the benefits and negatives of AM pertaining to antennas. AM dielectric substrates enable multiband approaches to AM phased arrays. We demonstrate our AM antenna by comparing experimental data to an identical antenna manufactured via legacy materials and techniques.

2. Role of additive manufacturing for antennas and radio frequency components

As governments, industries, and universities move toward multipurpose wireless platforms, engineers must integrate functionality of disparate frequency bands into single systems. This requires planar and vertical integration of apertures, substrates, and feed networks to enable multiple modes of operation. Now RF front ends must integrate several different antennas and their feed networks consisting of transmission lines, amplifiers, filters, and switches across increasing bandwidth. Integrated designs require both hybrid material and fully volumetric, as opposed to planar, approaches. The ability of AM to achieve geometries not possible by today's manufacturing processes makes AM a critical enabler of future of RF systems.

In the recent past, designers achieved robust mechanical and structural components through AM processes, but companies have not yet developed AM materials and feedstocks suitable for the design of antennas and other RF components. Engineers must also conduct research in the area of conductive inks for AM. Current silver inks yield metal layers with lower conductivity compared to their bulk metal counterparts. Increased conductivity of printable inks enhances the power efficiency of RF components.

Currently, low dielectric constants of commercially available AM feedstocks are limiting antenna designs. However, through the development of high-dielectric constant and low-loss electromagnetic materials, AM opens the RF design space to complex geometries and material gradients not currently achievable. One example is the Luneberg lens [1], which relies on a graded dielectric constant. By controlling the fill density of printed substrates, AM achieves a continuously graded slope in dielectric constant that is the enabling feature of the Luneberg lens design [2].

Increasing the RF versatility of AM requires research into feedstocks that achieve high dielectric constants. Recent research shows AM filaments with dielectric constants 0f 4 or greater can be extruded from polymer/ceramic composites [3–6]. The process loads a low-loss host polymer with a given volume fraction 0f ceramic nanoparticles with high dielectric constants. The host polymer will tend to have a low dielectric constant and the volume fraction of the dispersed high-dielectric ceramic particles will determine the macroscopic dielectric constant of the extruded filament [5, 7]. Since we know the presence of voids during the 3D printing process can cause deviations in the dielectric constant, electromagnetic characterization of the final substrate becomes very important.

AM for hybrid material antennas has additional obstacles other than limitations in dielectric constant. The layered nature of 3D printing causes the potential for inconsistencies in the bonding of interfaces between printed layers. Similarly, 3D printing processes yield anisotropy in the electromagnetic properties of printed layers that can yield RF properties that differ depending on direction within the printed substrate. Porosity, surface roughness, and repetitiveness are also concerns relating to AM technologies for antennas and RF devices [8–11].

Beyond dielectric feedstocks for AM, research into increasing the conductivity of printable inks is also of interest. The best conductivity of conductive inks is currently 5–10 times less than that of bulk metals, and even these reduced conductivities require sintering processes in excess of 175°C [12]. Reduced conductivity will cause decreased radiation efficiency in antennas and increased transport inefficiencies in transmission lines. Whereas high sintering temperatures required for 3D printed inks would degrade and melt the thermoplastic-based dielectric compounds discussed previously. There are alternate methods for sintering currently under investigation such as laser sintering and flash annealing, but these methods still require further research to demonstrate their viability.

2.1 High-dielectric filaments compatible with additive manufacturing

Current research investigates methods for extruding high-dielectric AM filaments through a robust and repeatable method that allows for the printing of AM substrates with a given value of dielectric constant. This will give engineers a continuum of achievable dielectric constants for AM substrates they can use in their models when designing RF components. A second challenge is developing a technique for sintering of conductive inks printed on an AM substrate without compromising the integrity of the dielectric substrate. A final hurdle is to produce a fully integrated AM antenna including ground plane, RF connectors, feed, multiple dielectric substrates, and aperture through a fully automated process.

Typically, composite materials are prepared using mixing techniques such as dissolving pellets of ABS using acetone [12, 13]. Afterward, mixers add plasticizers and surfactants in small concentrations. These additives are necessary, as the plasticizer acts as a lubricant between molecular chains in the polymer, enabling flexibility even with loading of ceramic powders; however, the addition of too much plasticizer negatively affects the composite by rendering it too elastic. Materials use surfactants to prevent aggregation between the ceramic particles [14].

Adding high-dielectric ceramic nanoparticles in in volume fractions between 15 and 30% to the mixtures allows to them to homogenize within the polymer. Once the acetone has fully evaporated, we cut the composite slab into pieces and then ground them into approximately 2-mm pellets. We can then extrude filaments at high temperature, but this process has its limitations. Adding more than 40% by volume of ceramic powder results in a filament too brittle for use, even with the addition of plasticizer. Furthermore, as the loading of ceramic inclusions increases, so too does the viscosity of the material. Designers must exercise care, as viscous materials are prone to printing defects, such as voids [14].

3. Hybrid material dual-band and dual-polarization antenna design

As the world continues to move to fully integrated multifunction antennas, the design of new multiband and multimode antenna geometries continues to yield structures that are more complicated to manufacture. However, current challenges in maintaining, upgrading, and networking an ever-increasing number of antenna nodes are alleviated through adapting these types of multifunction antennas and other RF devices. The low profile and lightweight designs of references [15–18] show great promise for multifunction antenna applications. This chapter compares the performance of a multifunction antenna prototyped using traditional off-the-shelf materials and fabrication techniques, to the same antenna geometry manufactured using AM materials and 3D printing manufacturing processes. The idea being that the versatility allowed via AM of antennas can help navigate the newfound complexities of emerging multifunction antenna designs.

3.1 Antenna performance using conventional materials and techniques

We base the hybrid substrate-shared aperture antenna on the shorted annular ring and concentric patch geometry explored by Dorsey and Zaghloul [15–17]. **Figure 1** (left) shows the geometry of the dual-band antenna on two nested dielectric substrates, and **Figure 1** (right) shows the layout of the nested substrates themselves. **Table 1** gives the values in millimeters for the antenna dimensions illustrated in **Figure 1**. We shrink the overall footprint of the dual-band antenna by 30% by increasing the dielectric constant of the substrate under the annular ring from ε_{r1} = 2.33 to ε_{r1} = 6.15.

Two pairs of orthogonal microstrip pin feeds control the different operational modes of the antenna by exciting either S- or X-band frequencies and either vertical or horizontal polarization. **Figure 2** gives the locations of the pin feeds with respect to the annular ring and concentric patch antennas respectively. **Figure 2** also shows a shorting wall grounding the inner perimeter of the annular ring to diminish surface waves. Suppression of surface waves on the dielectric substrates helps

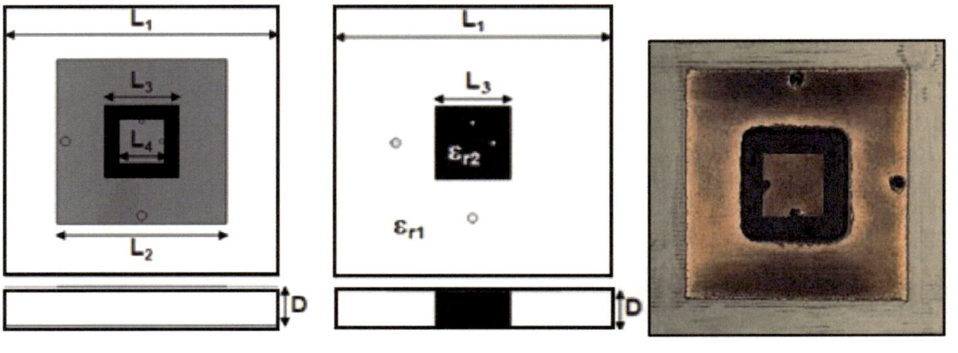

Figure 1.
Top view geometry of the dual-band antenna (left), layout of the nested substrates (center), and top view of the prototype dual-band antenna.

L_1	L_2	L_3	L_4	D	ε_{r1}	ε_{r2}
36.7	22.57	10.23	7.08	5.07	6.15	2.33

Table 1.
Antenna dimensions in millimeters of ***Figure 1.***

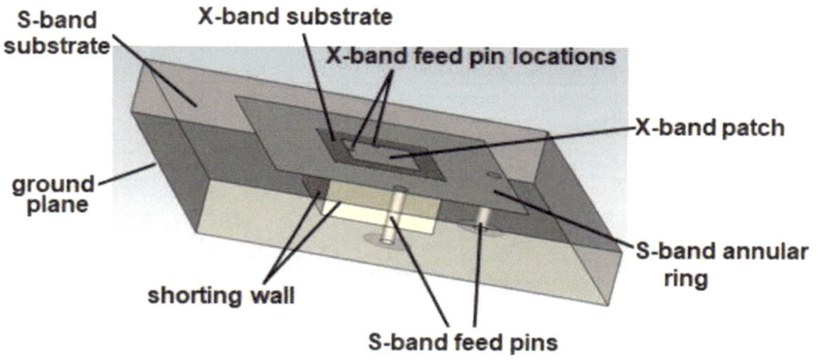

Figure 2.
3D view of the dual-band antenna and pin feed network. The outer dielectric layer is transparent for clarity.

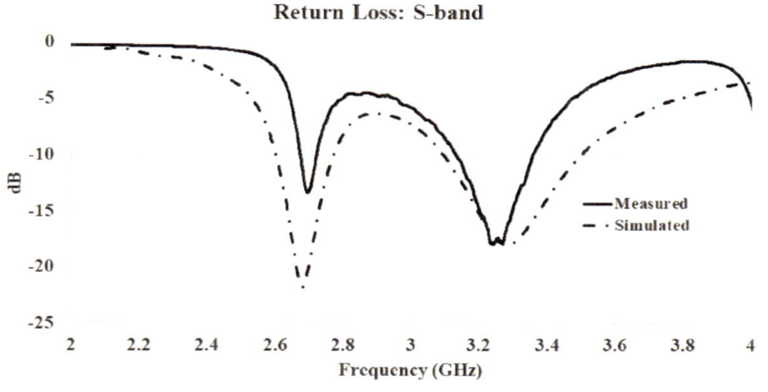

Figure 3.
Comparison of simulated return loss to measured return loss of the horizontal port at S-band.

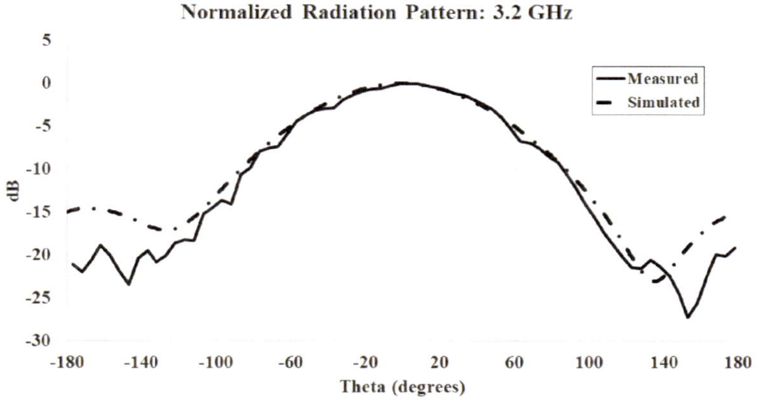

Figure 4.
Measured and simulated normalized radiation pattern at S-band horizontal port.

increases isolation between the excitation ports of the dual-band antenna. The antenna comprises a top metal layer, a hybrid Rogers 3006/Rogers 5870 substrate layer, and a bottom metal ground layer. All metal layers are 0.1 mm thick, and the dielectric substrate layer is 5.05 mm thick.

Figures 3 and **4** show the measured versus simulated return loss and normalized radiation patterns of the dual-band antenna at S-band. **Figures 5** and **6** show the

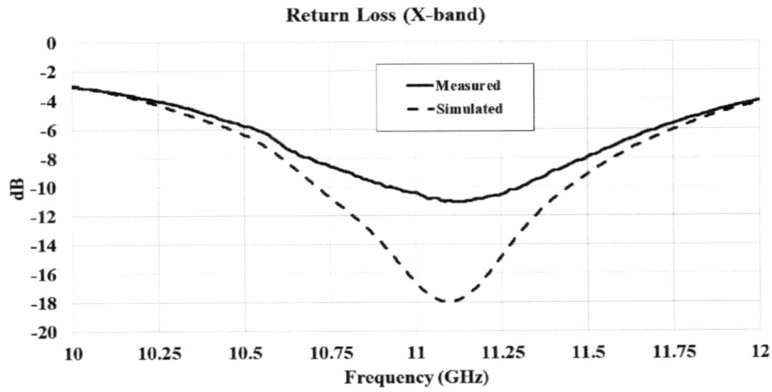

Figure 5.
Measured and simulated return loss at X-band horizontal port.

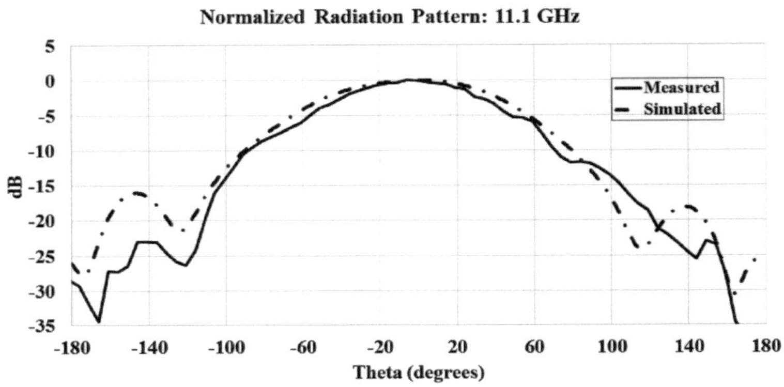

Figure 6.
Measured and simulated normalized radiation pattern at X-band horizontal port.

same for at X-band. The radiation shows good agreement at both S-band and X-band. **Figures 3** and **5** show a return loss measurement of −17.5 dB at the resonant frequency of 3.2 GHz and measurement of −11.0 dB at the resonant frequency of 11.1 GHz. There is a second S-band resonance seen at a frequency of 2.7 GHz in **Figure 3**, but the radiation characteristics in **Figure 4** show that this is not the dominant S-band resonance. We attribute discrepancies in the depth of the return loss curve in **Figure 5** to manufacturing tolerances in the placement of the pin feeds with respect to the edges of the concentric patch. Since electrical length decreases with wavelength at X-band, smaller tolerance errors can have larger effects than at S-band on the impedance match quality. However, both the simulated and measured resonant frequency at X-band is the same as expected.

3.2 Antenna performance using additive manufacturing

AM allows engineers to rethink the RF design space. AM facilitates complex designs that required properties not achievable by traditional manufacturing methods. Strides in AM produce robust structural and mechanical parts, but industry has yet to develop a full suite of electromagnetic properties for AM feedstocks. Low dielectric constants of commercial feedstocks limit current antenna designs, but recent research into the composition of high-dielectric feedstocks for AM opens the design space for antennas [12–14]. Additionally, space-filling algorithms paired with the extrusion of high-dielectric filaments, discussed in Section 2.1, allow customized dielectric constants not previously achievable [2].

We choose a slightly modified version of the antenna design presented in Section 3.1. Due to the hybrid material design and the utilization of a fully embedded vertical shorting wall, we believe this is a good design to test capabilities of AM as it pertains to antennas. For this investigation, we design the dual-band antenna for S-band and X-band respectively. We perform all simulations using the finite difference time domain (FDTD) solver of CST Studio Suite 2019.

Figure 7 shows a to-scale 3D printed prototype of the antenna shown in **Figures 1** and **2**. The antenna utilizes a microstrip stack of a printed copper layer, hybrid dielectric layer, and a copper ground layer. We printed all pieces separately and assembled by hand afterward. All conductive layers are 1.0 mm thick. The hybrid substrate layer is 5.05 mm thick. The total profile of the antenna is 6.05 mm due to printing thicker metal layers to prevent warping.

Table 2 shows the dimensions of the geometries given in **Figure 2** as they pertain to **Figure 7**. We chose two AM filaments compatible with a Fuse Deposition Modeling (FDM) 3D printer. For the high dielectric, we chose Preperm ABS 650 with a reported dielectric constant of $\varepsilon_r = 6.5$ to mimic the Rogers 6010 material with $\varepsilon_r = 6.15$. However, after conducting waveguide measurements on a 3D printed sample, we found anisotropic values of $\varepsilon_{rx} = \varepsilon_{ry} = 5.7$ and $\varepsilon_{rz} = 5.3$. We chose ABS to mimic the Rogers 5870 material with $\varepsilon_r = 2.33$. Measured values of the ABS were nearly isotropic with a value of $\varepsilon_r = 2.38$. The loss tangents of both materials were about 10^{-3}. We printed the metal layers via the selective laser sintering (SLS) method allowing us to print actual copper as opposed to using a lower conductivity ink. Using a substrate of $\varepsilon_{r1} = 5.3$ instead of $\varepsilon_{r1} = 6.15$ under the S-band element shifts the resonances from 3.2 to 3.8 GHz. Changing the dimensions of the concentric slot and X-band patch, shown in **Figure 1** and **Table 2**, shifts the X-band frequency from 11.1 to 9.35 GHz.

Figure 7.
Fully assembled AM dual-band antenna.

L_1	L_2	L_3	L_4	D	ε_{r1}	ε_{r2}
37.08	25.36	11.03	7.38	7.05	5.3	2.38

Table 2.
Antenna dimensions in millimeters of 3D printed antenna based on schematic in ***Figure 1***.

We show the measured versus simulated return loss and realized gain of the dual-band antenna at S-band in **Figures 8** and **9** as well as at X-band in **Figures 10** and **11**. We took all realized gain versus frequency measurements at boresight to the antenna. We see general agreement at both bands for all measurements. One

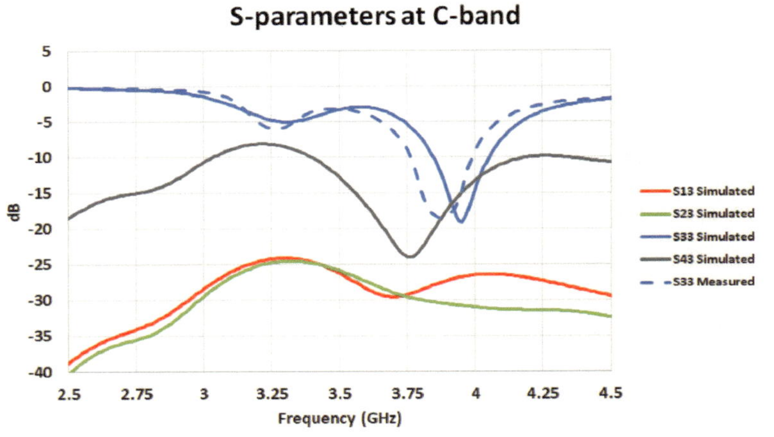

Figure 8.
S-parameters at S-band horizontal polarization port.

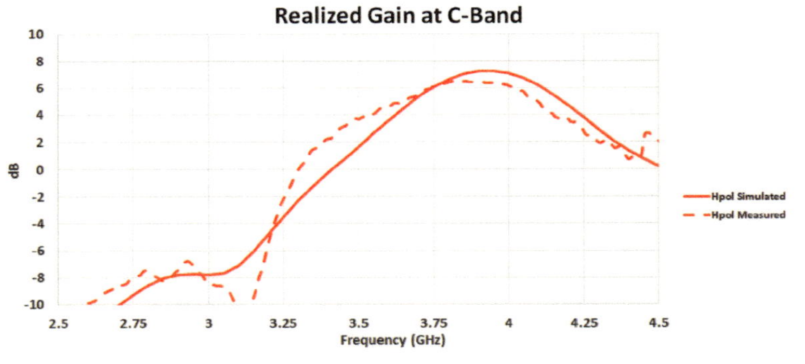

Figure 9.
Realized gain to boresight at S-band horizontal polarization port.

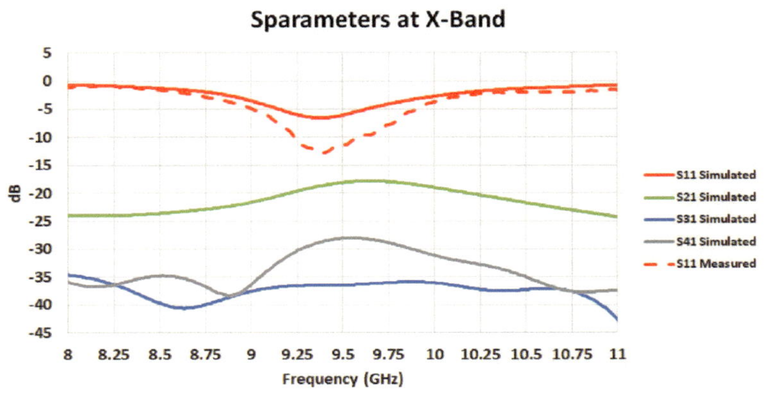

Figure 10.
S-parameters at X-band horizontal polarization port.

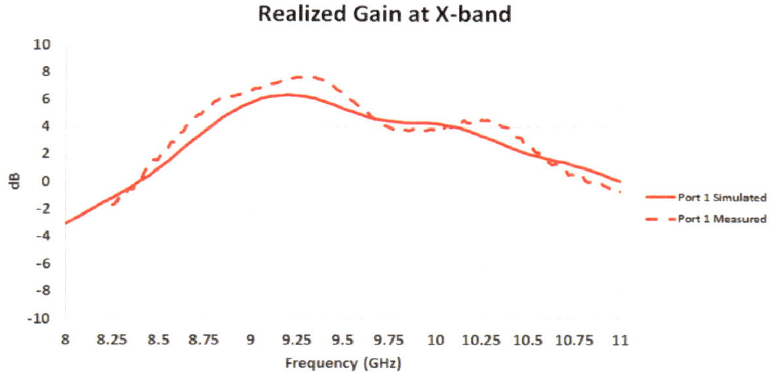

Figure 11.
Realized gain to boresight at S-band horizontal polarization port.

discrepancy is in the S-band realized gain where measurements are up to 2.0 dB lower than expected from 3.25 to 3.6 GHz. The resonance at S-band is also about 100 MHz or 2.6% off. The X-band resonance and realized gain curves show better agreement than those at S-band. The return loss at X-band is even better than that predicted by simulation and this shows up in **Figure 11** where the measured realized gain is higher than simulation at resonance.

We attribute measured differences in the return loss to manufacturing tolerances since the pin fed patch is an extremely resonant type of feed.

4. Conclusions

This chapter describes both the benefits and current challenges facing AM for antennas and RF devices. The forefront of which is a lack of commercially available high-dielectric materials that are compatible with filament fed 3D printers. As a comparison, we fabricated two similar antenna designs utilizing multiple nested dielectrics and an embedded shorting wall within the dielectrics to compare performance between traditional materials and manufacturing methods versus those of AM. Due to restricted access to an AM feedstock with the proper dielectric constant, direct comparisons of the two antennas is not the preferred way of comparison. However, experimental results for both prototypes agree well with the simulation data. There also seems to be no degradation in the performance of the AM prototype over the traditional prototype in terms of the agreement with the respective antenna models. With new methods of extruding higher dielectric filaments for FDM 3D printers, AM seems to be a good fit for future work in antenna design especially as antenna and RF front ends grow increasingly complex and more fully integrated.

Author details

Gregory Mitchell[1*] and David Turowski[2]

1 Army Research Laboratory, Adelphi, USA

2 Lockheed Martin, Moorestown, USA

*Address all correspondence to: gregory.a.mitchell1.civ@mail.mil

References

[1] Luneburg R, Herzberger M. Mint: A mathematical theory of optics. American Journal of Physics. 1996;**34** (80). DOI: 10.1119/1.1972799

[2] Roper GB, Yarlagadda S, Mirotznik M. Fabrication of a flat Luneburg lens using functional additive manufacturing. In: Proceedings of National Radio Science Meeting (Joint AP-S Symposium), USNC-URSI, Memphis, TN; 6–11 July 2014; DOI: 10.1109/USNC-URSI.2014.6955649

[3] Moulart A, Marrett C, Colton J. Mint: Polymeric composites for use in electronic and microwave devices. Journal of Polymer Engineering and Science. 2004;**44**:588-597. DOI: 10.1002/pen.20053

[4] Agarwala MK et al. Structural ceramics by fused deposition of ceramics. In: Proceedings of Solid Freeform Fabrication Symposium. Austin, TX. 1995. pp. 1-8

[5] Duncan B et al. 3D printing of millimeter wave RF devices. In: Workshop on Additive Manufacturing of Antennas and Electromagnetic Structures. McLean, VA: MITRE; 2017

[6] Castles F, et al. "Microwave Dielectric Characterization of 3D-Printed $BaTiO_3$/ABS Polymer Composites", US National Library of Medicine, PMC4778131; 2016. Available from: https://www.ncbi.nlm.nih.gov/pmc/articles/PMC4778131/#b20

[7] Rao Y et al. Novel polymer-ceramic nanocomposite based on high dielectric constant epoxy formula for embedded capacitor application. Journal of Applied Polymer Science. 28 Nov 2001;**83**(5): 1084-1090. Available from: https://doi.org/10.1002/app.10082

[8] Ahn D et al. Representation of surface roughness in fused deposition modeling. Journal of Materials Processing Technology. 2009;**209** (15-16):5593-5600

[9] Sood A et al. Improved dimensional accuracy of fused deposition modeling processed part using grey Taguchi method. Journal of Material and Design. 2009;**30**(10):4243-4252

[10] Ang K et al. Investigation of the mechanical properties and porosity relationships in fused deposition modeling-fabricated porous structures. Rapid Prototyping Journal. 2006;**12**(2): 100-105

[11] Roberson D, Wicker R, MacDonald E. Ohmic curing of printed silver conductive traces. Journal of Electronic Materials. 2012;**41**(9):2553-2566

[12] Parsons P, Larimore Z, Muhammed F, Mirotznik M. Fabrication of low dielectric constant composite filaments for use in fused filament 3D printing. Additive Manufacturing, Science Direct. December 2019;**30**. Available from: https://doi.org/10.1016/j.addma.2019.100888

[13] Castles F, Isakov D, Lui A, Lei Q, Dancer CEJ, Wang Y, et al. Microwave dielectric characterization of 3D-printed $BaTiO_3$/ABS polymer composites. Scientific Reports. 2016;**6**. Available from: https://doi.org/10.1038/srep22714

[14] Parsons P, Larimore Z, Mitchell G. Composite materials development for fused filament fabrication of RF systems. In: Proceedings of Applied Computational Electromagnetics Symposium (ACES). Monterey, CA; March 2020

[15] Dorsey W, Zaghloul A. Dual-polarized dual-band antenna element for ISM bands. In: Proceedings of IEEE

Antennas and Propagation Society. International Symposium. Charleston, SC: APS/URSI; 2009

[16] Zaghloul A, Dorsey W. Evolutionary development of a dual-band, dual-polarization, low-profile printed circuit antenna. In: Proceedings of International Conference on Electromagnetics in Advanced Applications. Torino, IT: ICEAA; 2009. pp. 994-997

[17] Dorsey W, Zaghloul A. Dual-band dual-circularly polarized antenna element. IET Microwaves, Antennas and Propagation. 2013;7:283-290

[18] Khan M, Yang Z, Warnick K. Dual-circular-polarized high-efficiency antenna for ku-band satellite communication. IEEE Antennas and Wireless Propagation Letters. 2014;**13**: 1624-1627

Section 4

Advanced Antenna Arrays

Chapter 10

A Review to Massive MIMO Detection Algorithms: Theory and Implementation

Bastien Trotobas, Amor Nafkha and Yves Louët

Abstract

Multiple-input multiple-output (MIMO) systems entered most major standards in the past decades, including IEEE 802.11n (Wi-Fi) and long-term evolution (LTE). Moreover, MIMO techniques will be used for 5G by increasing the number of antennas at the base station end. MIMO systems enable spatial multiplexing, which has the potential of increasing the capacity of the communication channel linearly with the minimum of the number of antennas installed at both sides without sacrificing any additional bandwidth or power. To handle the space-division multiplexing (SDM), receivers have to implement new algorithms to exploit the spatial information in order to distinguish the transmitted data streams. This chapter provides an overview of the most well-known and promising MIMO detectors, as well as some unusual-yet-interesting ones. We focus on the description of the different paradigms to highlight the different approaches that have been studied. For each paradigm, we describe the mathematical framework and give the underlying philosophy. When hardware implementations are available in the literature, we provide the results reported and give the according references.

Keywords: MIMO systems, MIMO detectors, space-division multiplexing, SDM-MIMO, linear detection, interference cancelation, tree-search

1. Introduction

In multiple-input multiple-output (MIMO) communication systems, both the transmitters and receivers are equipped with several antennas which will help in achieving high gains in spectral, power, and energy efficiency compared to conventional single-input single-output (SISO) systems where both the transmitters and receivers have only one antenna each. As a matter of fact, the MIMO systems have the ability to turn multipath propagation and multipath delay spread into a benefit for the receiver. The key advantage of MIMO systems is the many orders of magnitude of the signal-to-noise ratio (SNR) at no extra bandwidth. However, a non-negligible software and hardware processing complexity is added at both sides (transmitter and receiver). Present wireless communication standards including Wi-Fi standards like IEEE 802.11n/ac, long-term evolution (LTE), and WiMAX are considering MIMO technology as a key element. Moreover, in the next generation of wireless technology systems (i.e., 5G), massive MIMO is emerging as a new research field in which base stations are equipped with 100 or more antennas. At the receiver side, designing reliable and energy-efficient MIMO detectors is a very challenging task, because of the complexity of the implementation of the signal detection due to the interfering sub-streams. The signal detection problem refers to finding the most probable transmitted symbols based on the perfect channel state information (CSI) available at the receiver and the received signal.

The hardware implementation of massive MIMO detector is of particular interest to deal with 5G wireless technology. Optimal massive detectors such as the maximum likelihood detector (MLD) or the sphere decoding (SD) are considered infeasible given their high computational complexity. Hence, low computational complexity algorithm achieving near-optimal performance is required; many existing detection algorithms like zero forcing (ZF), minimum mean-square error

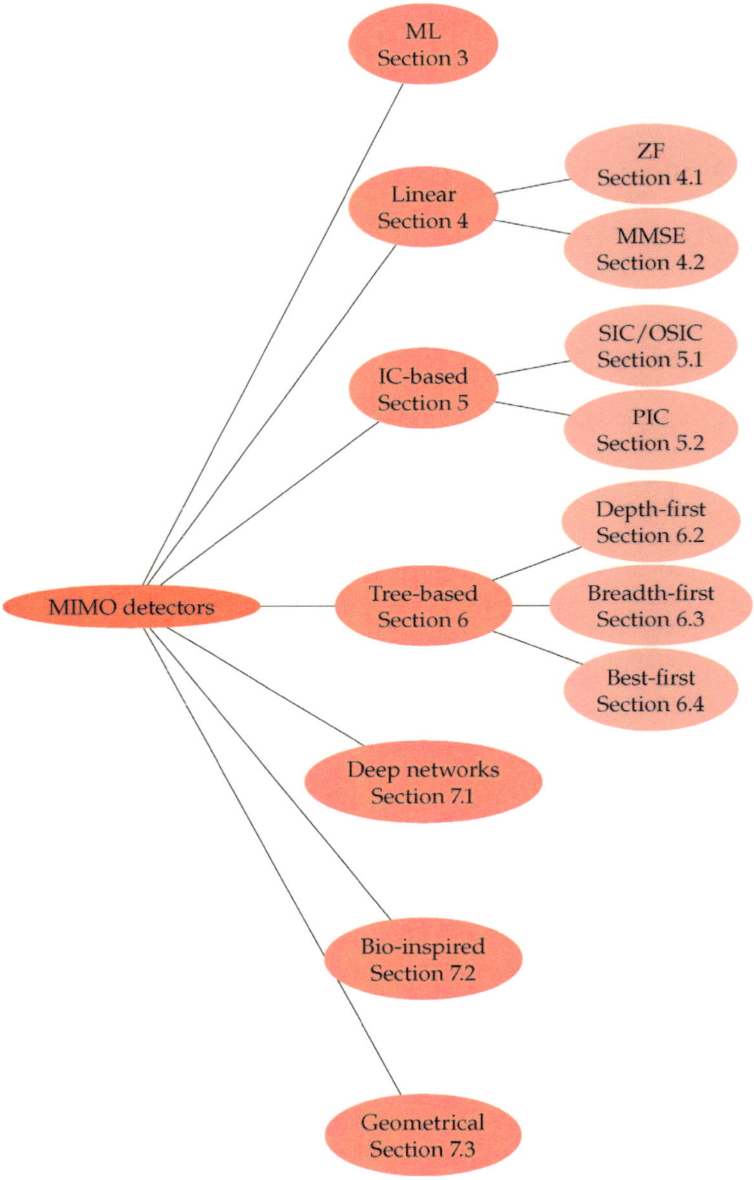

Figure 1.
Tree mind map of the detectors described in this chapter and the number or the corresponding section.

(MMSE), and successive interference cancelation (SIC) are used to deal with massive MIMO detection. In [1, 2], the authors presented surveys on various MIMO and massive MIMO detection techniques from algorithmic viewpoints. Although many classical massive MIMO detectors have been proposed in the literature, herein, new recent algorithms based on the application of machine learning, geometrical techniques, and bioinspired methods are presented and discussed.

In this chapter, we propose an overview of the SDM detection algorithms. We specifically stress out the different paradigms that are used to solve the detection problem and compare all of them. Thus, we describe the most well-known and promising MIMO detectors, as well as some unusual-yet-interesting ones. Section 2 presents the framework and the assumptions that are used in the remainder section. Section 3 introduces the maximum likelihood (ML) optimal detector, and then Section 4 describes the linear ones. Section 5 details algorithms based on the interference cancelation, and Section 6 discusses the one based on tree-search. Finally, Section 7 highlights unusual-yet-interesting detectors before Section 8 concludes the chapter. **Figure 1** provides an overview of all the detectors described in this chapter as a tree mind map.

2. Introduction to MIMO detection algorithms

In the SDM framework, data streams are transmitted at the same time and at the same frequency, and the receiver relies on spatial consideration to distinguish the streams. Herein, we assume that the MIMO transmitter does not use any spatial coding and that all data streams are independent. To give the reader a unified mathematical description through this chapter, we adopt the following notation: scalars, vectors, and matrices are denoted by lower-case, bold-face lower-case, and bold-face higher-case letters, respectively. We call $\mathbf{v}(i)$ the ith coefficient of the vector $\mathbf{v}$, and $\mathbf{H}(i,j)$ is the element of the ith row and jth column in the $\mathbf{H}$ matrix.

In the linear input–output MIMO model where data are transmitted as the symbols of a constellation Φ, the received vector $\mathbf{y} \in \mathbb{C}^M$ is the result of the emitted symbols $\mathbf{x} \in \Phi^N$ propagated through the channel $\mathbf{H}$ and added to an additive noise $\mathbf{w}$. This model leads to the following equation:

$$\mathbf{y} = \mathbf{H}\mathbf{x} + \mathbf{w} \tag{1}$$

and the MIMO detection problem then refers to the combinatorial optimization problem:

$$\arg\min_{\mathbf{x} \in \Phi^N} \|\mathbf{y} - \mathbf{H}\mathbf{x}\|^2. \tag{2}$$

Assuming a circularly symmetric Gaussian noise, solving Eq. (2) is equivalent to searching the most probable emitted symbol vector based on the signal on each receive antennas and the channel state. Even if $\|\mathbf{y} - \mathbf{H}\mathbf{x}\|^2$ is a convex function with respect to x, the detection problem is not a convex optimization problem due to the discrete feasible solution set Φ^N. As a result, a special algorithm has to be used, and this chapter will describe the most common ones.

Let us start by outlining the traditional assumptions that we will use in the present chapter. Although many constellation types could be used in MIMO systems, we limit the discussion to the square quadrature amplitude modulations (QAMs) that are most commonly investigated. Besides, the channel is considered memoryless, linear, and flat and with a block fading[1]. In this chapter, we assume that channel state information (CSI) is correctly estimated at the receiver side but not at the transmitter side. The impact of imperfect CSI at the receiver on the performance of detection algorithms is not addressed in the present chapter. Some known training symbols are sent from the transmitter, based on which the receiver estimates the channel before proceeding to the detection of the transmitted data symbols.

The channel matrix is modeled as a complex matrix $\mathbf{H} \in \mathbb{C}^{(M,N)}$. In that case, the element $\mathbf{H}(i,j)$ refers to the complex channel gain between the jth transmit antenna to the ith receive antenna. Many channel models can fit in this framework, and we stick to the most popular one: the uncorrelated Rayleigh fading channel [3, 4]. The uncorrelated channel model provides a good approximation of propagating environments with rich scattering where the signals between the transmitter and the receiver experience many different paths and no strong line of sight between the transmitter and the receiver. This situation occurs, for instance, in urban and indoor conditions. In these conditions, each receiver antenna receives a sum of a large number of signal paths, and the channel transfer functions can be modeled as the realization of a circularly symmetric normal distribution.

3. Maximum likelihood detector

Obtaining the optimal result requires, in the most straightforward approach, the use of the ML detector that solves Eq. (2) using an exhaustive search. Even if this method gives the best result since all $x \in \Phi^N$ are evaluated, it is not suitable for real implementation. Indeed, the number of vectors to be tested grows exponentially with the number of transmit antenna and the constellation size. Thus, the computational cost of evaluating Eq. (2) requires an unrealistic quantity of resources to detect the transmitted vector $\mathbf{x}$. That is why a variety of detection algorithm has been developed throughout the past year to achieve the same detection performance of ML detectors while having a tractable complexity.

From a computational theory perspective, the detection problem is an instance of the closest lattice-point search (CLPS) problem with a specified lattice [5]. It has been proved that regardless of the preprocessing on the lattice (i.e., the channel matrix), the problem is always NP-hard [5]. The NP-hardness implies that it is not possible, at the moment, to find any detector that is sure to have both an optimal performance and a polynomial complexity[2]. For that reason, all the following detectors have suboptimal performance (which can be very close to optimal) or a non-polynomial worst-case complexity (which can be polynomial is the average case).

4. Linear detectors

4.1 Zero forcing (ZF) detector

Linear detectors are the most simple algorithms to solve the detection problem. The most basic one is the ZF algorithm that follows a two-step process. First, the ZF

---

[1] The nomenclature section provides definitions.

[2] It is widely believed that it does not exist any solution to solve NP-hard problems in polynomial time. Yet, this assumption is not proven so that there is still a possibility that such an algorithm exists and is just not discovered nowadays.

detector solves Eq. (2) transforming the constraint from $\mathbf{x} \in \Phi^N$ to $\mathbf{x} \in \mathbb{C}^N$ such that the problem become an easy-to-solve convex optimization with a known mathematical solution:

$$\mathbf{x}_0 = \mathbf{H}^+ \mathbf{y} \tag{3}$$

with $\mathbf{H}^+ = \left(\mathbf{H}^H \mathbf{H}\right)^{-1} \mathbf{H}^H$ being the left Moore-Penrose pseudoinverse. Then, the constraint on x is reintroduced by quantizing the vector accordingly to the constellation in use. This quantization should lead to a good estimation as after the application of the detection matrix $\mathbf{T}_{\text{ZF}} = \mathbf{H}^+$, Eq. (1) becomes

$$\mathbf{T}_{\text{ZF}} . \mathbf{y} = \mathbf{x} + \mathbf{H}^+ \mathbf{w} \tag{4}$$

highlighting that all the interference are canceled. The previous equation is also the proof that the ZF detector is the optimal linear one regarding the signal-to-interference ratio (SIR) criteria. Indeed, one can see that the vector $\mathbf{T}_{\text{ZF}} . \mathbf{y}$ contains each stream independently plus some noise but without any interference.

4.2 Minimum mean-square error (MMSE) detector

By only focusing on the interference, the ZF detector performance suffers from not taking the noise into account. Indeed, if the noise level is known to the receiver, a Bayesian estimator including this information can provide a better detection. A linear Bayesian estimator minimizing the mean-square error can be derived using the orthogonality principle [6] leading to

$$\mathbf{T}_{\text{MMSE}} = \left(\mathbf{H}^H \mathbf{H} + 2\sigma^2 \mathbf{I}\right)^{-1} \mathbf{H}^H \tag{5}$$

with σ^2 being the noise variance per real direction. The detector based on this detection matrix, followed by the quantization, is called the minimum mean-square error (MMSE), and it is known to maximize the signal-to-noise-plus-interference ratio (SINR). When the signal-to-noise ratio (SNR) is low (i.e., σ^2 is high), the MMSE detector provides better results, jointly minimizing the interference and the noise. Otherwise, when σ^2 is very low, the corrective term becomes negligible, and the ZF and MMSE detectors overlap.

5. Interference cancellation detectors

To improve further the performance, it is necessary to drop the linear detector approach and look for more elaborate decoding algorithms. Historically, the first nonlinear detector type is still based on the principle of canceling signal interference. This concept leads to two approaches: an iterative one named successive interference cancelation (SIC) and a simultaneous version named parallel interference cancelation (PIC).

5.1 Successive interference cancellation (SIC) detector

The SIC detectors opt for a two-step iterative process: first, a decision is taken on the first position $\mathbf{x}_1$, and then assuming that the decision was right, the detector corrects $\mathbf{y}$ by removing the interference that would have been generated by $\mathbf{x}_1$.

Then, SIC detectors repeat this process on the next **x**'s entry until the whole vector is received.

Even if the performance is better than with the linear detectors, the SIC process is very prone to error, given that the assumption at an iteration has an impact on all the following ones. For this reason, the simple SIC detector has quickly been replaced by a variant seeking for an optimal iteration order [7]. This variant called ordered successive interference cancelation (OSIC) aims to make the first assumption on the position that leads to the better SNR or SINR.

To select the best symbol to detect at each iteration, the OSIC detector computes the post-SNR or post-SINR for each symbol, assuming that the kth element is canceled using the detection matrix **T**. Most of the time, the detection matrix is chosen to be $\mathbf{T}_{ZF}$ or $\mathbf{T}_{MMSE}$ optionally updated after each iteration. When using the SNR criterion, the value of the kth post-SNR is computed as in [7, 8].

$$\mathrm{SNR}_k = \frac{<\mathbf{x}(k)^2>_\Phi |\mathbf{T}_k \mathbf{h}_k|^2}{\sigma^2 \|\mathbf{T}_k\|^2} \tag{6}$$

with $\mathbf{T}_k$ being the kth row of **T**, $\mathbf{h}_k$ the kth column of **H**, and $<\mathbf{x}_k^2>_\Phi$ the expected value over the constellation set. The latter term is the average signal power of the kth data stream that can be computed, assuming that each symbol is equiprobable, as

$$<\mathbf{x}_k^2>_\Phi = \frac{1}{\varphi} \sum_{\mathbf{x}\in\Phi} \mathbf{x}(k)^2 \tag{7}$$

with φ the number of symbols in constellation Φ.

When using the SINR criterion, the post-SINR expression becomes slightly more complex as the post-processed power of each other channel appears in the expression

$$\mathrm{SINR}_k = \frac{<\mathbf{x}(k)^2>_\Phi |\mathbf{T}_k \mathbf{h}_k|^2}{\sum_{l\neq k} <\mathbf{x}(l)^2>_\Phi |\mathbf{T}_l \mathbf{h}_l|^2 + \sigma^2 \|\mathbf{T}_k\|^2}. \tag{8}$$

For clarity sake, **Figure 2** sums up the OSIC detection algorithm introducing a process $t : \mathbf{H} \mapsto T$ to build a detection matrix from a channel one. This process is most of the time the Moore-Penrose pseudoinverse or the process described in Eq. (5). We also denote by $\mathcal{D}$ the set of the symbol index to be decoded and by $\leftarrow$ the affectation. One must note that an instruction is optional and may be skipped. If this instruction is applied, performance is increased by canceling the interference in the post-criterion computation and so is the complexity.

5.2 Parallel interference cancellation (PIC) detector

The main drawback of the OSIC algorithm is that the number of iterations grows linearly with the number of antennas. The number of stage becomes an issue for large MIMO system since each stage adds a reception delay. For that reason, a detector capable of canceling the interference for all antennas at once was developed.

The first application of such an algorithm to SDM systems dates from the early 2000s, and it is based on a few basic steps summed up in **Figure 3** as published in [9]. The main point of the PIC algorithm is to start by using a simple detector with poor performance, most of the time a linear one, and cancel the interference on all antennas at once based on the assumption. If better performance is required, it is possible to iterate the last three instructions as many times as needed by using the new detected symbol as the new assumption.

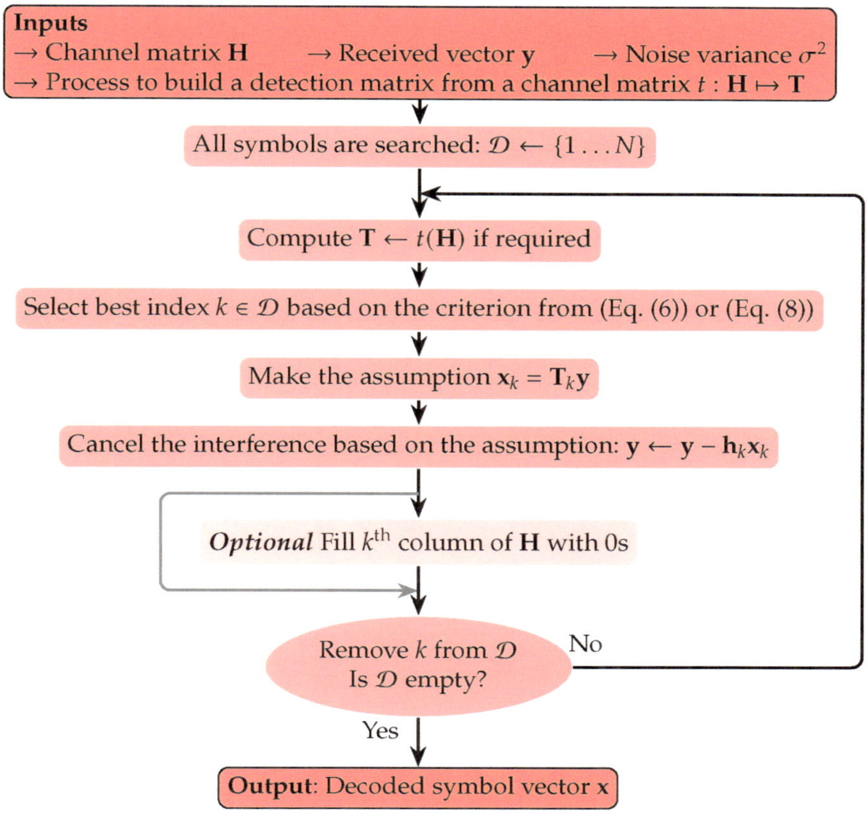

Figure 2.
OSIC algorithm outline.

Simultaneously, the iterative reception techniques developed for turbo codes and single-input single-output channels are adapted to MIMO systems. The goal is to receive a coded message by alternating between soft-input soft-output detector and decoder. Each algorithm uses a priori information from the other to improve its performance [10]. This method leads to one of the current most accomplished version of the PIC family: a soft-input soft-output detector to be used in iterative decoding with any message coding [11].

This version adds several improvements to the basic algorithm described in **Figure 3**. First, it uses the soft symbols from [12] that are defined as the expected value of the symbols knowing the a priori. The reliability of a soft symbol is computed as its variance. Then, the parallel cancelation (see the third instruction in **Figure 3**) is performed using the soft symbols in place of the rough estimation. Finally, a last MMSE filtering is performed before the computation of the log-likelihood ratios (LLRs). Further reductions in complexity are also used, such as the max-log approximation [10, 13, 14] or the channel Gram matrix [15]. Thanks to all of these improvements, an application-specific integrated circuit (ASIC) is reported to achieve a throughput greater than 750 Mb/s with good BER performance [11].

5.3 Selecting between SIC and PIC detectors

The key idea to select between SIC and PIC detectors is to compare the relative quality of data streams. As stated earlier, SIC algorithms guess the best data stream and then process the other one based on this assumption. This process makes the

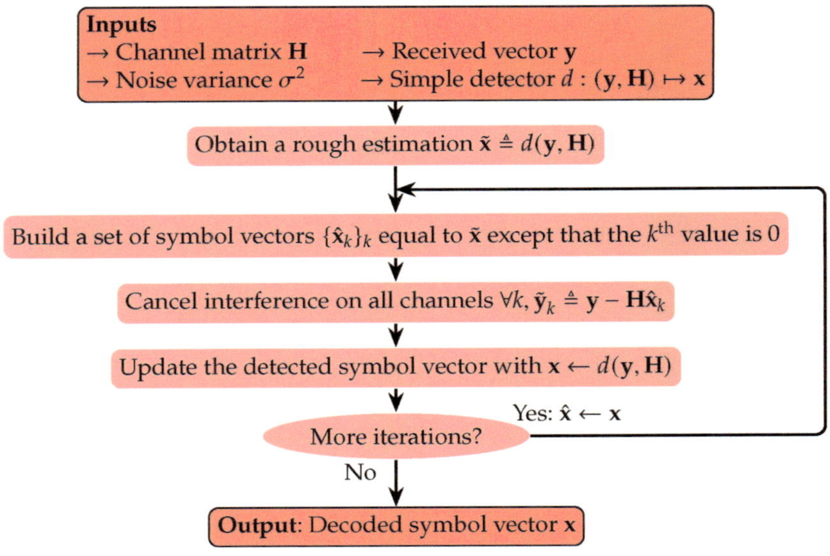

Figure 3.
PIC algorithm outline as published in [9].

SIC algorithms very prone to error propagation. Indeed, if an assumption is wrong, the error has consequences on all the data streams to be detected. Hence, SIC detectors should be used when there is a net ranking in the quality of each data stream. A basic scenario for this would be a MIMO system receiving data from several users with different channel qualities.

On the contrary, PIC detectors process all the data streams at once so that they are more resilient to interstream error propagation. However, the parallel computations assume that every data stream is as reliable as the other. Due to this assumption, a poor-quality data stream propagates its error to the whole system. For that reason, PIC detectors are well suited when all data streams have the same quality level.

6. Tree-search-based detectors

Tree-search-based detectors are the current most investigated algorithms. They use a different framework than the linear and the interference cancelation detector. As the name suggests, the tree-search detector interprets the detection problem from Eq. (2) as the search for the best path in a tree. Tree-search-based detectors can either be optimal with a non-polynomial yet small complexity or quasi-optimal yet not optimal with a polynomial convexity.

Figure 4 gives an example of the tree interpretation for a constellation with four symbols and two data streams. In this configuration, solving the detection problem is equivalent to find two symbols in the set $\Phi = \{s_1, s_2, s_3, s_4\}$ that minimize the objective function. This process can be seen as finding the path in the tree that leads to the best objective function. The first tree level corresponds to the first symbol and so on for each level.

In this paradigm, the exhaustive search detector described in Section 3 computes the objective function for each leaf node and then selects the best path. Tree-search-based detectors search for the best leaf without trying every path. This leads to three enumeration paradigms: depth-first, breadth-first, and best-first. These

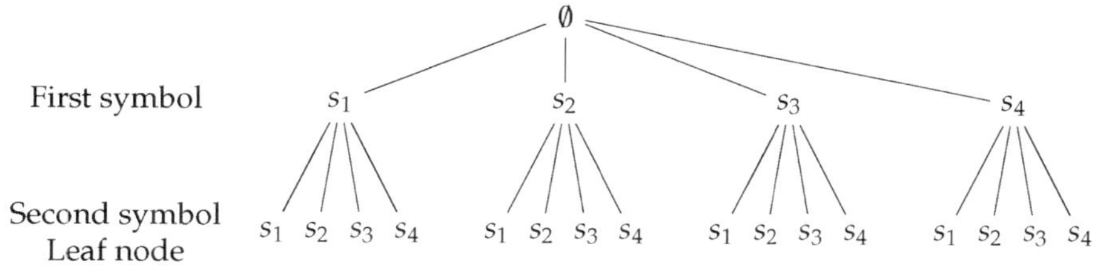

Figure 4.
Tree view of the detection problem: Example for $\Phi = \{s_1, s_2, s_3, s_4\}$ *and two data streams.*

paradigms will be detailed after the description of the preprocessing used by all the variants.

6.1 Preprocessing using QR decomposition

All tree-search-based detectors use the same preprocessing. Let be $\mathbf{H} = \mathbf{QR}$ the QR decomposition of the channel matrix with $\mathbf{Q}$ a unitary matrix and $\mathbf{R}$ an upper triangular one. The decomposition is computed only once per coherence block leading to a negligible overhead of the complexity per received symbol. Using the QR decomposition, we have

$$\|\mathbf{y} - \mathbf{Hx}\| = \|\mathbf{y} - \mathbf{QRx}\| = \|\mathbf{Q}^H\mathbf{y} - \mathbf{Q}^H\mathbf{QRx}\| \tag{9}$$

as unitary matrices act as isometries. Thus, by exploiting the property of unitary matrices $\mathbf{Q}^H = \mathbf{Q}^{-1}$, this norm can be rewritten as

$$\|\mathbf{y} - \mathbf{Hx}\| = \|\tilde{\mathbf{y}} - \mathbf{Rx}\| \tag{10}$$

with $\tilde{\mathbf{y}} \triangleq \mathbf{Q}^H\mathbf{y}$. Computing $\tilde{\mathbf{y}}$ is the only overhead in complexity that is required on a symbol basis as it cannot be preprocessed for the whole coherence block.

The point of this QR preprocessing is that the triangularity of $\mathbf{R}$ allows to compute the objective function iteratively. Indeed, we can introduce for all $k \in \{1, \dots, N\}$, $\mathbf{d}(k) \triangleq \tilde{\mathbf{y}}(k) - (\mathbf{Rx})(k)$ with $(\mathbf{Rx})(k)$ being the kth coefficient of the product Rx. Given this definition, the objective function is written as

$$\|\tilde{\mathbf{y}} - \mathbf{Rx}\|^2 = \sum_{k=1}^{n} \mathbf{d}(k)^2. \tag{11}$$

Moreover, the triangularity of $\mathbf{R}$ gives

$$\forall k \in \{1, \dots, N\}, (\mathbf{Rx})(k) = \sum_{j=1}^{n} \mathbf{R}(k,j)\mathbf{x}(j) = \sum_{j=k}^{n} \mathbf{R}(k,j)\mathbf{x}(j) \tag{12}$$

that leads to

$$\forall k \in \{1, \dots, N\}, \mathbf{d}(k) = \tilde{\mathbf{y}}(k) - \sum_{j=k}^{n} \mathbf{R}(k,j)\mathbf{x}(j). \tag{13}$$

With this expression, it is clear that we can compute partial estimations of the objective function and $\mathbf{d}(k)$ coefficients while the symbol vector is built. Indeed, starting from the last component, Eq. (13) is fully evaluated for the jth position as soon as a hypothesis is made on $\mathbf{x}(j)$. Thus, it is possible to add a new operand in Eq. (11) and to have an idea of how promising the partial symbol vector is. The partial objective function from Eq. (11) is traditionally called the partial Euclidean distance (PED).

6.2 Depth-first tree-search detection: sphere decoding

The depth-first paradigm is the oldest one, and it is commonly known in the communication field as the sphere decoding (SD). SD is the transposition of the mathematical Fincke-Pohst algorithm in the telecommunication field [16]. The basic principle of this algorithm is to define an upper-bound for the objective function named the radius r^2 and then to use it to prune paths as early as possible. Reintroducing x_0 from Eq. (3), the upper-bound constraint gives

$$\|\mathbf{y} - \mathbf{Hx}\|^2 = \|\mathbf{H}(\mathbf{x}_0 - \mathbf{x})\| \leq r^2. \tag{14}$$

This inequality highlights that constraining the objective function may be interpreted as looking for solution no so far from $\mathbf{x}_0$. As stated in Eq. (4), the only deviation from $\mathbf{x}_0$ is due to the noise so that the choice of r must be adapted to the SNR. Thus, if the SNR is high, the radius can be small, while in the contrary scenario, the radius should be increased so that there is at least one vector $\mathbf{x}$ satisfying the constraint from Eq. (14).

In the remainder of this section, we assume that r^2 is adequately chosen and that there is at least a solution. As stated in Section 6.1, the QR decomposition allows us to compute a PED at each level. As the PED is a sum of squares, it can only increase during the decoding process. Thus, if at some point, a PED violates the constraint from Eq. (14), then all the vectors build upon this partial solution are bound to be infeasible. From a tree-search perspective, this means that if a node already breaks the constraint, all its children will do the same. Thus, all paths starting from this node can be pruned without performance loss.

The SD is referred to as a depth-first detector as starting from the root node, it goes as depth as possible until it reaches a leaf or violates Eq. (14). If a leaf is reached, it is compared to the best leaf so far and saved if it is the new best leaf. If Eq. (14) is violated, the SD algorithm backtracks and explores a new path. When all paths are either completed or pruned, the result is the best leaf reached.

The Schnorr-Euchner (SE) enumeration is another depth-first enumeration known to perform better by using a lattice reduction method [17, 18]. The basic idea is to explore the node's children by the increasing order of their PED. This is particularly useful when using the radius reduction technique that sets an infinite r^2 at the beginning and then updates it to the best objective function for a leaf encounter so far.

The SD algorithm and its SE version are optimal as they ensure to find the exact solution of the detection problem. Indeed, the best leaf is obviously the best leaf among all the completed paths, and the pruned paths cannot lead to a better point due to their already worst PED. The NP-hardness argument detailed in Section 3 implies that SD has a non-polynomial worst-case complexity. Moreover, SD expected complexity is also non-polynomial even if the exponential growth is slow enough to compete with polynomial detectors under certain circumstances [19].

A very efficient soft-input soft-output depth-first algorithm is the single tree-search sphere decoding (STS-SD) [20]. To produce its soft-output, it uses the max-log approximation [10, 13, 14] and makes some changes on the pruning criterion. The max-log approximation avoids the computation of the exact LLRs by claiming that

$$L_i \approx \frac{1}{2\sigma^2}\left(\min_{x\in\chi_i^0}\|\mathbf{y}-\mathbf{H}\mathbf{x}\|^2 - \min_{x\in\chi_i^1}\|\mathbf{y}-\mathbf{H}\mathbf{x}\|^2\right) \tag{15}$$

where $\chi_i^k = \{x \in \Phi^N : b_i = k\}$ is the set of all symbols with the ith bit set to k. Thus, to compute the max-log approximation, one must know the objective function of the best leaf (i.e., one of the minimum in Eq. (15)) but also the objective function of each best counter-hypotheses (the other minimum). A path should then be pruned only if its PED is greater than the current radius r^2 and if this path cannot lead to a better counter-hypothesis. This can be implemented by adding another radius called the hypothetical radius constraint.

One of the most advanced ASIC-implemented depth-first reported so far uses a two-dimensional Schnorr-Euchner enumeration. This implementation reaches a throughput higher than 600 Mb/s for the soft-output version and exceeds 1.2 Gb/s for the hard-output one while keeping an excellent energy efficiency [21]. The high throughput is achieved by using several SD cores in parallel to decode several vectors simultaneously.

6.3 Breadth-first tree-search detectors: K-best and M-algorithm

Breadth-first detectors drop out of optimality for better implementability. Indeed, they address the two main issues of the depth-first paradigm: the unpredictable complexity that depends on the SNR through r^2 and the depth-and-backtrack enumeration that prevents the use of hardware pipelines. Breadth-first detectors achieve this goal by removing the pruning criterion and always keep the same number of paths instead. At each level, the detector compares all the current paths' PED and keeps only the best ones. This number is traditionally called K for the K-best algorithm [22] or M for the M-algorithm [23]. A recent work mixed this approach with the upper-bound radius from death-first to prune even more path per level and reduce the complexity further [24]. This method converges to the breadth-first if all PED are always under the upper-bound, but if some PEDs overgrow, it can reduce the number of surviving paths.

The restricted number of surviving paths induces that the right one can be pruned early if its PED has grown too quickly in the early levels. This is the reason for the optimality loss. Then, some detectors implement a post-detection SNR criterion to reorder the tree levels such that the most certain one is at the top. Thus, the right path is less likely to be pruned by mistake in the early stages. Thanks to this reordering, the K-best algorithm performs very well yet, not optimally in a mathematical sense.

From a hardware implementation perspective, breadth-first tree-search detectors are very efficient. They do not require any backtrack so that an expanded node can be safely deleted as it will never be revisited. Moreover, the number of visited nodes per level is fixed. Thus, ASIC can embed the exact amount of resources required. These two characteristics allow the construction of efficient hardware pipelines that substantially increase the throughput. K-best can achieve at least the same throughput as depth-first without the need for parallel cores. Hard-output implementations exceeding 2.5 Gb/s are reported using the breadth-first paradigm [25]. Another study focused on energy efficiency designed a breadth-first variant that can handle both the channel noise and the hardware noise generated by the voltage over the scaling method in memories [26].

Breadth-first detectors can provide soft-output using the max-log approximation and a list. This list approach is used by several detectors, including some other tree-search algorithm and detectors from other families. The point of this approach is to produce a list Γ of point associated with their objective function and to approximate Eq. (15) as

$$L_i \approx \frac{1}{2\sigma^2}\left(\min_{x\in\Gamma\cap\chi_i^0}\|\mathbf{y}-\mathbf{Hx}\|^2 - \min_{x\in\Gamma\cap\chi_i^1}\|\mathbf{y}-\mathbf{Hx}\|^2\right). \tag{16}$$

For most breadth-first algorithms, Γ is the list of all completed paths [22, 25].

6.4 Best-first tree-search detector: fast descent tree-search and parallel tree-search

Best-first detectors are also sometimes called metric-first. The basic idea besides this enumeration is to not favor depth or breadth over each other. Instead, the node with the best PED is expanded, regardless of its level. The best-first algorithm keeps a node pool with nodes to be expanded. First, the pool is initialized with the root node. Then, at each iteration, the node with the lower PED is popped out from the pool, and all its children are computed and pushed in the pool unless they are leaves. If they are, a comparison with the best leaf so far allows us to keep track of the best result. When a leaf is reached, its objective function may be used as an upper-bound to prune the pool for each node with a PED higher than the new reference. The detection ends when the pool is empty.

This simple method quickly overfills the pool as several nodes are added when only one is popped out. Rather than providing a huge pool to contain all the nodes, improvement is to convert the φ-ary tree (with φ the constellation size) to a first-child next-sibling binary tree [27]. This method is called the modified best-first algorithm (MBF). With this variant, the only nodes added in the pool after an expansion are the best child and the best yet-to-visit siblings. Then, the growth rate of the pool size is controlled. However, this method struggles to provide a full path solution quickly as the popped out node is the one with the lower PED that is often close to the root. To solve this issue, a variant implementation called MBF fast descent (MBF-FD) changes the expansion rule. When a node is expanded, the process goes through the best child until reaching a leaf, pushing in all best siblings along the way [28].

Recently, a best-first algorithm ASIC is reported to reach 800 Mb/s in a soft-input soft-output framework [29]. The modified algorithm, called cross-level parallel tree-search, splits the pool node into several pools, one per level. At each iteration, a node from each pool in popped out expanded using the best-child/best-sibling framework, and the new nodes are pushed in the according pool. Moreover, the presented detector prune nodes in each pool using the upper-bound to keep only the one that can improve the result or the counter-hypothesis (see Section 6.2 for details). The slit pool helps the parallelization process so that this algorithm variant is very suitable for hardware implementation.

7. Other unusual detectors: bioinspired and geometrical detectors

7.1 Deep neural MIMO detection: learning to detect

The rise of deep learning leads to the search for an efficient neural network to solve the detection problem such as DetNet [30]. This network is inspired by the projected gradient descent algorithm that is a major option to solve convex optimization. It is trained for both static channel (**H** is fixed) and on a time-varying channel (the same condition as previously). As the errors are sometimes unavoidable due to a bad channel realization, the loss function should not be the objective function. Thus, the DetNet designers opt for a

$$\sum_{k=1}^{L} \log(k) \frac{\|\mathbf{x} - \mathbf{x}_k\|}{\|\mathbf{x} - \mathbf{x}_{\mathrm{ZF}}\|} \tag{17}$$

with $\mathbf{x}_{\mathrm{ZF}}$ the result of ZF detection and $\mathbf{x}_k$ the detected symbol of the kth layer. Then, the normalization with the ZF result avoids over-penalizing the situation with bad **H** realization. Moreover, the logarithm weights the result from each layer to give more credit to the final ones.

7.2 Bioinspired detectors

The most studied bioinspired decoders fall into two categories: ant colony optimizations (ACO) and particle swarm optimizations (PSO) that include the firefly algorithm (FA). These techniques are often very complex compared to the previously described algorithms, but they claim to be resilient to challenging conditions. Bioinspired algorithms should be able to decode messages with imperfect CSI, or the data streams are correlated.

ACO-based detectors simulate several ants that choose a path randomly to follow with a nonuniform probability function [31]. Each antenna is processed independently. At each iteration, an ant selects the symbol $s \in \Phi$ with the probability

$$p_s = \frac{\tau_s^{\alpha} \eta_s^{\beta}}{\sum_{s \in \Phi} \tau_s^{\alpha} \eta_s^{\beta}} \tag{18}$$

with τ_s the pheromone level on the path, η_s an image of the objective function, most of the time through a log-sigmoid function, and (α, β) the two parameters that balance the relative importance of each term. After each iteration, the pheromone level is updated according to the following principle: the better the objective function the ant achieves, the more pheromone it dropped off. Thus, the ant selects more often the path that seems more promising regarding the objective function and the previous runs while preserving some chance of exploring a new path.

FA-based detectors simulate several fireflies that try to find the best mating partner. The objective function determines the attractiveness of a firefly. Then a firefly goes toward more attractive congeners biased with a random influence to promote exploration [32]. Some FA variant implements a memory effect that makes it even closer to a PSO-based algorithm [33]. This framework is applied to MIMO detection using the QR decomposition described in Section 6.1. The FA represents each symbol to decode as a nest containing as many fireflies as the constellation size. Thus, the fireflies have to select a partner in each nest from the last symbol to the first based on the biased attractiveness. When the firefly population searched all the nests, the best path represents the decoded symbol vector. FA-based detectors can be related to tree-search-based algorithm with a randomness exploration and a fixed number of path allowed.

7.3 Geometrical detectors

Geometrical detectors are based on a two-step process: an exploration to find a small set of promising solutions and an exploitation to improve this set at a small cost. It follows the traditional approach in nonconvex optimization to perform simple descents that can be stuck in local optimums from several starting points. Geometrical detectors use a real-valued model and the singular value decomposition (SVD) rather than the QR one [34]. Let us rewrite the objective function by introducing the SVD of $\mathbf{H} = \mathbf{U}\mathbf{D}\mathbf{V}^T$ with $\mathbf{U}$ and $\mathbf{V}$ two orthogonal matrices and $\mathbf{D}$ the diagonal matrix containing the singular values $\{\lambda_i : 1 \le i \le n\}$ in ascending order.

Consequently, the objective function can be rewritten as

$$\|\mathbf{y} - \mathbf{H}\mathbf{x}\|^2 = \left(\mathbf{V}^T(\mathbf{x} - \mathbf{x}_0)\right)^T \mathbf{D}\mathbf{U}^T\mathbf{U}\mathbf{D}\left(\mathbf{V}^T(\mathbf{x} - \mathbf{x}_0)\right) \tag{19}$$

using $\mathbf{x}_0$ from Eq. (3). As the vectors of $\mathbf{V}$, named $\{\mathbf{v}_i : 1 \le i \le n\}$, constitute a basis, we can define $\{\alpha_i : 1 \le i \le n\}$ the coordinates of $\mathbf{x} - \mathbf{x}_0$ on this basis. Using the orthogonality of $\mathbf{U}$ and $\mathbf{V}$ and the diagonality of $\mathbf{D}$, Eq. (19) leads to

$$\|\mathbf{y} - \mathbf{H}\mathbf{x}\|^2 = \sum_{i=1}^{n} \alpha_i^2 \lambda_i^2. \tag{20}$$

Let Δ_i be the straight line passing through x_0 and directed by $\mathbf{v}_i$. One can note that Eq. (20) highlights that the objective function grows more slowly along the first Δ_i rather than along the last ones so that promising points must be around these first straight lines. Then the solution is most likely to be found along this line. The geometrical exploration step is then performed, selecting some points near the first Δ_i. Then a straightforward descent algorithm is performed by looking for the best point in the close neighborhood until convergence.

A soft-output version of this algorithm is possible using the max-log approximation and the list approach detailed in [35], Section 5.2. A field-programmable gate array (FPGA) implementation has recently been proposed. This groundwork points out that geometrical detectors may achieve good performance in the future yet being far from mature at that point [36].

Detector	BER	Complexity	Comment
ML	Optimal	Dramatically complex	
ZF	Very poor	Very simple	Best linear detector regarding SNR criterion
MMSE	Poor	Simple	Best linear detector regarding SINR criterion
SIC/OSIC	Good	Rather complex	Best when there is a clear ranking in the quality of each data stream
PIC	Good	Rather complex	Best when all data streams have the same quality level
Depth-first	Optimal	Very complex	
Breadth-first	Good	Rather complex	Possible trade-off between BER and complexity via the number of surviving paths
Best-first	Good	Less complex	
Deep neural	Good	Rather complex	Possible trade-off between BER and complexity via the number of layers
Bioinspired	Good	Very complex	Resilient to imperfect CSI and channel correlation
Geometrical	Rather good	Rather complex	Possible trade-off between BER and complexity via the number of descents

Table 1.
Summary of all detectors described in this chapter.

8. Conclusions and summary

MIMO detection is a well-studied problem that has been tackled from several perspectives. The mathematical interpretation, as a combinatorial optimization problem, leads to the optimal and linear detectors. From the signal processing perspective, detecting a signal means improving the SNR or SINR so that the direct answer is to cancel the interference and to remove the noise. From an algorithmic perspective, the detection problem is the search for the best path in a weighted tree that relies on some well-known algorithms. Other sources of inspiration, such as nature or geometry, provide some interesting perspectives. These paradigms and the associated detectors are summed up in **Table 1**, and we compare all of them according to the BER-complexity trade-off.

All these perspectives shed a different light on the problem, leading to fruitful experimentation. Indeed, some methods take inspiration from others to keep on improving. Therefore, some improvement axes remain open, for instance, the permanent decrease of complexity with equal performance, the development for efficient hardware implementations, or the optimization of the interaction with decoders to exploit channel codings better.

Conflict of interest

The authors declare no conflict of interest but to research and publish in this area, in particular on geometrical detectors.

Nomenclature

Memoryless	the channel outputs only depend on its state and its inputs.
Linear	the channel outputs are linear combinations of the inputs.
Flat	the channel coherence bandwidth is large enough to consider that all frequencies experience the same transfer function.
Block fading	the channel states vary slow enough to be considered constant over many symbols named coherence block.

Abbreviations

ACO	ant colony optimization
ASIC	application-specific integrated circuit
CDMA	code-division multiple access
CLPS	closest lattice-point search
CSI	channel state information
FA	firefly algorithm
FPGA	field-programmable gate array
LLR	log-likelihood ratio
MBF	modified best-first
MBF-FD	modified best-first fast descent
ML	maximum likelihood
MMSE	minimum mean-square error
OSIC	ordered successive interference cancelation
PED	partial Euclidean distance
PIC	parallel interference cancelation

PSO	particle swarm optimization
QAM	quadrature amplitude modulation
SD	sphere decoding
SE	Schnorr-Euchner
SDM	space-division multiplexing
SIC	successive interference cancelation
SIR	signal-to-interference ratio
SINR	signal-to-noise-plus-interference ratio
SNR	signal-to-noise ratio
SVD	singular value decomposition
ZF	zero forcing

Author details

Bastien Trotobas, Amor Nafkha*† and Yves Louët†
SCEE/IETR UMR CNRS 6164, CentraleSupélec, Cesson Sévigné, France

*Address all correspondence to: amor.nafkha@centralesupelec.com

† These authors contributed equally.

References

[1] Yang S, Hanzo L. Fifty years of MIMO detection: The road to large-scale MIMOs. IEEE Communication Surveys and Tutorials. 2015;**17**(4):1941-1988. ArXiv: 1507.05138

[2] Idowu-Bismark OB, Kennedy O, Idachaba F, Atayero AA. A primer on MIMO detection algorithms for 5G communication network. International Journal on Communications Antenna and Propagation. 2018;**8**(3):194

[3] Saleh AAM, Valenzuela R. A statistical model for indoor multipath propagation. IEEE Journal on Selected Areas in Communications. 1987;**5**(2): 128-137

[4] Debbah M, Muller RR. MIMO Channel modeling and the principle of maximum entropy. IEEE Transactions on Information Theory. 2005;**51**(5): 1667-1690

[5] Micciancio D. The hardness of the closest vector problem with preprocessing. IEEE Transactions on Information Theory. 2001;**47**(3): 1212-1215

[6] Kay SM. Fundamentals of Statistical Signal Processing. Englewood Cliffs, NJ: Prentice-Hall; 1993

[7] Wolniansky PW, Foschini GJ, Golden GD, Valenzuela RA. V-BLAST: An architecture for realizing very high data rates over the rich-scattering wireless channel. In: International Symposium on Signals, Systems, and Electronics; 1998. pp. 295-300

[8] Riadi A, Boulouird M, Hassani MM. ZF/MMSE and OSIC detectors for UpLink OFDM massive MIMO systems. In: IEEE Jordan International Joint Conference on Electrical Engineering and Information Technology; 2019. pp. 767-772

[9] Chin WH, Constantinides AG, Ward DB. Parallel multistage detection for multiple antenna wireless systems. Electronics Letters. 2002; **38**(12):597

[10] Hochwald BM, ten Brink S. Achieving near-capacity on a multiple-antenna channel. IEEE Transactions on Communications. 2003;**51**(3):389-399

[11] Studer C, Fateh S, Seethaler D. ASIC implementation of soft-input soft-output MIMO detection using MMSE parallel interference cancellation. IEEE Journal of Solid-State Circuits. 2011; **46**(7):1754-1765

[12] Tuchler M, Singer AC, Koetter R. Minimum mean squared error equalization using a priori information. IEEE Transactions on Signal Processing. 2002;**50**(3):673-683

[13] Koch W, Baier A. Optimum and sub-optimum detection of coded data disturbed by time-varying intersymbol interference. In: IEEE Global Telecommunications Conference and Exhibition; 1990. pp. 1679-1684

[14] Robertson P, Villebrun E, Hoeher P. A comparison of optimal and sub-optimal MAP decoding algorithms operating in the log domain. In: IEEE International Conference on Communications, Vol. 2; 1995. pp. 1009-1013

[15] Tomasoni A, Ferrari M, Gatti D, Osnato F, Bellini S. A low complexity turbo MMSE receiver for W-LAN MIMO systems. In: IEEE International Conference on Communications, Vol. 9; 2006. pp. 4119-4124. ISSN: 1938-1883

[16] Fincke U, Pohst M. Improved methods for calculating vectors of short length in a lattice, including a complexity analysis. Mathematics of Computation. 1985;**44**(170):463-471

[17] Schnorr CP, Euchner M. Lattice basis reduction: Improved practical algorithms and solving subset sum problems. Mathematical Programming. 1994;**66**(1–3):181-199

[18] Agrell E, Eriksson T, Vardy A, Zeger K. Closest point search in lattices. IEEE Transactions on Information Theory. 2002;**48**(8):2201-2214

[19] Jalden J, Ottersten B. On the complexity of sphere decoding in digital communications. IEEE Transactions on Signal Processing. 2005;**53**(4): 1474-1484

[20] Studer C, Bolcskei H. Soft–input soft–output single tree-search sphere decoding. IEEE Transactions on Information Theory. 2010;**56**(10): 4827-4842

[21] Chuang PIJ, Sachdev M, Gaudet VC. VLSI implementation of high-throughput, low-energy, configurable MIMO detector. In: IEEE International Conference on Computer Design; New York, USA; 2015. pp. 335-342

[22] Guo Z, Nilsson P. Algorithm and implementation of the K-best sphere decoding for MIMO detection. IEEE Journal on Selected Areas in Communications. 2006;**24**(3):491-503

[23] Wong KKY, McLane PJ. A low-complexity iterative MIMO detection scheme using the soft-output M-Algorithm. In: IST Mobile Summit; 2005. p. 5

[24] Choi SJ, Shim SJ, You YH, Cha J, Song HK. Novel MIMO detection with improved complexity for near-ML detection in MIMO-OFDM systems. IEEE Access. 2019;7:60389-60398

[25] Yan Z, He G, Ren Y, He W, Jiang J, Mao Z. Design and implementation of flexible dual-mode soft-output MIMO detector with channel preprocessing. IEEE Transactions on Circuits and Systems I: Regular Papers. 2015;**62**(11): 2706-2717

[26] Khairy MS, Shen CA, Eltawil AM, Kurdahi FJ. Algorithms and architectures of energy-efficient error-resilient MIMO detectors for memory-dominated wireless communication systems. IEEE Transactions on Circuits and Systems I: Regular Papers. 2014; **61**(7):2159-2171

[27] Knuth DE. The Art of Computer Programming: Volume 1: Fundamental Algorithms. 3rd ed. Boston: Addison-Wesley Professional; 1997

[28] Liao CH, Wang TP, Chiueh TD. A 74.8 mW soft-output detector IC for 8 x 8 spatial-multiplexing MIMO communications. IEEE Journal of Solid-State Circuits. 2010;**45**(2):411-421

[29] He G, Zhang X, Liang Z. Algorithm and architecture of an efficient MIMO detector with cross-level parallel tree-search. IEEE Transactions on Very Large Scale Integration Systems. 2020; **28**(2):467-479

[30] Samuel N, Diskin T, Wiesel A. Deep MIMO detection. In: International Workshop on Signal Processing Advances in Wireless Communications; 2017. pp. 1-5. ISSN: 1948-3252

[31] Marinello JC, Abrão T. Lattice reduction aided detector for MIMO communication via ant colony optimisation. Wireless Personal Communications. 2014;77(1):63-85

[32] Datta A, Bhatia V. A near maximum likelihood performance modified firefly algorithm for large MIMO detection. Swarm and Evolutionary Computation. 2019;**44**:828-839

[33] Nasiri B, Meybodi MR. History-driven firefly algorithm for optimisation in dynamic and uncertain environments. International Journal of

Bio-Inspired Computation. 2016; **8**(5):326

[34] Nafkha A, Boutillon E, Roland C. Quasi-maximum-likelihood detector based on geometrical diversification greedy intensification. IEEE Transactions on Communications. 2009; **57**(4):926-929

[35] Trotobas B, Nafkha A. Fixed complexity soft-output detection algorithm through exploration and exploitation processes. In: AICT 2018; Barcelona, Spain; 2018. pp. 89-93

[36] Trotobas B, Akourim Y, Nafkha A, Loüet Y, Weiss J. Evaluation of the complexity, performance and implementability of geometrical MIMO detectors: The example of the exploration and exploitation list detector. International Journal on Advances in Telecommunications. June 2020;**13**(1&2)

Chapter 11

Reconfigurable Fabry-Pérot Cavity Antenna Basing on Phase Controllable Metasurfaces

Peng Xie, Guangming Wang, Haipeng Li, Yawei Wang and Xiangjun Gao

Abstract

Fabry-Pérot cavity (FPC) antenna is a kind of high-gain antenna. Compared with other high-gain antennas, such as array antenna and reflector antenna, the FPC antenna enjoys the advantages of simple structure and high efficiency. So it has attracts many attention since proposed. However, it also suffers the disadvantages of narrow band and fixed radiation patterns, due to its resonance structure. In order to overcome these disadvantages, we proposed novel strategies to realize reconfigurable FPC antennas using the phase controllable metasurfaces (MSs). Through adding PIN diodes into every unit cell of the MS, the reflection phase of the MS can be controlled by tuning the states of the diodes. Then the designed phase controllable MSs are used as the partially reflection surfaces (PRS) to realize frequency or radiation pattern reconfigurable FPC antennas. In this chapter, we analyze the basic theory of the FPC antenna and describe its radiation principle firstly. Then, reflection phase controllable MSs are designed and applied to the FPC antennas. Thus frequency and radiation pattern reconfigurable FPC antennas are formed. The design processes are described in details, and the proposed antennas are fabricated and measured. The measured results verify the correctness of the designs. Through this chapter, the readers can form a comprehensive understanding of reconfigurable FPC antenna design.

Keywords: metasurface, Fabry-Pérot cavity, reconfigurable, phase controllable, PIN diode

1. Introduction

The leaky-wave antennas have attracted much attention since proposed and found wide applications in wireless communication systems. As one kind of the leaky-wave antenna, Fabry-Pérot cavity (FPC) antennas are preferred by scientists due to their broadside pattern and high-gain performance. However, the FPC antenna usually suffer some disadvantages due to its inherent limitations of the resonate structure, such as narrow band, fixing beam, and high profile. So many researches have been done on wideband and beam tilted FPC antenna. To overcome these disadvantages of the conventional FPC antenna, we proposed novel strategies that realize reconfigurable FPC antennas using the PIN diodes. In this chapter, we designed three reconfigurable FPC antennas by metasurfaces (MSs).

Through adding PIN diodes on the MS, the reflection phase of the MS can be controlled by tuning the states of the diodes. The different reflect phase distributions of the MS can make the FPC antenna present different frequency and radiation performance. The reconfigurable FPC antenna can improve the performance of antenna, so that the antenna can be widely used.

2. Basic theory of the Fabry-Pérot cavity antenna

Due to its well frequency selection characteristics, the Fabry-Pérot (FP) resonate cavity are widely used in many applications, such as spectral analyzer, interference filter, and so on. The FP resonate cavity can also be used in the design of high-gain antennas [1–4]. Adding a feeder into the cavity, the electromagnetic wave emanating from the feeder experiences multiple reflections and transmissions in the cavity [5, 6]. When the resonance condition is satisfied, the wave coming out of the cavity will be in phase, and then bidirectional high-gain radiation is achieved. However, most applications require unidirectional antenna radiation. So the FP resonate cavity should be changed slightly to design high-gain FPC antenna [7–9].

In practice, the artificial partially reflection surfaces (PRS) are usually used to design FPC antenna. In order to realize unidirectional radiation, one of the PRS of the FP resonate cavity should be replaced by the metal ground plane [10–12]. The PRS and the metal ground plane form the resonate cavity of the FPC antenna. The schematic model of the FPC antenna is shown in **Figure 1**. The distance between the PRS and the ground plane is h. The radiator locates in the middle of the cavity above the ground plane. The electromagnetic wave emanating from the radiator is incident on the PRS with an angle of θ. One part of the electromagnetic wave transmits through the PRS and the other part is reflected into the cavity by the PRS. The reflected wave is totally reflected by the ground plane and is incident to the PRS again with the same angle, forming the secondary transmission and reflection. The electromagnetic wave experiences multiple reflections and transmissions between the PRS and the ground plane and finally all get through the PRS [11, 13, 14].

Assuming that the transmission coefficient and reflection coefficient of the PRS are

$$r = Re^{j\varphi_1}, \quad t = Te^{j\theta_1} \tag{1}$$

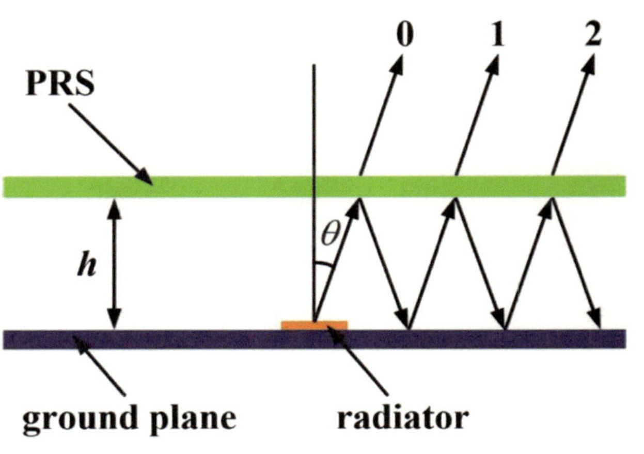

Figure 1.
The schematic model of the FPC antenna.

then the first transmission coefficient is

$$t_0 = t \tag{2}$$

After once reflected by the PRS, the transmission coefficient is

$$t_1 = tre^{-2jkh/\cos\theta + 2jkh\tan\theta\sin\theta + j\varphi_2} = tre^{-2jkh\cos\theta + j\varphi_2} \tag{3}$$

where φ_2 is the reflection phase of the metal ground plane, λ is the wavelength in the free space, and $K = 2\pi/\lambda$ is the propagation constant. According to the above equation, it can be deduced that the transmission coefficient of the resonate cavity after n times of reflection is

$$t_n = tr^n e^{-2jnkh\cos\theta + in\varphi_2} \tag{4}$$

So, the total transmission coefficient along θ direction is

$$t_{total} = \sum_{n=0}^{\infty} t_n = t\sum_{n=0}^{\infty} r^n e^{-2jnkh\cos\theta + jn\varphi_2} = \frac{t}{1 - re^{-2jkh\cos\theta + j\varphi_2}} \tag{5}$$

According to the energy conservation law, $T^2 = 1 - R^2$. Therefore, the total power transmission coefficient is

$$D = t_{total}t_{total}^{*} = \frac{1 - R^2}{1 + R^2 - 2R\cos\Phi} \tag{6}$$

where

$$\Phi = \frac{4\pi}{\lambda}h\cos\theta - \varphi_1 - \varphi_2 \tag{7}$$

φ_1 is the reflection phase of the PRS. According to Eq. (6), when $\cos\Phi = 1$, the maximum value of the power transmission coefficient is

$$D = \frac{1 - R^2}{1 + R^2 - 2R} = \frac{1 + R}{1 - R} \tag{8}$$

It can be seen that the power transmission coefficient is mainly relative to the reflection magnitude of the PRS. The power transmission coefficient increases with the increase of the reflection magnitude of the PRS, so does the gain of the antenna. The maximum radiation direction of the antenna is $\theta = 0°$. And the resonate condition of the antenna is

$$h = \frac{\lambda}{4\pi}(\varphi_1 + \varphi_2) + \frac{\lambda}{2}N, \quad N = 0, 1, 2 \ldots \tag{9}$$

3. Frequency reconfigurable Fabry-Pérot antenna

3.1 Design of the reconfigurable PRS

In order to realize the reflection phase configuration of the PRS, the unit cell of the PRS must be configurable [15]. The structure of the unit cell is shown in **Figure 2a**. The unit cell is printed on a substrate of FR4 with a thickness of 1.6 mm and

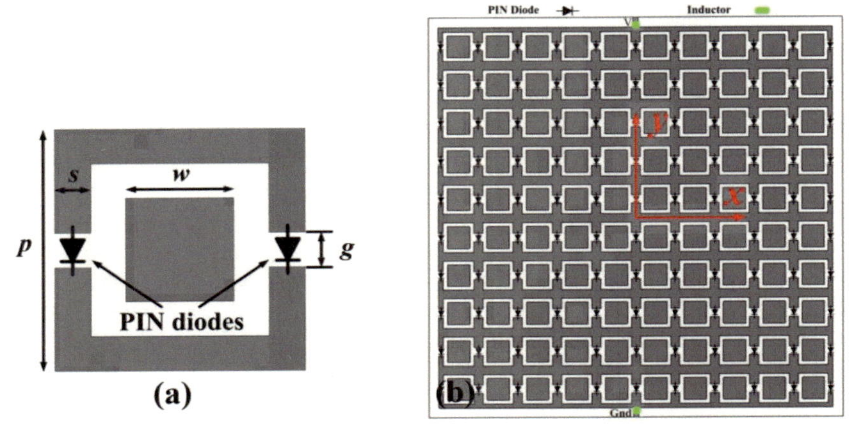

Figure 2.
(a) The structure of the unit cell and (b) the view of reconfigurable PRS.

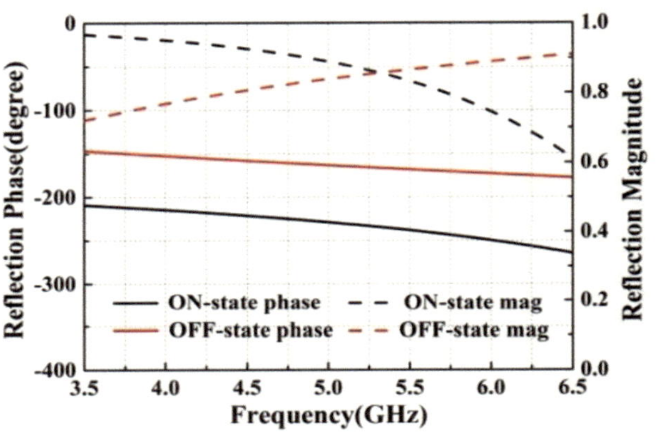

Figure 3.
Reflection coefficients of the unit cell.

permittivity of 4.4. It consists of a square patch and a square ring. Two slots are added on the square ring, and two PIN diodes are inserted into the slots. The two PIN diodes are controlled simultaneously. When diodes are in different states, the unit cell presents different reflection phases. **Figure 3** shows the simulation reflection coefficients of the unit cell with $w = 7$ mm. We can see that, when the diode state changes from OFF to ON, the reflection phase of the unit cell is delayed. According to Eq. (9), the λ increases with the decrease of φ_1. So the antenna will operate at low frequency when the diodes are ON and high frequency when the diodes are OFF. Besides, the reflection phase of the unit cell with the variation of w is plotted in **Figure 4**. It can be seen that the reflection phase of the unit cell decreases with the increase of w no matter if the diodes are open or closed. So the w can be used in tuning the reflection phase of the unit cell to meet the requirement of the antenna.

Figure 2b shows the structure of the configurable PRS. It consists of 10 × 10 unit cells. All diodes on the PRS are set along the same orientation, so all diodes can be controlled by only one DC source. The biasing point V on the top of the PRS connects the anode of the DC source, and the biasing point Gnd on the bottom of the PRS connects the cathode of the DC source. Inductors are added between the feeding point and the metal structure on the PRS to prevent the RF signal from entering the DC source. The periodic boundary is adopted in the simulation of the unit cell, so the

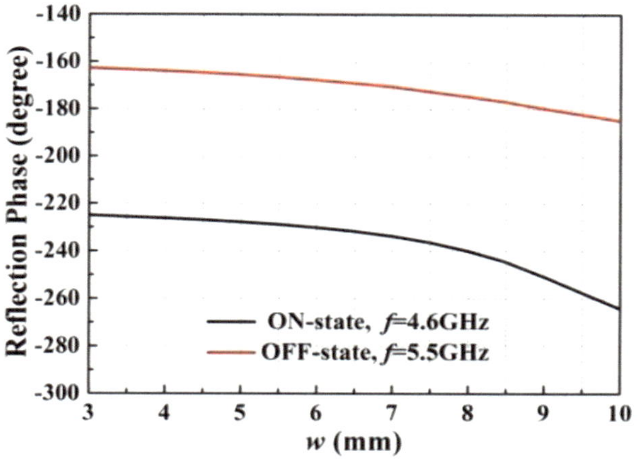

Figure 4.
Effect of w *on reflection phase of the unit cell.*

reflection phase of the unit cell can be considered as the reflection phase of the PRS. The diodes are the BAR50-03 W from Infineon. Because there is no model of the diode in the CST, it is replaced by the equivalent circuit in the simulation. When the diode is closed, they can be replaced by resistors of 1 Ω. When the diode is open, it is replaced by a circuit that consists of a 10 kΩ resistor in parallel with a 0.15 pF capacitor.

3.2 Antenna design

In order to realize frequency reconfiguration, the FPC antenna must adopt dual-band feeder. So we design a dual-band microwave antenna as the feeder of the FPC antenna. The structure of the feeder antenna is shown in **Figure 5a**. The antenna is printed on a substrate of FR4 with the permittivity of 4.4 and thickness of 1.6 mm. It consists of two different metal patches which are connected by a microwave line. Two patches have different lengths and widths, so they can resonate at different frequencies. The bottom of the substrate is a metal ground plane. A SMA connecter is used to feed the antenna from the bottom side, and the feeding point is on the microstrip line. The antenna can realize impedance matching through tuning the position of the feeding point and the microstrip line. The feed antenna is designed to work at 4.6 and 5.5 GHz.

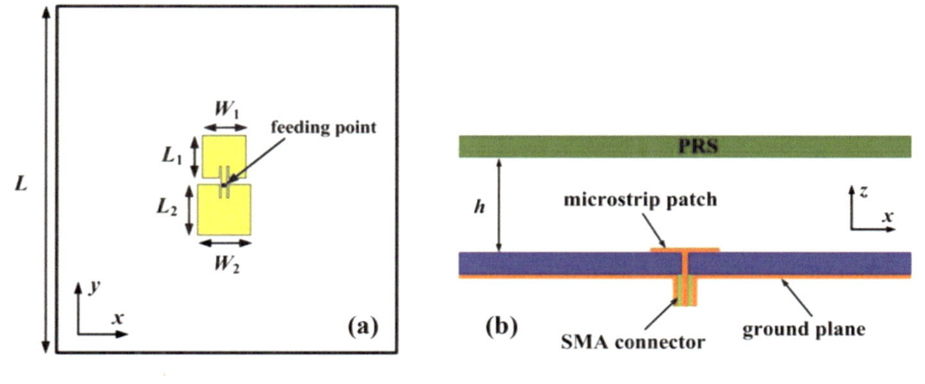

Figure 5.
Reflection coefficients of the unit cell. Geometry of the antenna: (a) top view of the feed antenna and (b) side view of the FPC antenna.

Next, we will introduce how to realize the dual-band working of the antenna through tuning the reflection phase of the PRS. **Figure 5b** shows the geometry of the FPC antenna. We consume that the height of the cavity is h_L when the antenna work at low frequency and h_H when the antenna work at high frequency. According Eq. (9), they can be expressed as.

$$h_L = \frac{(\varphi_L + \varphi_2)\lambda_L}{4\pi} + \frac{N_L \lambda_L}{2} \tag{10}$$

$$h_H = \frac{(\varphi_H + \varphi_2)\lambda_H}{4\pi} + \frac{N_H \lambda_H}{2} \tag{11}$$

When $h_L = h_H$, it can be considered as the antenna operates at two different frequencies without changing the antenna structure. In Eqs. (10) and (11), the N is set to 1 to obtain low profile of the antenna. The antenna is designed to work at 4.6 and 5.5 GHz, so λ_L = 65.2 mm and λ_H = 54.5 mm. The reflection phase of the metal ground plane φ_2 is always π; thus the h is only related to the reflection phase of the PRS. The φ is related to the width of the patch in the unit cell. So we can deduce the relationship between h and w, as shown in **Figure 6**. The black line is the relationship

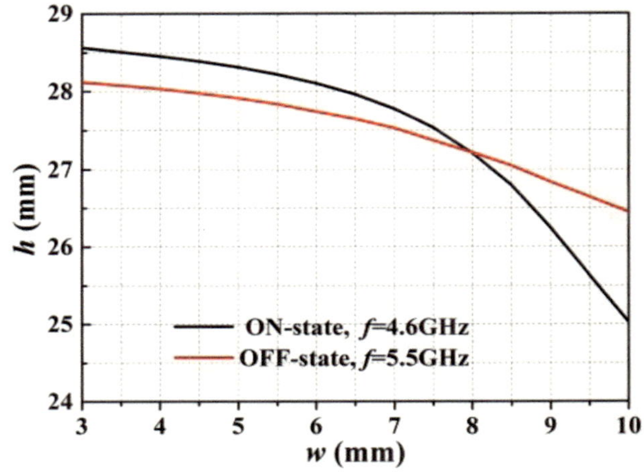

Figure 6.
The relationship between w *and* h.

Figure 7.
The photographs of the fabricated antenna.

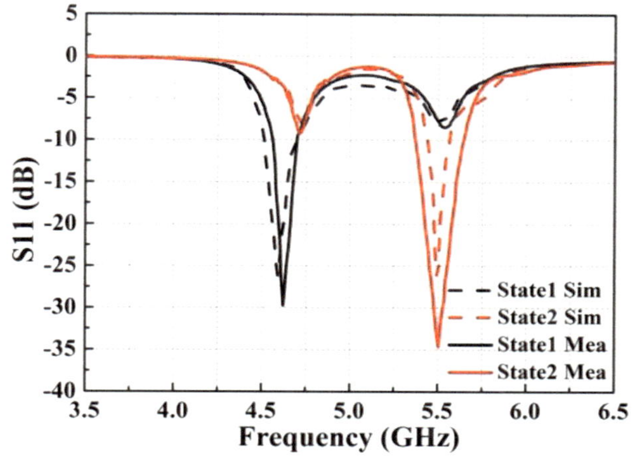

Figure 8.
Simulated and measured S11 of the antenna.

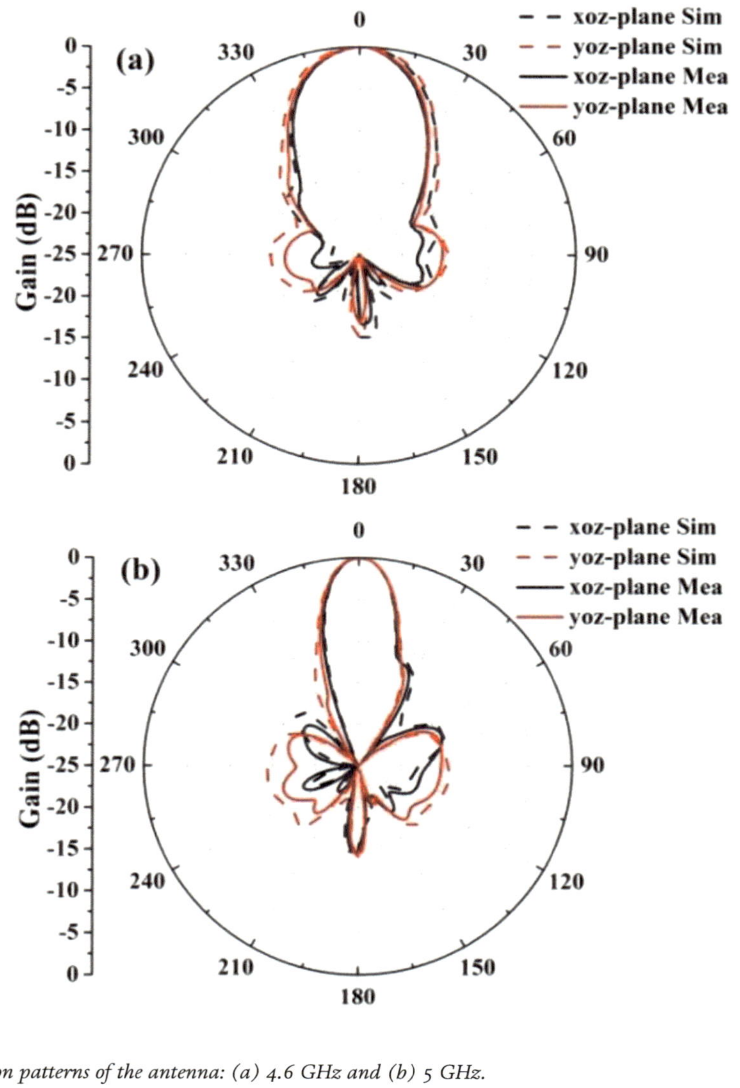

Figure 9.
The radiation patterns of the antenna: (a) 4.6 GHz and (b) 5 GHz.

when f = 4.6 GHz and the diodes are ON, and the red line is the relationship when f = 5.5 GHz and the diodes are OFF. The two lines intersect at the point w = 8 mm and h = 27.2 mm. This indicates that when h = 27.2 mm and w = 8 mm, the antenna can work at 4.6 GHz if the diodes are ON and at 5.5 GHz if the diodes are OFF. So the antenna realizes frequency reconfiguration through tuning the states of the diodes.

3.3 Fabrication and measurement of the antenna

The photographs of the fabricated antenna are shown in **Figure 7**. The red lines connect the anode of the DC source, and the black lines connect the cathode. The S_{11} and radiation patterns of the antenna at two frequencies are measured. Good agreement between the simulated and measured results is obtained. The simulated and measured S11 of the antenna is shown in **Figure 8**. The −10 dB impedance band of the antenna is 4.55–4.7 GHz (3.3%) when diodes are ON and 5.37–5.63 GHz (4.7%) when diodes are OFF. The measured center of the low-frequency band is higher than the simulated result. This may be caused by the fabrication errors. During the welding of SMA joints, some unnecessary metals may be introduced. This results in the deviation of the resonant frequency of the antenna.

Figure 9 presents the simulated and measured radiation patterns of the antenna. **Figure 9a** shows the radiation patterns at 4.6 GHz and **Figure 9b** shows at 5.5 GHz. The measured maximum gain of the antenna is 13.1 dB at 4.6 GHz and 17.1 dB at 5.5 GHz. The 3 dB gain bandwidth of the antenna is 11.9 and 8.2% at two frequency bands, respectively. The gain bandwidth is wider than the impedance bandwidth, so the impedance bandwidth is the working bandwidth of the antenna. The simulated and measured results show that the antenna can realize frequency reconfiguration through tuning the states of the diodes on the PRS. Meanwhile, the antenna obtains well radiation characteristic in both frequencies.

4. Pattern reconfigurable Fabry-Pérot cavity antenna

4.1 Design of the configurable PRS

Firstly, we design a reflection phase reconfigurable unit cell, as shown in **Figure 10**. The unit cell is printed on a substrate of FR4.4 with a thickness of 1.6 mm and permittivity of 4.4. It consists of a square patch and a square ring. A slot is inserted into the unit cell to separate the patch and the ring. Three PIN diodes are

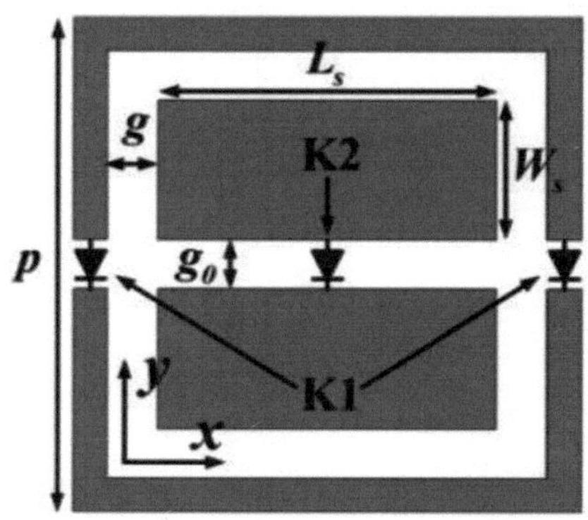

Figure 10.
The structure of the unit cell.

inserted in the slot. The diodes are divided into two groups. The diodes on the ring are named K1 and the diode in the patch is named K2. The different states of two group diodes give the four cell states, as shown in **Table 1**.

The unit cell is simulated by the CST. The simulation setting of the unit cell is the same as the unit cell presented in Section 2. The reflection coefficients of the unit cell under y-polarized wave are shown in **Figure 11**. The simulation results show that the reflection phase of the unit cell at 5 GHz ranged from −165 to −262°. This suggests that the unit cell has a large range of reflection phase variation. Meanwhile, the reflection magnitudes of the unit cell in four states are always greater than 0.7. This can ensure a high gain of the FPC antenna.

A reconfigurable PRS is formed by the unit cell shown in **Figure 10**. The structure of the PRS is shown in **Figure 12**. It is composed by 6 × 6 unit cells and divided

Diodes	S1	S2	S3	S4
K1	OFF	OFF	ON	ON
K2	OFF	ON	OFF	ON

Table 1.
The four states of the unit cell.

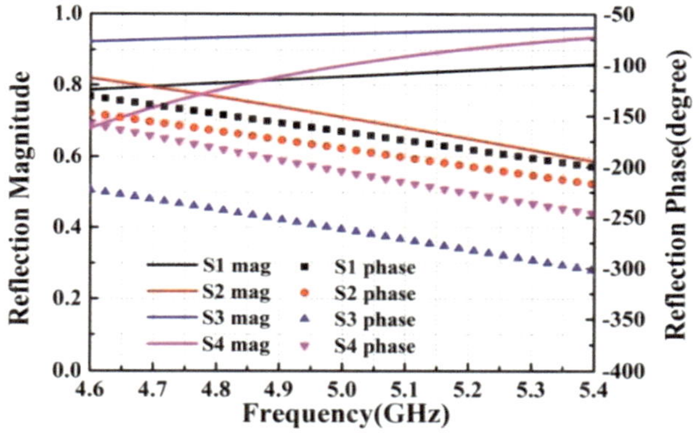

Figure 11.
The reflection phase of the unit cell.

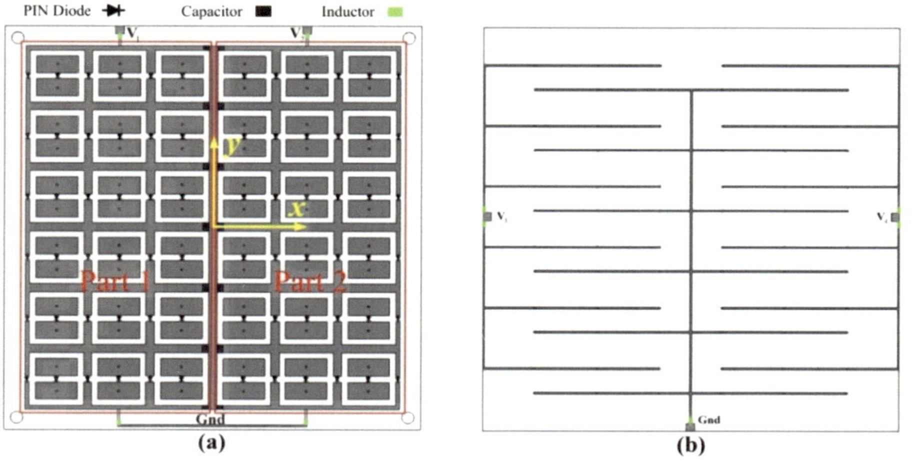

Figure 12.
The structure of the PRS: (a) top view and (b) bottom view.

into two parts. The reflection phase of two parts of the PRS can be controlled by tuning the states of the PIN diodes. Some DC-biased circuit is added on the PRS to bias the PIN diodes. Firstly, since the square rings in the unit cells form a net structure when combing the PRS and the K1 is on the net structure, so two lines of capacitors (220 pF) are added between the two parts to bias the K1 in different parts of the PRS independently. In **Figure 12a**, the points V1 and V2 are used to bias the K1 in different parts. Then, in order to bias the K2, two metalized via hole are added into each unit cell to connect two parts of the patch, and the biasing lines on the bottom side of the PRS which are shown in **Figure 12b**. The points V3 and V4 are used to bias the K2 in different parts of the PRS. The Gnd on the top and bottom side of the PRS are connected by a metalized via hole. Some inductors (20 nH) are inserted into the lines to prevent the RF signal from going into the DC sources. With the designed biasing lines, the K1 and K2 in different parts of the PRS can be controlled independently. The biasing lines are very thin (0.2 mm), and the diameter of the metalized via holes is tiny enough (0.3 mm). So the simulated results show that the DC biasing circuits have barely influence on the performance of the metasurface.

4.2 Antenna design

From the theoretical analysis in Section 1, when the PRS has a uniform reflection phase, the antenna radiates toward the broadside direction [16]. And the main beam of the antenna is perpendicular to the antenna surface. However, when the PRS have reflection phase gradient, the main beam of the antenna will be tilted. So when the two parts of the PRS have different reflection phases, the beam of the antenna will tilt to the direction that has lagging phase.

The structure of the FPC antenna is shown in **Figure 12**, and the values of the parameters of the antenna are shown in **Table 2**. The FPC antenna is feed by a slot coupled patch antenna, as shown in **Figure 13a**. The feed antenna is set under the PRS with a distance of h. The side view of the FPC antenna is shown in **Figure 13b**.

p	Ls	Ws	g	g_0	Lp	L	h
15	10	4.25	1.5	1.5	10.4	100	30.5

Table 2.
Values of the parameters (mm).

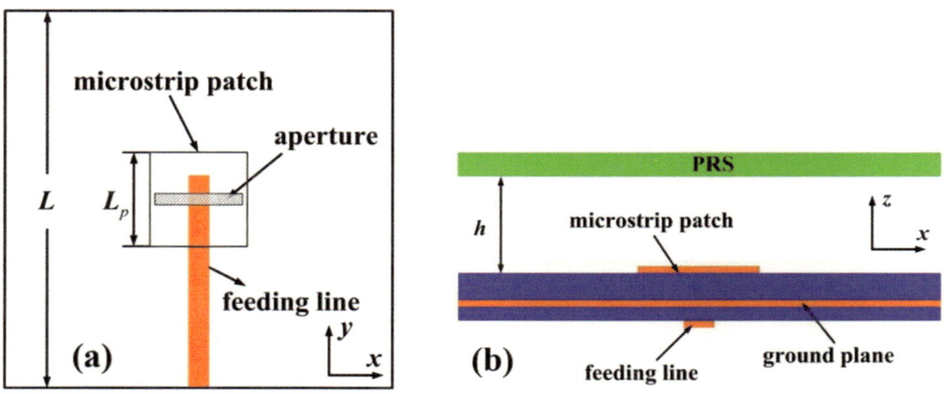

Figure 13.
Structure of the antenna: (a) top view of the feeding antenna and (b) side view of the FPC antenna.

Through controlling the states of the diodes, the PRS will present different reflection phase distributions, so the beam of the antenna will tilt toward different directions. Due to the large reflection phase variation range, the antenna obtains a large steering angle. The antenna operates at 5 GHz and the reflection phase gradient of the PRS is along x- direction, so the beam of the antenna will tilt in xoz plane. We calculate the height of the cavity when the PRS is in the state of S1. After being optimized by the CST, the h is set to 30.5 mm finally.

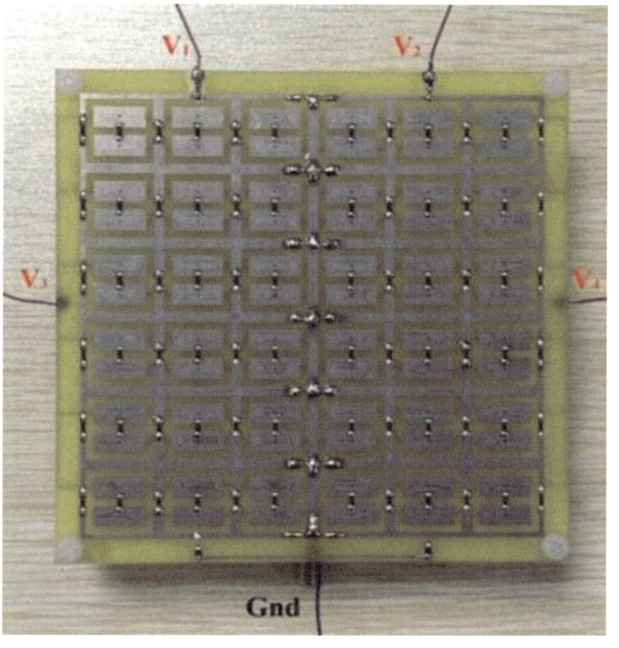

Figure 14.
Photograph of the fabricated antenna.

States	Part 1		Part 2		Beam direction (°)	Gain (dB)
	K1	K2	K1	K2		
1	OFF	OFF	OFF	OFF	0	11.1
2	OFF	ON	OFF	OFF	−6	9.8
3	OFF	OFF	OFF	ON	6	9.8
4	ON	ON	OFF	OFF	−18	8.7
5	OFF	OFF	ON	ON	18	8.7
6	ON	OFF	ON	ON	−24	7.6
7	ON	ON	ON	OFF	24	7.6
8	ON	OFF	OFF	ON	−46	6.6
9	OFF	ON	ON	OFF	46	6.6
10	ON	OFF	OFF	OFF	−54	6.3
11	OFF	OFF	ON	OFF	54	6.3

Table 3.
Detail information of the antenna at different states.

4.3 Fabrication and measurement of the antenna

The photograph of the fabricated antenna is shown in **Figure 14**. After simulation, we find that the antenna can work in at kinds of states. We measured the S_{11} and radiation patterns of the antenna in these 11 states. **Table 3** gives the detail performance of the antenna in different states, including the states of the diodes, beam directions, and maximum gain. Because the antenna structure at state 2 (4, 6, 8, 10) is completely symmetrical to that at state 3 (5, 7, 9, 11), so they have same the S_{11}. For simplicity, we just show the simulated and measured S11 in states of 1, 2, 4, 6, 8, and 10 in **Figure 15**. The measured results agree well with the simulated ones. The measured impedance bandwidth of the antenna is 4.92–5.08 GHz (3.2%). The S_{11} of the antenna only has little difference between different states.

The radiation patterns of the antenna are measured in anechoic chamber. The simulated and measured radiation patterns of the antenna are presented in **Figure 16**. Good agreement is obtained between the measured results and the simulated ones. We know that the main beam of the antenna tilts in *xoz* plane,

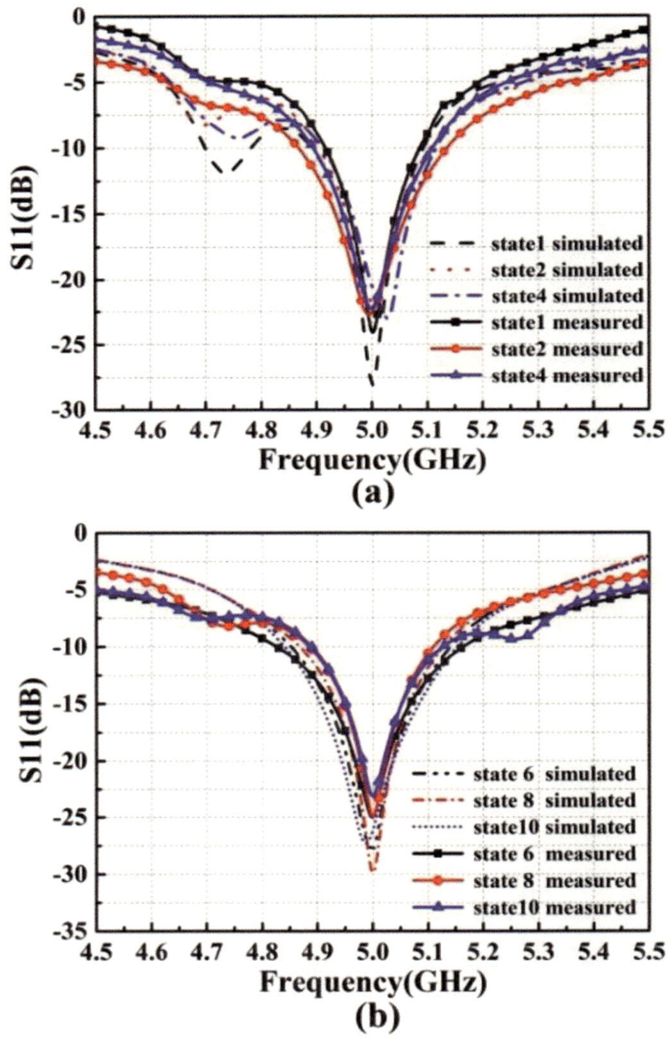

Figure 15.
Simulated and measured S11: (a) states 1, 2, and 4 and (b) states 6, 8, and 10.

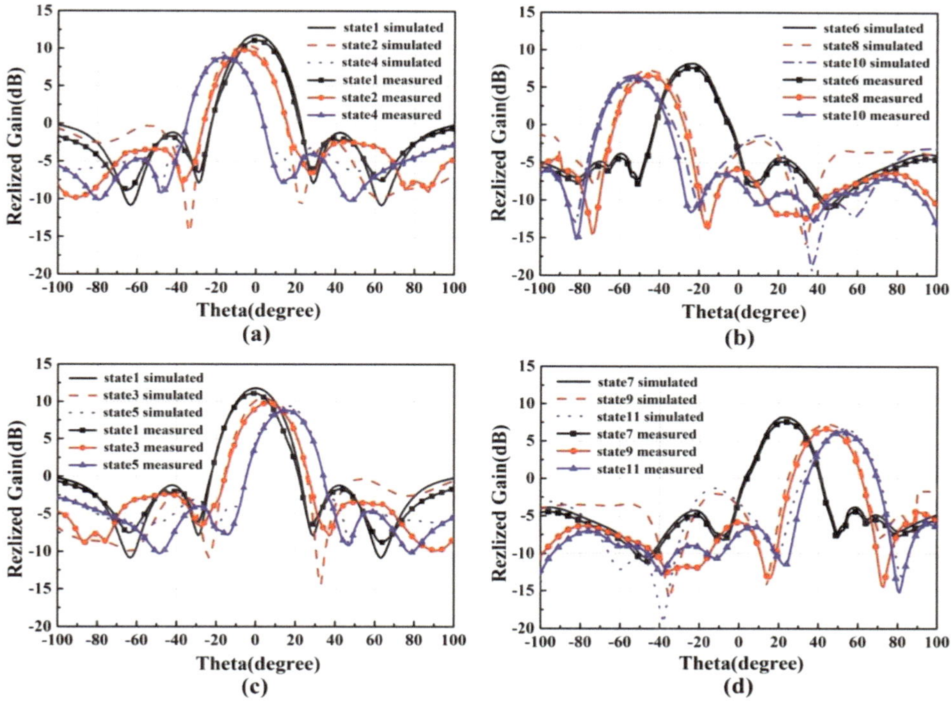

Figure 16.
Radiation patterns of the antenna: (a) states 1, 2, and 4; (b) states 6, 8, and 10; (c) states 1, 3, and 5; and (d) states 7, 9, and 11.

so the radiation patterns of the antennas are all in *xoz* plane. The main beam of the antenna in state 1 is toward the +*z* direction. However, it tilts to –*y* direction in states 2, 4, 6, 8, and 10 and to +*y* direction in states 3, 5, 7, 9, and 11. And the same tilted angle is obtained at states 2 (4, 6, 8, 10) and 3 (5, 7, 9, 11).

The simulated maximum gain of the antenna at state 1 is 11.8 dB, and the measured maximum gain is 11.1 dB. With the increase of the beam tilted angle, the maximum gain of the antenna decreases. The antenna obtains maximum beam tilted angle (about 54°) at states 10 and 11, and the maximum gain of the antenna decreases to 6.3 dB. The antenna remains to have high-gain performance while realizing beam tilting. Besides, the gain floating of the antenna at all states in the operating band are less than 2.8%. So the impedance bandwidth of the antenna is the working band of the antenna.

5. Polarization and pattern reconfigurable FPC antenna

5.1 Design of the reconfigurable PRS

In the last section, we design an FPC antenna with large beam steering angle. However, it can only realize beam tilting in one plane. In practice application, we prefer the antenna to be able to steer the beam in two directions. So in this Section, we will design a configurable PRS to realize two-dimensional beam steering of the FPC antenna [17].

The unit cell of the PRS is shown in **Figure 17**. It is also printed on the substrate of FR4. Each unit cell consists of four square patches. Four PIN diodes are inserted between two adjacent patches. The four diodes are controlled simultaneously, and the

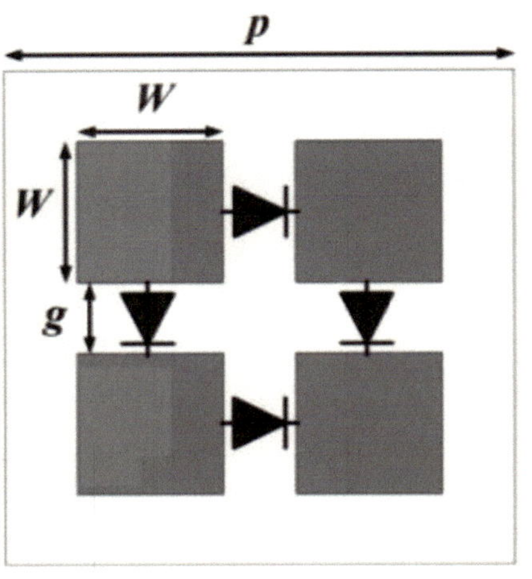

Figure 17.
The unit cell of the PRS.

unit cell can present two states. The unit cell has different reflection coefficients in different states. The reflection coefficients of the unit cell are simulated by the CST, and the results are shown in **Figure 18**. We can see that the unit cell has the same reflection coefficients for both polarization waves. The reflection phases of the unit cell in two states at 5.5 GHz are −187° and −217°, respectively. And the reflection magnitude is always larger than 0.75. This suggests that the unit cell remains to have high reflectivity when tuning the reflection phase through controlling the state of the diodes. The high reflectivity ensures the high gain of the antenna.

We compose a PRS with 6 × 6 unit cells by the unit cell shown in **Figure 17**. The top view of the PRS is shown in **Figure 19**. The PRS is divided into four parts. The diodes in different parts can be controlled independently. Through tuning the state of the diodes in different parts, the PRS presents different reflection phase distributions. A DC biasing circuit is designed on the PRS to realize independent control of the diodes in different parts. Two patches of each unit cell are connected to the biasing lines on the bottom side of the PRS through metalized via holes. By reasonably arranging the orientation of the diodes in each unit cell, the diodes can be controlled simultaneously. The biasing lines on the bottom of the PRS are shown in **Figure 20b**. The biasing point V1, V2, V3, and V4 are connected to the anode of the

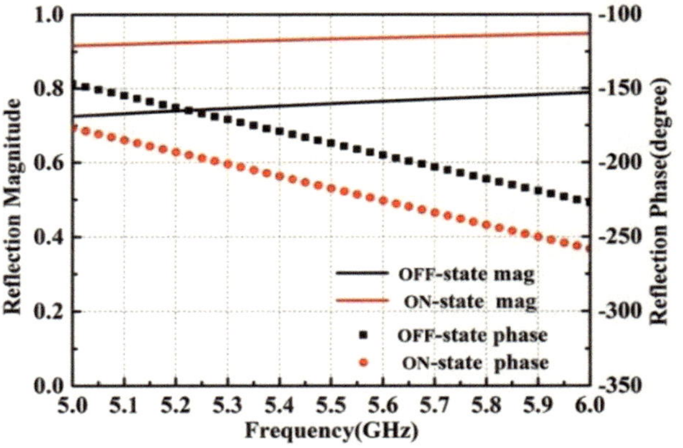

Figure 18.
The reflection coefficients of the unit cell.

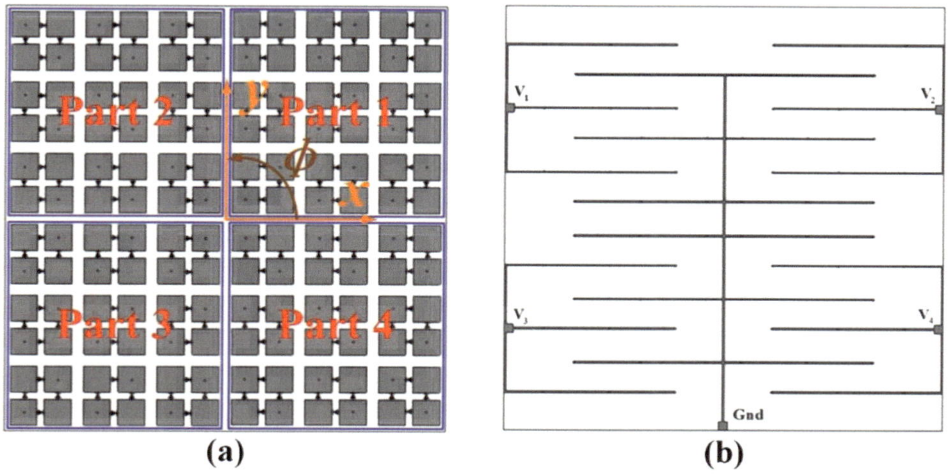

Figure 19.
The structure of the PRS: (a) top view and (b) bottom view.

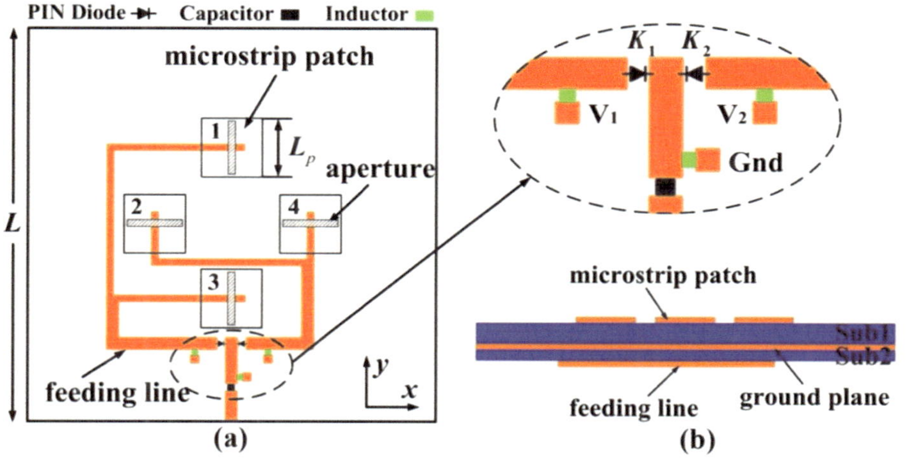

Figure 20.
Polarization configurable feeder antenna: (a) top view and (b) side view.

DC source, and the Gnd is connected to the cathode. The width of the biasing line is 0.2 mm, and the diameter of the metalized via holes is 0.3 mm to reduce the influence of biasing lines on the RF signal.

5.2 Design of the polarization configurable feeder antenna

In order to realize the polarization configuration of the antenna, a polarization configurable feeder antenna is needed. So we designed a polarization configurable slot coupled patch antenna, as shown in **Figure 20**. The feeder antenna is composed by two substrates with different thickness (1.6 mm for Sub1 and 0.8 mm for Sub2). Four identical patches are chosen as the radiators of the antenna. They are printed on the top side of the Sub1 and are symmetric with respect to the center of the antenna. Between the two substrates is a metal ground plane, and four slots are etched on the plane under four patches. The slots 1 and 3 are along the y direction, and the slots 2 and 4 are along the x direction. The feeding line is printed on the bottom side of the Sub2. The four stubs are used to feed four patches through slots.

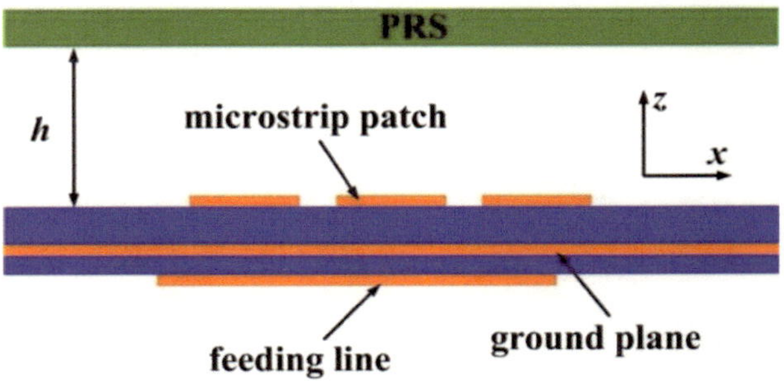

Figure 21.
Geometry of the FPC antenna.

States	Part 1	Part 2	Part 3	Part 4	Beam direction
1	OFF	OFF	OFF	OFF	Broadside
2	ON	OFF	OFF	ON	$\varphi = 0°$
3	ON	OFF	OFF	OFF	$\varphi = 45°$
4	ON	ON	OFF	OFF	$\varphi = 90°$
5	OFF	ON	OFF	OFF	$\varphi = 135°$
6	OFF	ON	ON	OFF	$\varphi = 180°$
7	OFF	OFF	ON	OFF	$\varphi = 225°$
8	OFF	OFF	ON	ON	$\varphi = 270°$
9	OFF	OFF	OFF	ON	$\varphi = 315°$

Table 4.
Detail information of the antenna at different states.

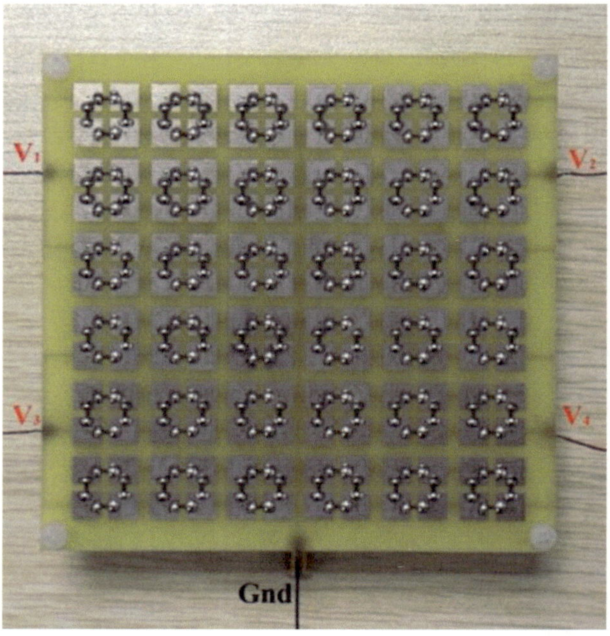

Figure 22.
The photograph of the fabricated antenna.

The lengths of the four stubs are properly designed to make the patches receive the excited with the same phase. Two PIN diodes K1 and K2 are inserted into the feeding line. The biasing points V1, V2, and Gnd are added near the microstrip line to bias the PIN diodes. These points are connected to the microstrip line through inductors. The inductors can allow DC to pass through and block the RF signals. When V1 connects the anode of the DC source, the K1 is turned ON, and when V2 connects the anode of the DC source, the K2 is turned ON. When K1 is ON and K2 is OFF, patches 1 and 3 are excited and the antenna radiates x-polarized wave. Whereas when K1 is OFF and K2 is ON, patches 2 and 4 are excited and the antenna radiates y-polarized wave. Besides, in order to prevent the DC from going into the RF sources, a slot is added in the microstrip line, and a capacitor is inserted into the slot. The capacitor can allow RF signal to pass through and block the DC.

5.3 Design of the antenna

Using the proposed configurable PRS and the polarization configurable patch antenna, a polarization and pattern reconfigurable FPC antenna are formed, as shown in **Figure 21**. The PRS is set above the feeder antenna with a distance of h. The antenna is designed to work at 5 GHz. The height of the cavity h is calculated

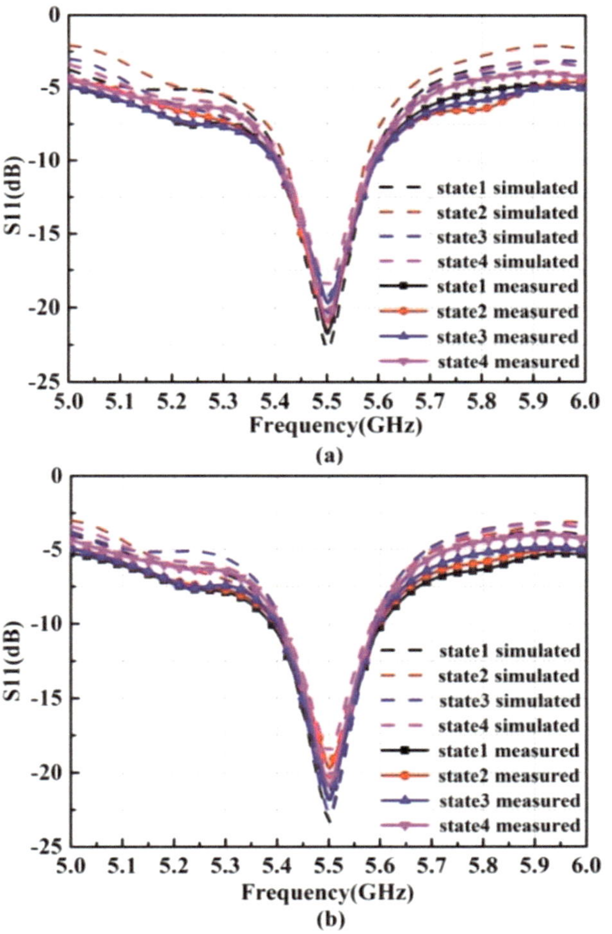

Figure 23.
Simulated and measured S11 of the antenna: (a) x-polarized and (b) y-polarized.

according to Eq. (9) with the reflection phase of the unit cell when diodes are OFF. After being optimized, the h is set to 28 mm.

Through controlling the diodes in four parts of the PRS, the PRS can present nine different states, and the antenna obtains nine beam directions in both polarizations. **Table 4** gives the detailed information of the antenna at different states. Due to the PRS having the same function on different polarized waves, the antenna has the same beam direction at the same diode states for different polarizations.

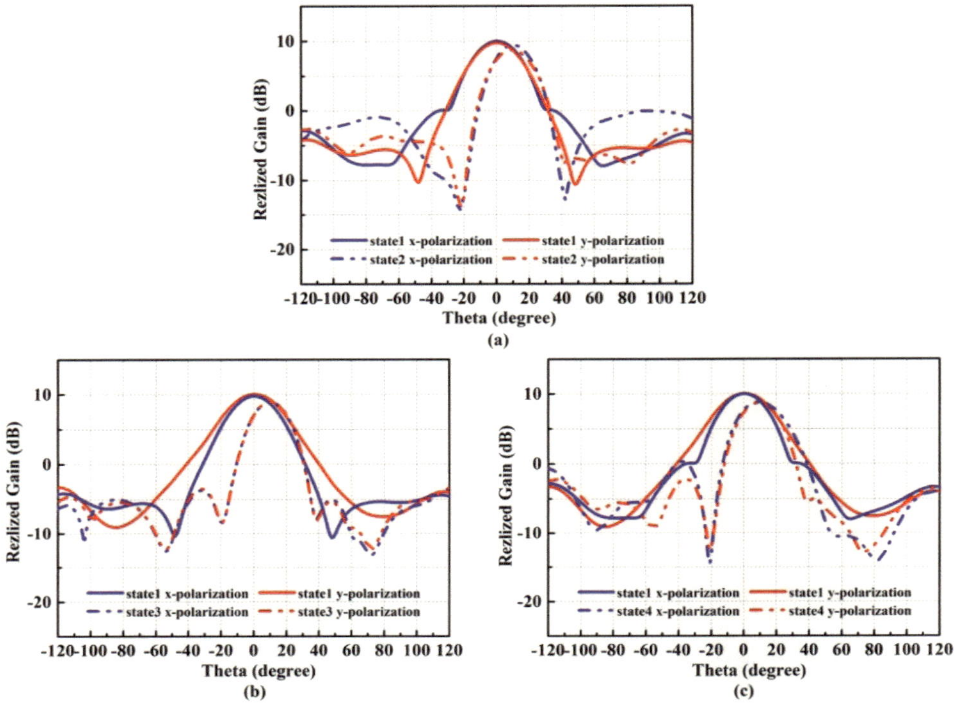

Figure 24.
The radiation patterns of the antenna: (a) φ = 0°, (b) φ = 45°, and (c) φ = 90°.

States	Beam tilted angle (°)	Gain (dB)	Direction (dBi)	Aperture efficiency (%)	Radiation efficiency (%)
1	0	9.7	12.5	27.3	52.5
2	10	8.7	12	21.7	46.8
3	8	8.9	11.9	22.7	50.1
4	10	8.7	12	21.7	46.8
5	8	8.9	11.9	22.7	50.1
6	10	8.7	12	21.7	46.8
7	8	8.9	11.9	22.7	50.1
8	10	8.7	12	21.7	46.8
9	8	8.9	11.9	22.7	50.1

Table 5.
The antenna performance.

5.4 Fabrication and measurement of the antenna

Figure 22 shows the photograph of the fabricated antenna. We measured the reflection coefficients and the radiation patterns of the antenna. Because of the symmetry of the PRS, the simulated and measured results of the antenna in states 2 and 6 (states 4 and 8 and states 3, 5, 7, and 9) are almost the same. So we just show the results in states 1, 2, 3, and 4.

Figure 23 plots the simulated and measured S_{11} of the antenna at states 1, 2, 3, and 4. **Figure 23a** is the result under x- polarization and **Figure 23b** is under y-polarization. The measured impedance bandwidth of the antenna is 5.4–5.6 GHz (3.6%). Because the feeder of the antenna remains constant in different states, the S_{11} of the antenna in different states varies very little. The measured results agree well with the simulated ones.

The radiation patterns of the antenna are measured in anechoic chamber. **Figure 24** plots the measured results of the antenna. **Figure 24a** is the radiation pattern of the antenna at states 1 and 3 in the plane of $\varphi = 0°$, **Figure 24b** is at states 1 and 3 in the plane of $\varphi = 45°$, and **Figure 24c** is at states 1 and 4 in the plane of $\varphi = 90°$. The results show that the main beam of the antenna is in the $+z$ direction at state 1 and tilts in the plane of $\varphi = 0°$, $\varphi = 45°$, and $\varphi = 90°$ at states 2, 3, and 4, respectively. The tilted angle at states 2 and 4 is 10° and that at state 3 is 8°. The maximum gains of the antenna at different states are also measured. The antenna achieves maximum gain of 9.7 dB at state 1, 8.9 dB at state 3, and 8.7 dB at states 2 and 4. Meanwhile, the antenna obtains stable gain at all states. The gain floating of the antenna at all states are less than 2.3% within the impedance bandwidth. **Table 5** summarizes the antenna performance. The beam tilted angle and gain are measured results and the directivity is simulated results. The aperture efficiency is calculated by the gain and size of the antenna, as shown in Eq. (12). And the radiation efficiency is calculated by the gain and direction, as shown in Eq. (13).

$$\eta_1 = G\frac{{\lambda_0}^2}{4\pi A} \tag{12}$$

$$\eta_2 = G/D \tag{13}$$

6. Conclusion

In this chapter, we summarize our recent efforts in realizing reconfigurable FPC antenna basing on the reconfigurable MS. PIN diodes are added on the MS to realize the reflection phase control. Using the flexible phase control capability of the MS, it is easy to tune the frequency and radiation pattern of the FPC antenna, so as to realize reconfigurable FP antenna. The reconfigurable FPC antenna is a good way to improve the performance of antenna. This method can make the FPC antenna more widely used in the field of wireless communication systems.

Abbreviations

FP	Fabry-Pérot
FPC	Fabry-Pérot cavity
MS	metasurface
PRS	partially reflection surfaces

Author details

Peng Xie, Guangming Wang*, Haipeng Li, Yawei Wang and Xiangjun Gao
Air Force Engineering University, Air and Missile Defense College, Xi'an, Shaanxi Province, China

*Address all correspondence to: wgming01@sina.com

References

[1] Trentini GV. Partially reflecting sheet arrays. IRE Transactions on Antennas and Propagation. 1956;**4**(4): 666-671

[2] Zeb BA, Hashmi RM, Esselle KP, Ge Y. The use of reflection and transmission models to design wideband and dual-band Fabry-Perot cavity antennas. In: 2013 International Symposium on Electromagnetic Theory; 2013. pp. 1084-1087

[3] Abdelghani ML, Attia H, Denidni TA. Dual- and wideband Fabry–Perot resonator antenna for WLAN applications. IEEE Antennas and Wireless Propagation Letters. 2017;**16**: 473-476

[4] Kim D, Jeongho J, Choi J. A broadband Fabry-Perot cavity antenna designed using an improved resonance prediction method. Microwave and Optical Technology Letters. 2011;**53**(5): 1065-1069

[5] Wang N, Liu Q, Wu C, Talbi L, Zeng Q, Xu J. Wideband Fabry-Perot resonator antenna with two complementary FSS layers. IEEE Transactions on Antennas and Propagation. 2014;**62**(5):2463-2471

[6] Konstantinidis K, Feresidis AP, Hall PS. Multilayer partially reflective surfaces for broadband Fabry-Perot cavity antennas. IEEE Transactions on Antennas and Propagation. 2014;**62**(7): 3474-3481

[7] Kim D. Noval dual-band Fabry-Pérot cavity antenna with low frequency separation ratio. Microwave and Optical Technology Letters. 2009;**51**(8): 1869-1872

[8] Meng F, Sharma SK. A dual-band high-gain resonant cavity antenna with a single layer superstrate. IEEE Transactions on Antennas and Propagation. 2015;**63**(5):2320-2325

[9] Debogović T et al. Dual-polarized partially reflective surface antenna with MEMS-based beamwidth reconfiguration. IEEE Transactions on Antennas and Propagation. 2014;**62**(1): 228-236

[10] Sultan F, Mitu SSI. Superstrate-based beam scanning of a Fabry–Perot cavity antenna. IEEE Antennas and Wireless Propagation Letters. 2016;**15**: 1187-1190

[11] Guzmán-Quirós R, Weily AR, Gómez-Tornero JL, Guo YJ. A Fabry–Pérot antenna with two-dimensional electronic beam scanning. IEEE Transactions on Antennas and Propagation. 2016;**64**(4):1536-1541

[12] Ji LY, Guo YJ, Qin PY, Gong SX, Mittra R. A reconfigurable partially reflective surface (PRS) antenna for beam steering. IEEE Transactions on Antennas and Propagation. 2015;**63**: 2387-2395

[13] Guzmán-Quirós R, Gómez-Tornero JL, Weily AR, Guo YJ. Electronically steerable 1-D Fabry-Perot leaky-wave antenna employing a tunable high impedance surface. IEEE Transactions on Antennas and Propagation. 2012;**60**: 5046-5055

[14] Ourir A et al. Directive metamaterial-based subwavelength resonant cavity antennas—Applications for beam steering. Comptes Rendus Physique. 2009;**10**(5):415-422

[15] Xie P, Wang G. Design of a frequency reconfigurable Fabry-Pérot cavity antenna with single layer partially reflecting surface. Progress in Electromagnetics Research. 2017;**70**: 115-121

[16] Xie P, Wang G, Cai T, Li H, Liang J. A dual-polarized two-dimensional beam-steering Febry-Pérot cavity antenna with a reconfigurable partially reflecting surface. IEEE Antennas and Wireless Propagation Letters. 2017;**16**: 2370-2374

[17] Xie P, Wang G, Cai T, Li H, Liang J. Novel Fabry-Pérot cavity antenna with enhanced beam steering property using reconfigurable meta-surface. Applied Physics A: Materials Science and Processing. 2017;**123**(7):462

Chapter 12

Time Modulated Linear Array (TMLA) Design

Oussama Gassab, Arab Azrar and Sara Bouguerra

Abstract

In this chapter, time modulated linear array (TMLA) is presented and discussed in detail where all its theoretical backgrounds are derived. The difference between single and multiple time modulation frequencies of TMLA is shown, where different examples in designing them are presented. In addition, the power and directivity of TMLAs are derived in their closed form. Moreover, the relation between the steering angle of each sideband with respect to the first sideband angle is developed analytically. Also, an efficient mathematical method is presented to design TMLA with desired sidelobe (SLL) and sideband levels (SBLs) with maximum attainable directivity. It is shown that the TMLA can be designed by only controlling its time sequence distributions which is a very good advantage as compared to the conventional antenna array.

Keywords: antenna array, time modulated linear array (TMLA), time modulation, power radiation, directivity, sidelobe level (SLL), sideband level (SBL), electronic beam steering, single time modulation frequency, multiple time modulation frequency

1. Introduction

The antenna array performance can be improved by decreasing its sidelobe level (SLL) and increasing its directivity. To do that, many different methods and techniques were proposed such as genetic algorithm (GA), particle swarm optimization (PSO), and hybridization between different arrays [1–3]. However, these methods provide very satisfactory results in the designed array; the realization of the designed excitations by using conventional approaches, such as tapered amplitude distributions and amplitude attenuators, is very challenging due to the fact that any small inaccuracy in the design will cause unwanted deviations in the SLL [4]. In order to overcome this problem, the time modulated linear array (TMLA), also called 4-D antenna array, was proposed. The main concept of this idea was used in [5] and applied to antenna array in order to achieve ultralow sidelobe level by Kummer et al in [6]. The idea of TMLA is to use the time as an additional degree of freedom in the design by using radio-frequency switches that periodically modulate the elements. The concept of TMLA is to use switching modulation (on, off) in order to reduce the effects of errors because the on-off switching can be controlled at a very high accuracy level.

IntechOpen

2. Time modulated linear array

Suppose an N-isotropic element 4-D linear array aligned along the z-axis and centered on its origin as shown in **Figure 1**.

The array factor of time modulated array is given by [7]

$$\mathrm{AF}(\theta, t) = e^{j2\pi f_0 t} \sum_n I_n g_n(t) e^{jkz_n \cos(\theta)} \tag{1}$$

where f_0 is the center frequency; θ is the elevation angle of the usual spherical coordinate; I_n are the time-independent static excitation amplitude; $k = 2\pi/\lambda$ is the wavenumber, in which λ is the wavelength; and z_n is the position of the nth element of the array along the z-axis.

2.1 TMLA with single time modulation frequency (STMF)

$g_n(t)$ are the periodic switch-on time sequence functions, and they are written for the case of STMF as [7] (see **Figure 2**)

$$g_n(t) = \begin{cases} 1, & t_{\mathrm{on},n} < t \le t_{\mathrm{off},n} \\ 0, & 0 < t \le t_{\mathrm{on},n} \text{or } t_{\mathrm{off},n} < t \le T_p \end{cases} \tag{2}$$

It should be indicated that in the STMF, the switching period T_p is the same for all the antenna elements.

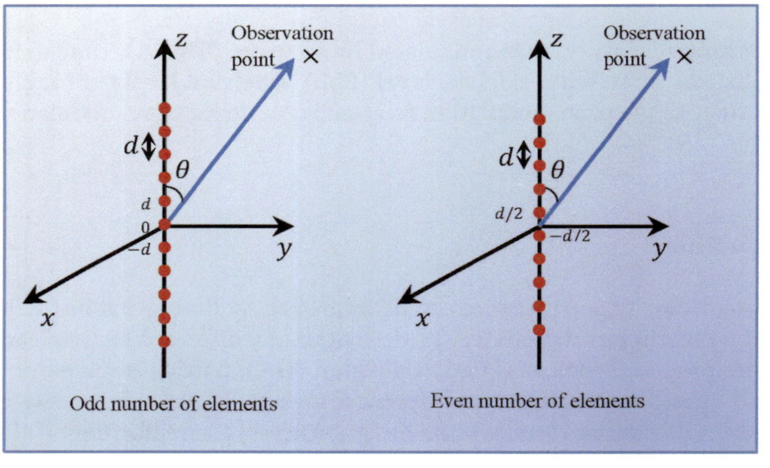

Figure 1.
Time modulated array elements positions aligned along the z-axis and centered on its origin.

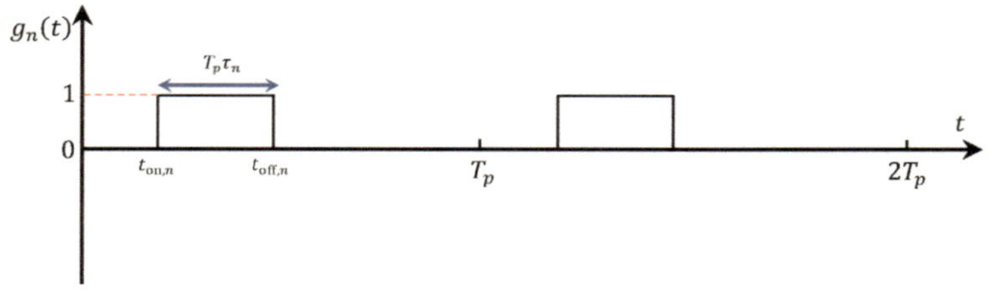

Figure 2.
The periodic time sequence graph.

The topology of TMLA with STMF is shown in **Figure 3**, where single-throw switches are connected to each antenna so that to control the switching between the two states: on and off.

Since $g_n(t)$ are periodic functions, they can be expanded by the Fourier series as

$$g_n(t) = \sum_{m=-\infty}^{+\infty} G_{nm} e^{j2\pi m f_p t} \tag{3}$$

where $f_p = 1/T_\mathrm{p}$ is the modulation frequency, $f_0 \gg f_p$, and G_{nm} is the mth Fourier coefficient of the nth radio-frequency switch, and they are represented as

$$G_{nm} = \frac{1}{T_p} \int_0^{T_p} g_n(t) e^{-j2\pi m f_p t} dt \tag{4}$$

$$G_{nm} = \tau_n \mathrm{sinc}(\tau_n m) e^{-jm\pi\left(2\xi_{\mathrm{on},n} + \tau_n\right)} \tag{5}$$

where $\tau_n = (t_{\mathrm{off},n} - t_{\mathrm{on},n})/T_p$ and $\xi_{\mathrm{on},n} = t_{\mathrm{on},n}/T_p$ are the normalized switch-on duration and the normalized switch-on instant, respectively, and $\mathrm{sinc}(x) = \sin(\pi x)/\pi x$, wherein $\mathrm{sinc}(0) = 1$. By using (3), equation (1) can be written as

$$\mathrm{AF}(\theta, t) = \sum_{m=-\infty}^{+\infty} \sum_n I_n G_{nm} e^{j2\pi\left(f_0 + m f_p\right)t} e^{jkz_n \cos(\theta)} \tag{6}$$

$$\mathrm{AF}(\theta, t) = \sum_{m=-\infty}^{+\infty} AF_m(\theta, t) \tag{7}$$

where

$$\mathrm{AF}_m(\theta, t) = \sum_n I_n G_{nm} e^{j2\pi\left(f_0 + m f_p\right)t} e^{jkz_n \cos(\theta)} \tag{8}$$

Note that $\mathrm{AF}_0(\theta, t)$ is the array factor at the desired frequency f_0 and $\mathrm{AF}_m(\theta, t)$ is the sideband array factor for the case of STMF.

The array factor at the desired frequency f_0 and for the case $I_n = 1$ is given as

$$\mathrm{AF}_0(\theta, t) = e^{j2\pi f_0 t} \sum_n \tau_n e^{jkz_n \cos(\theta)}, 0 \le \tau_n \le 1 \tag{9}$$

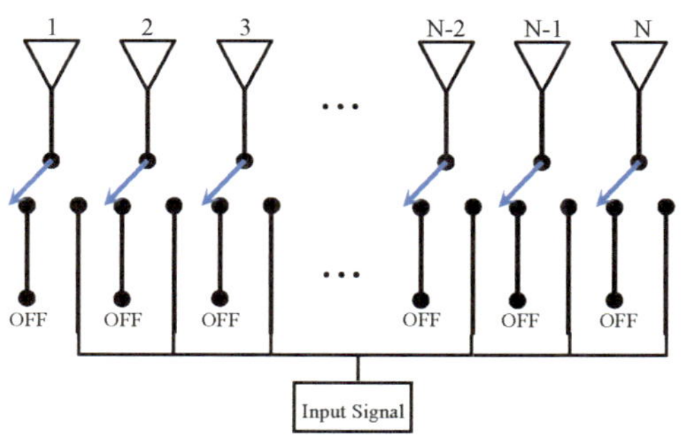

Figure 3.
Configuration of N-element TMLA with STMF using single-throw switches.

It can be concluded that by controlling the normalized switch-on durations τ_n, any array pattern AF_0 can be generated with very high accuracy.

2.1.1 Power radiations in time domain

In this section, we outline how to obtain the generalized power expression of the TMLA. By aligning the array along the z-axis and considering spherical coordinate with θ_p and ϕ_p, the elevation and azimuth angles, respectively, Eq. (1) can be represented in terms of θ_p as

$$AF(\theta_p, t) = e^{j2\pi f_0 t} \sum_{n=1}^{N} I_n g_n(t) e^{jkz_n \cos(\theta_p)} = e^{j2\pi f_0 t} \mathcal{AF}(\theta_p, t) \tag{10}$$

where z_n denotes the positions of the TMLA elements along the z-axis.

Let's consider $\vec{E}_0$ and $\vec{H}_0$ be the intensities of the electric and magnetic field (complex spatial form) radiated by a given antenna used in the TMLA, hence, the electric field and magnetic field generated by the array are given as

$$\vec{E}(t) = e^{j2\pi f_0 t} \mathcal{AF}(\theta_p, t) \vec{E}_0 \tag{11}$$

$$\vec{H}(t) = e^{j2\pi f_0 t} \mathcal{AF}(\theta_p, t) \vec{H}_0 \tag{12}$$

The instantaneous Poynting vector is given as [8]

$$\vec{\mathcal{P}} = \vec{\mathcal{E}} \times \vec{\mathcal{H}} \tag{13}$$

where $\vec{\mathcal{E}}$ and $\vec{\mathcal{H}}$ represent the instantaneous field vectors, and they are given as

$$\vec{\mathcal{E}}(t) = \text{Re}\left\{ e^{j2\pi f_0 t} \mathcal{AF}(\theta_p, t) \vec{E}_0 \right\} \tag{14}$$

$$\vec{\mathcal{H}}(t) = \text{Re}\left\{ e^{j2\pi f_0 t} \mathcal{AF}(\theta_p, t) \vec{H}_0 \right\} \tag{15}$$

By using $\text{Re}\{X\} = (X + X^*)/2$ and inserting (14) and (15) in (13) and proceeding with the same analysis performed in [8], the following result is obtained:

$$\mathcal{P} = \frac{1}{2} \text{Re}\left\{ |\mathcal{AF}|^2 \vec{E}_0 \times \vec{H}_0^{\,*} \right\} + \frac{1}{2} \text{Re}\left\{ \mathcal{AF}^2 \vec{E}_0 \times \vec{H}_0 e^{j4\pi f_0 t} \right\} \tag{16}$$

Note that $\vec{E}_0$ and $\vec{H}_0$ are not functions of time and $\mathcal{AF}$ is a periodic function with period $T_p \gg T_0$, and then the second term has zero average power; hence, the average power density is equal to

$$P_{\text{density}} = \frac{W_{\text{avg}}}{T_p} \int_0^{T_p} |\mathcal{AF}|^2 dt \tag{17}$$

where $W_{\text{avg}} = \frac{1}{2} \text{Re}\left\{ \vec{E}_0 \times \vec{H}_0^{\,*} \right\}$ is the power density of each antenna element in the TMLA. By considering isotropic antenna elements, $W_{\text{avg}}\, r^2$ is the radiation intensity, and it is constant over all the space, and it can be taken as unity (r is the radial distance from the TMLA to the observation point at the far-field region). By using (1), (17) is written as

$$P_{\text{density}} = \frac{W_{\text{avg}}}{T_p} \sum_{n,k=1}^{N} I_n I_k^* \int_0^{T_p} g_n(t) g_k^*(t) dt e^{jk(z_n - z_k)\cos(\theta_p)} \tag{18}$$

The total power is given

$$P_T = \int_0^{2\pi} \int_0^{\pi} P_{\text{density}} r^2 \sin(\theta_p) d\theta_p d\phi_p = 4\pi \sum_{n,k=1}^{N} I_n I_k^* \frac{1}{T_p} \int_0^{T_p} g_n(t) g_k^*(t) dt \operatorname{sinc} \frac{2}{\lambda}(z_n - z_k) \tag{19}$$

We should indicate that the expression (19) is a very simple formula to determine the total power radiated by the TMLA.

For the case $d = \lambda/2$ and equal spacing distance between the elements, the total power can be written as

$$P_{\text{T}} = 4\pi \sum_{n=1}^{N} |I_n|^2 \frac{1}{T_p} \int_0^{T_p} g_n(t)^2 dt = 4\pi \sum_{n=1}^{N} |I_n|^2 \tau_n \tag{20}$$

2.1.2 Power radiations in frequency domain

In this section, the power radiation is represented in the frequency domain. By taking the Fourier series (3) of $g_n(t)$ and using the Parseval's theorem then

$$\frac{1}{T_p} \int_0^{T_p} g_n(t) g_k^*(t) dt = \sum_{m=-\infty}^{+\infty} G_{nm} G_{km}^* \tag{21}$$

and the total power is given as

$$P_T = 4\pi \sum_{n,k=1}^{N} I_n I_k^* \sum_{m=-\infty}^{+\infty} G_{nm} G_{km}^* \operatorname{sinc} \frac{2}{\lambda}(z_n - z_k) \tag{22}$$

It is worth noticing that the total power P_T can be written as

$$P_{\text{T}} = P_{f_0} + P_{\text{SB}} \tag{23}$$

and

$$P_{f_0} = 4\pi \sum_{n=1}^{N} |I_n|^2 {\tau_n}^2 + 4\pi \sum_{\substack{n,k=1 \\ k \neq n}}^{N} I_n I_k^* \operatorname{sinc} \frac{2}{\lambda}(z_n - z_k) \tau_n \tau_k \tag{24}$$

$$P_{\text{SB}} = 4\pi \sum_{n=1}^{N} |I_n|^2 \sum_{\substack{m=-\infty \\ m \neq 0}}^{+\infty} |G_{nm}|^2 + 4\pi \sum_{\substack{n,k=1 \\ k \neq n}}^{N} I_n I_k^* \operatorname{sinc} \frac{2}{\lambda}(z_n - z_k) \sum_{\substack{m=-\infty \\ m \neq 0}}^{+\infty} G_{nm} G_{km}^* \tag{25}$$

where P_{f_0} is the power radiated at the fundamental frequency f_0 and P_{SB} is the power of sidebands.

The complex Fourier coefficient G_{nm} in (5) can be written as

$$G_{nm} = \frac{j}{2\pi m}\left(e^{-j2\pi m\xi_{\text{off},n}} - e^{-j2\pi m\xi_{\text{on},n}}\right) \tag{26}$$

Then

$$G_{nm}G_{km}^{*} = \frac{1}{4\pi^2 m^2}\left[e^{j2\pi m\left(\xi_{\text{off},k}-\xi_{\text{off},n}\right)} - e^{j2\pi m\left(\xi_{\text{on},k}-\xi_{\text{off},n}\right)} - e^{j2\pi m\left(\xi_{\text{off},k}-\xi_{\text{on},n}\right)} + e^{j2\pi m\left(\xi_{\text{on},k}-\xi_{\text{on},n}\right)}\right] \tag{27}$$

Then, we have

$$\sum_{\substack{m=-\infty \\ m\neq 0}}^{+\infty} G_{nm}G_{km}^{*} = \left[2\sum_{m=1}^{+\infty}\frac{\cos 2\pi m\left(\xi_{\text{off},k}-\xi_{\text{off},n}\right)}{(2\pi m)^2} - 2\sum_{m=1}^{+\infty}\frac{\cos 2\pi m\left(\xi_{\text{on},k}-\xi_{\text{off},n}\right)}{(2\pi m)^2}\right.$$
$$\left. - 2\sum_{m=1}^{+\infty}\frac{\cos 2\pi m\left(\xi_{\text{off},k}-\xi_{\text{on},n}\right)}{(2\pi m)^2} + 2\sum_{m=1}^{+\infty}\frac{\cos 2\pi m\left(\xi_{\text{on},k}-\xi_{\text{on},n}\right)}{(2\pi m)^2}\right] \tag{28}$$

By using the results given in [9], then

$$\sum_{\substack{m=-\infty \\ m\neq 0}}^{+\infty} G_{nm}G_{km}^{*} = \overline{\tau}_{nk} - \tau_n\tau_k \tag{29}$$

where

$$\overline{\tau}_{nk} = \frac{1}{2}\left(-\left|\xi_{\text{off},k}-\xi_{\text{off},n}\right| + \left|\xi_{\text{on},k}-\xi_{\text{off},n}\right| + \left|\xi_{\text{off},k}-\xi_{\text{on},n}\right| - \left|\xi_{\text{on},k}-\xi_{\text{on},n}\right|\right) \tag{30}$$

It should be indicated that $\overline{\tau}_{nk}$ can be interpreted as the overlapped duration between the corresponding switch-on durations τ_n and τ_k.

At the case $n = k$, the Parseval's theorem (21) can be used, then

$$\sum_{\substack{m=-\infty \\ m\neq 0}}^{+\infty} |G_{nm}|^2 = \frac{1}{T_p}\int_0^{T_p} |g_n(t)|^2 dt - |G_{n0}|^2 \tag{31}$$

and

$$\sum_{\substack{m=-\infty \\ m\neq 0}}^{+\infty} |G_{nm}|^2 = \tau_n(1-\tau_n) \tag{32}$$

It should be indicated that (29) and (32) can be used in (25) in order to obtain the closed-form expression for the sideband power.

It is worth noticing that the total power expression (22) can be written as

$$P_T = 4\pi \sum_{n,k=1}^{N} I_n I_k^* \overline{\tau}_{nk} \mathrm{sinc}\frac{2}{\lambda}(z_n - z_k) \tag{33}$$

2.1.3 Directivity

The directivity at the fundamental frequency f_0 of TMLA with STMF is presented as [10]

$$D_{f_0} = \frac{4\pi \left| AF_0(\theta, t)_{\max} \right|^2}{P_\mathrm{T}} \tag{34}$$

By considering excitations with the same amplitude, i.e., $I_n = 1$ as shown in Eq. (9), the directivity D_{f_0} can be written as

$$D_{f_0} = \frac{4\pi \left| \sum_n \tau_n \right|^2}{P_{f_0}} \frac{P_{f_0}}{P_{f_0} + P_\mathrm{SB}} \tag{35}$$

It can be written in the following form [10]:

$$D_{f_0} = D_\mathrm{conv} \eta_{f_0} \tag{36}$$

where $D_\mathrm{conv} = 4\pi \left| \sum_n \tau_n \right|^2 / P_{f_0}$ is the directivity of the conventional antenna array, i.e., without modulation switches, and η_{f_0} is the dynamic factor. It is worth noticing that $\eta_\mathrm{SB} = 1$ denotes no sideband radiations (SBRs).

2.1.4 Simulation and computed results

To understand the benefits of TMLA with STMF, simulation examples should be analyzed in detail. Let's consider 30-element Chebyshev weighting with 30 dB SLL, where $I_n = 1$ is considered. The normalized array pattern at the fundamental frequency and the first four sidebands frequencies are shown in **Figure 4**, whereas the periodic time sequences for each element are shown in **Figure 5**. It is evident that the array pattern at the fundamental frequency has a Chebyshev array pattern. However, there are other array patterns at the multiple of the time modulation frequency f_p due to the modulating switches, which cause power losses at the SBRs. The power distribution over the sidebands is shown in **Figure 6**, where only the positive sidebands are shown because the negative sidebands have the same power distribution as the positive ones.

It is evident that most of the power resides at the fundamental frequency f_0 with 75.94%, where the remaining sidebands have only 24.06% of the total power. It should be indicated that the directivity of the array pattern at the fundamental frequency f_0 is equal to 19.93 dBi which is less than the conventional array directivity 26.24 dBi because of the SBR losses. Many wondering facts may arise now; since the conventional array has no sideband power losses and higher directivity than the TMLA, why is the TMLA really needed? Why do we need modulating switches that generate infinite sideband at the multiple of the modulation frequency? The answer is simple; the conventional array needs tapered amplitude

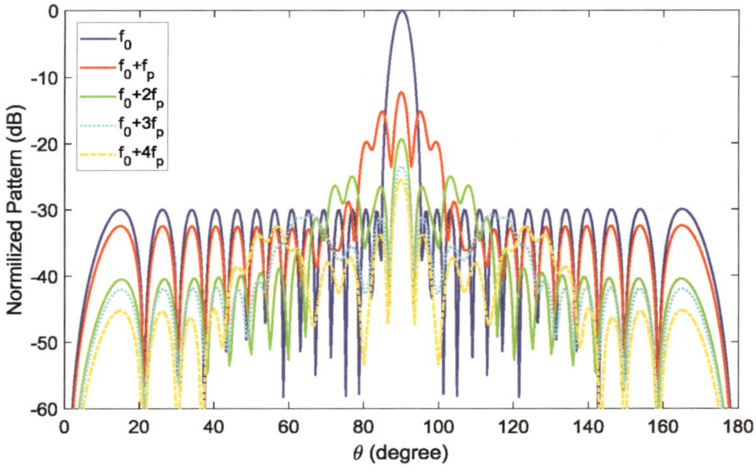

Figure 4.
The array patterns of TMLA with Chebyshev weighting at the fundamental frequency and the four positive first sidebands.

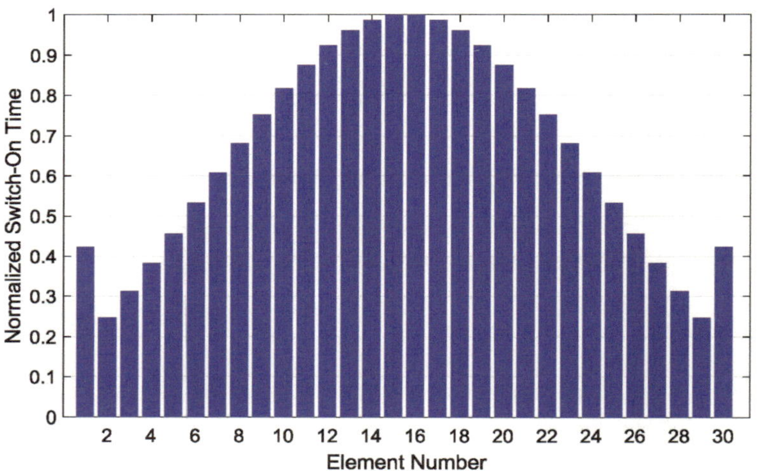

Figure 5.
The periodic time sequences of each element of TMLA, where all the switch-on instants are equal to zero.

distributions and amplitude attenuators [4] in order to realize the excitation amplitudes. However, this method is not so accurate; hence, it causes deviations in the desired array pattern. Therefore, high SLLs are generated. In this case, the realistic directivity is dramatically reduced. Whereas the excitation amplitudes in TMLA can be easily generated with very high accuracy by using modulating switches, therefore, the desired array pattern is totally preserved (its SLLs are kept at their designed level).

It is worth noticing that the sideband levels (SBLs) are high at the main lobe of the fundamental array pattern as shown in **Figure 4**. This kind of problem can be solved by shifting the sideband arrays by controlling the normalized switch-on instants $\xi_{\text{on},n}$.

The array factor at the mth sideband for static excitation amplitude $I_n = 1$ and equal spacing distance between the elements $(z_n = (n-1)d)$ is given by (8)

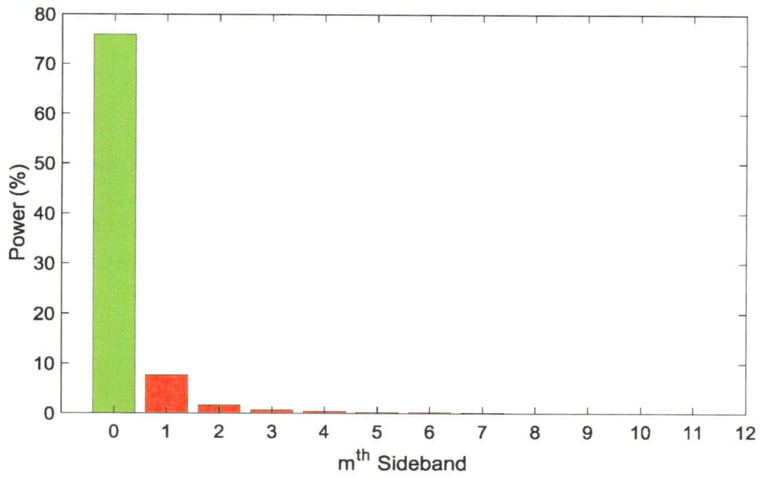

Figure 6.
Power percentage spectrum of TMLA (Chebyshev weighting with N = 30 and SLL= −30 dB).

$$\mathrm{AF}_m(\theta, t) = e^{j2\pi(f_0 + mf_p)t} \sum_{n=1}^{N} \tau_n \mathrm{sinc}(\tau_n m) e^{j\left(-m\pi\left(2\xi_{\mathrm{on},n} + \tau_n\right) + k(n-1)d\cos(\theta)\right)} \tag{37}$$

Without steering the mth sideband, the switch-on instant is zero $\xi_{\mathrm{on},n} = 0$ then $-m\pi\left(2\xi_{\mathrm{on},n} + \tau_n\right) = -m\pi\tau_n$, which represent an original phase shift at the mth sideband. Hence, (37) can be written as

$$\mathrm{AF}_m(\theta, t) = e^{j2\pi(f_0 + mf_p)t} \sum_{n=1}^{N} \tau_n \mathrm{sinc}(\tau_n m) e^{-m\pi\tau_n} e^{j\left(-2m\pi\xi_{\mathrm{on},n} + k(n-1)d\cos(\theta)\right)} \tag{38}$$

To steer the mth sideband toward θ_0, the following condition should be taken

$$e^{j\left(-2m\pi\xi_{\mathrm{on},n} + k(n-1)d\cos(\theta)\right)} = 1 \tag{39}$$

The general solution of Eq. (39) is given as

$$\xi_{\mathrm{on},n} = \frac{k(n-1)d}{2\pi m}\cos(\theta_0) + \frac{\mathcal{K}}{m}, \qquad \mathcal{K} \in \mathbb{Z} \tag{40}$$

We should indicate that $\mathcal{K}$ is an integer number which is chosen in order to maintain $\xi_{\mathrm{on},n}$ in the region $[0, 1]$. It should be indicated that when the first sideband is steered toward θ_0, i.e., $m = 1$ at relation (40), the mth sideband is self-steered toward a specific angle θ_m that can be determined as

By substituting (40) with $m = 1$ in (38), the following relation is obtained

$$\mathrm{AF}_m(\theta, t) = e^{j2\pi(f_0 + mf_p)t} \sum_{n} \tau_n \mathrm{sinc}(\tau_n m) e^{-m\pi\tau_n} e^{j(n-1)(-kmd\cos(\theta_0) + kd\cos(\theta))} \tag{41}$$

To find θ_m, the following equation should be solved

$$e^{j(n-1)(-kmd\cos(\theta_0) + kd\cos(\theta))} = 1 \tag{42}$$

Its general solution is given as

$$\theta_m = \arccos\left(m\cos\left(\theta_0\right) + 2\mathcal{K}\right), \qquad \mathcal{K} \in \mathbb{Z} \tag{43}$$

It should be indicated that if $|m\cos(\theta_0)| > 1$, then $\mathcal{K}$ is chosen from $\mathbb{Z}$ (integer numbers set) so that $|m\cos(\theta_0) + 2\mathcal{K}| \leq 1$. For example, let steer the first sideband toward $\theta_0 = 120^\circ$, then the 2nd, 3rd, and 4th sidebands are steered toward $180^\circ, 60^\circ$, and 90°, respectively. The results are shown in **Figure 7** (a), where the corresponding switch-on time is shown in **Figure** 7 (b). When the first sideband is steered toward $\theta_0 = 180^\circ$, then the 2nd, 3rd, and 4th sidebands are steered toward $90^\circ, 180^\circ$, and 90°, respectively. The normalized array pattern is shown in **Figure 8** (a), and the corresponding switch-on time for each antenna element is presented in **Figure 8** (b). We should indicate that the even sidebands, i.e., m is even, are not

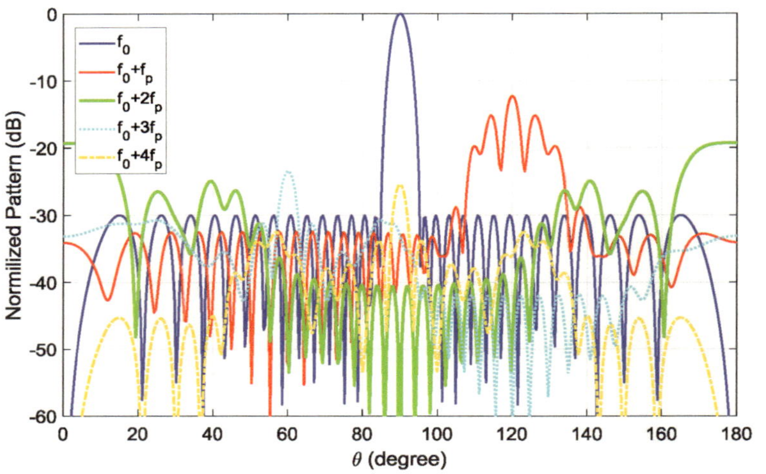

(a)

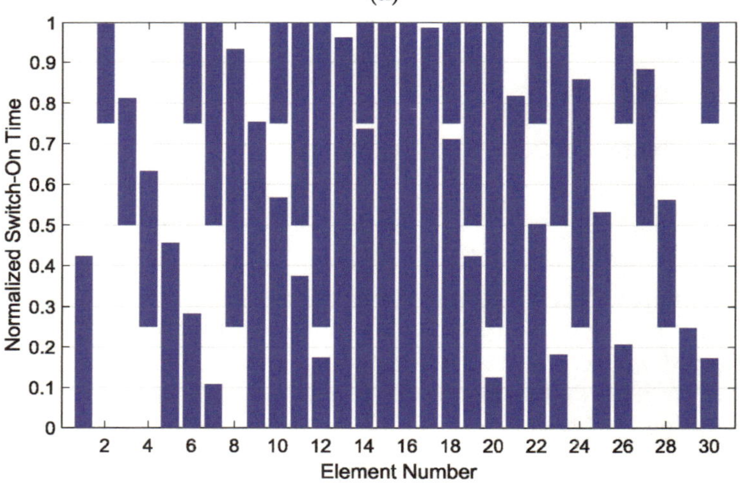

(b)

Figure 7.
(a) The array patterns of TMLA with Chebyshev weighting at the fundamental frequency and steered four positive sidebands where the first positive side is steered toward 120°. (b) The periodic time sequences of each element of TMLA.

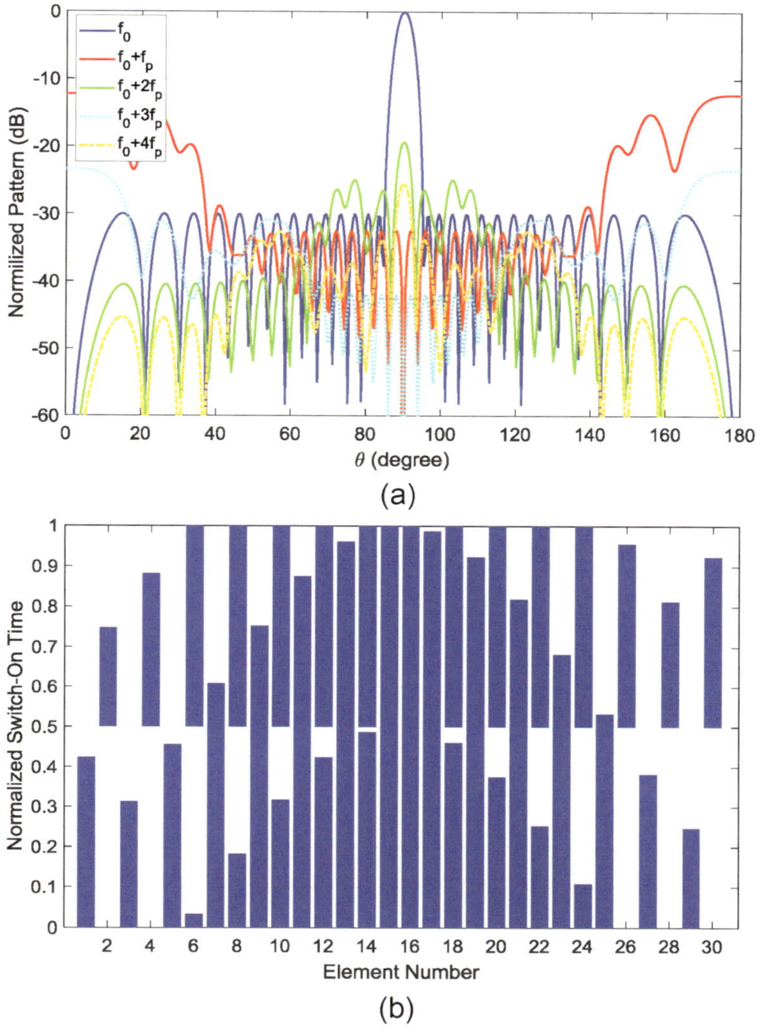

Figure 8.
(a) The array patterns of TMLA with Chebyshev weighting at the fundamental frequency and steered four positive sidebands where the first positive side is steered toward 180°. (b) The periodic time sequences of each element of TMLA.

steered at the case $\theta_0 = 180^\circ$ because there exists an integer number $\mathcal{K} \in \mathbb{Z}$ so that $-m + 2\mathcal{K} = 0$ in Eq. (43). It is worth noticing that the power distribution over the sidebands as given in **Figure 6** remains unchanged for all the steering angles.

From the above results, it can be deduced that the use of periodic switches to modulate the signal generates SBRs at the multiples of the time modulation frequency, which causes power loss and low directivity. To overcome the SBR problem, the optimization techniques, such as differential evolution (DE), GA, PSO, and the simulated annealing (SA), were used to reduce the SBL as well as maintain SLL at a certain low level [11–14]. In [14], the PSO technique was used in order to minimize the power losses and maintain the SLL and SBL at the desired level; therefore the time sequences generated by the PSO are given in **Figure 9**, and the corresponding array pattern is presented in **Figure 10**.

It can be observed that the SLLs are maintained at -20 dB and the maximum SBL is -30.2 dB, where only four elements are time modulated and the elements 1,

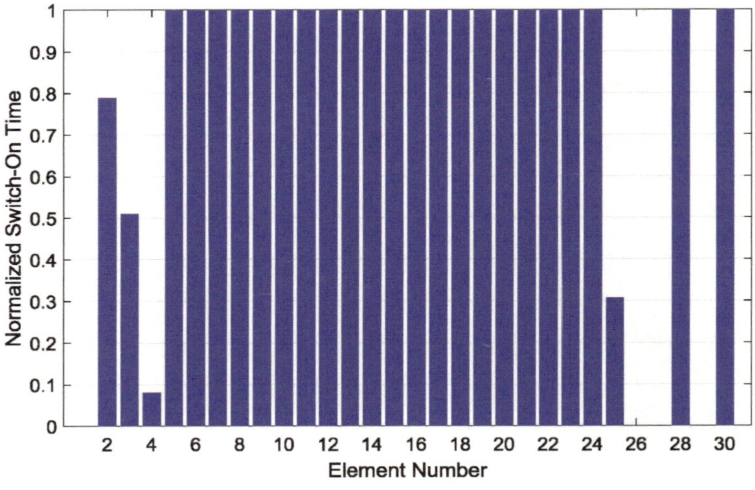

Figure 9.
Switch-on time sequences optimized by PSO technique in order to reduce the SBR ($N = 30, d = 0.7\lambda$) [13].

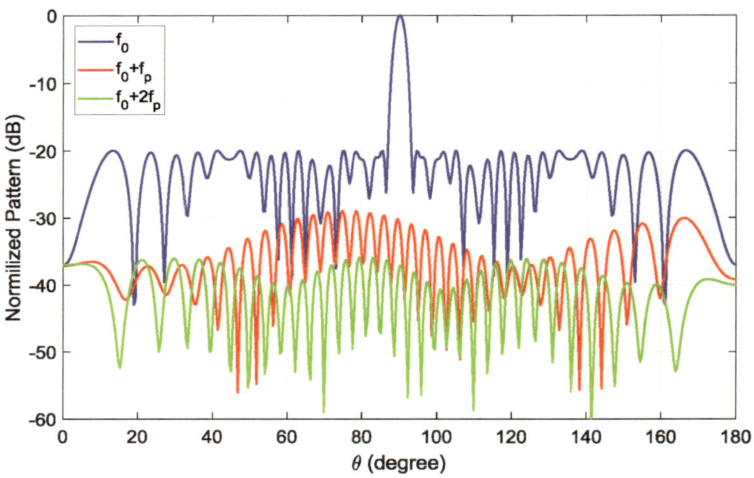

Figure 10.
Normalized array patterns (at the fundamental frequency and the two first sidebands) of the optimized TMLA by the PSO technique [13].

26, 27, and 29 are always turned off. This TMLA can be considered 26-element array with nonuniform spacing because the 4-turned off elements can be ignored. The power distribution over the sidebands is shown in **Figure 11**. It can be observed that most of the power resides at the fundamental frequency with 96.43% of the total power, where the remaining sidebands have only 3.57% of the total power. However, in this case, the SLL is only −20 dB.

In [14], the SA method was used in order to maintain the SLL at a certain level and minimize the SBL under −30 dB. Therefore, the obtained switch-on time sequences are shown in **Figure 12**, and the corresponding array patterns of the optimized TMLA are shown in **Figure 13**, where the power spectrum percentage is presented in **Figure 14**. It can be observed that only 9 elements are time modulated, where the remaining 21 elements are always turned on. In this case, the SBLs are minimized greatly, where they have only 3.96% of the total power. The PSO is more

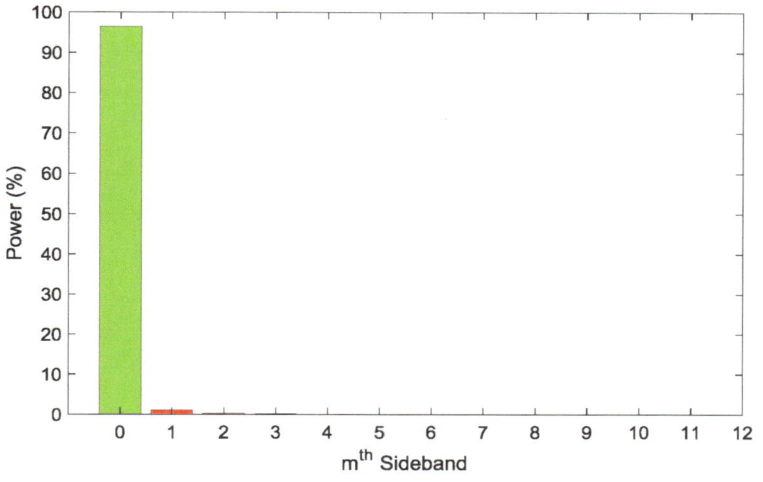

Figure 11.
Power percentage spectrum of the optimized TMLA by the PSO ($N = 30, d = 0.7\lambda$).

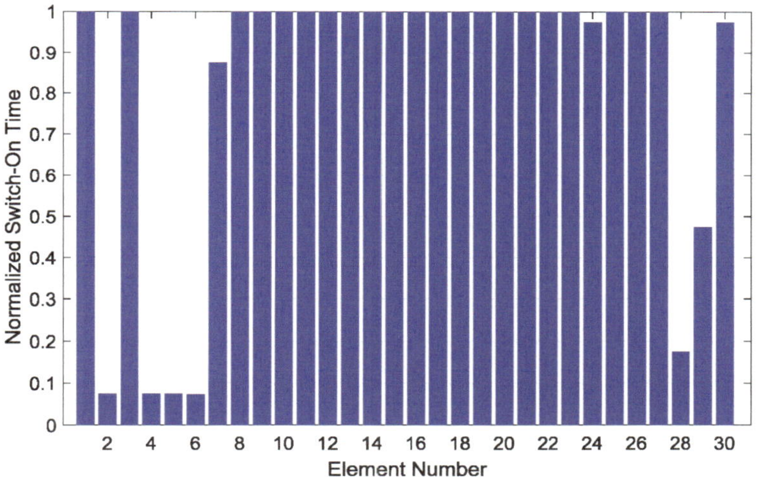

Figure 12.
Switch-on time sequences optimized by the SA technique in order to reduce the SBR ($N = 30, d = 0.7\lambda$) [14].

efficient than the SA one as indicated in [14]. For more details about the optimization methods, the reader should be referred to [13, 14].

The multiple time modulation frequency (MTMF) was proposed to reduce SBL of TMLA because of avoiding the accumulation of the sidebands in the space [15]; however, the SBR power was not decreased by using MTMF. In [16], the DE was used with MTMF to suppress SLL, SBL, and SBR power, and very good results were obtained. In the following section, the MTMF is investigated in detail.

2.2 TMLA with multiple time modulation frequency (MTMF)

In TMLA with MTMF, each antenna element has its time modulating switching period $T_{p,n}$. Hence, the periodic function $g_n(t)$ in (1) is expressed as [15]

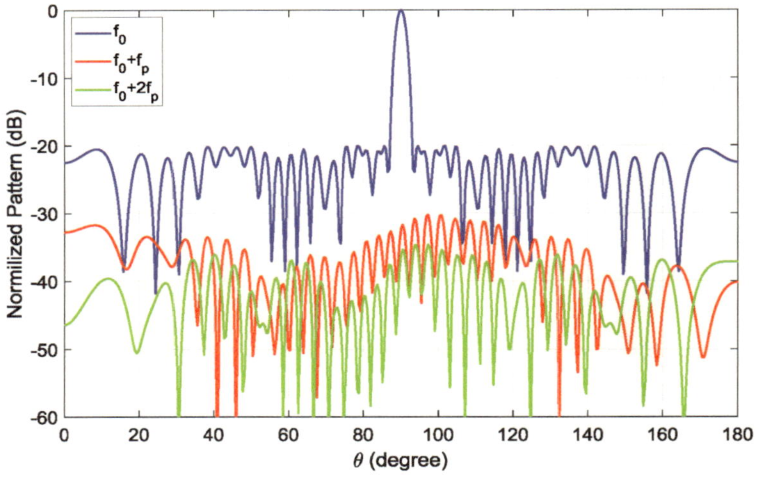

Figure 13.
Normalized array patterns (at the fundamental frequency and the two first sidebands) of the optimized TMLA by the SA technique [14].

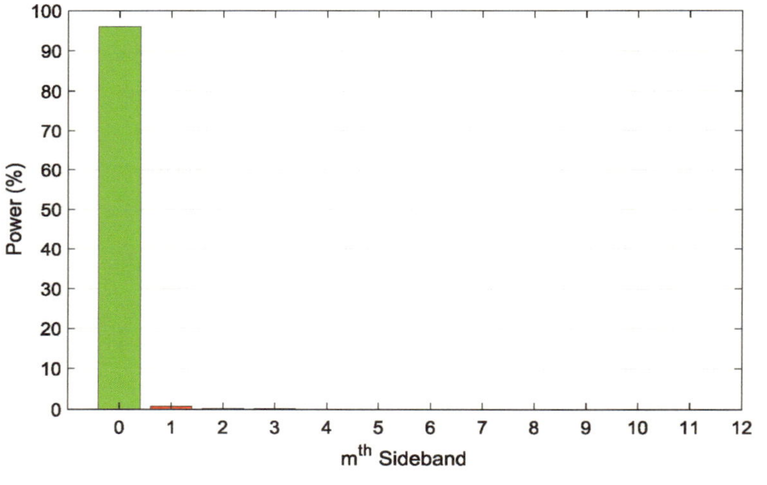

Figure 14.
Power percentage spectrum of the optimized TMLA by the SA method ($N = 30, d = 0.7\lambda$).

$$g_n(t) = \begin{cases} 1, & t_{\text{on},n} < t \le t_{\text{off},n} \\ 0, & 0 < t \le t_{\text{on},n} \text{or } t_{\text{off},n} < t \le T_{\text{p},n} \end{cases} \tag{44}$$

Since $g_n(t)$ are periodic functions, they can be expanded by the Fourier series as

$$g_n(t) = \sum_{m=-\infty}^{+\infty} G_{nm} e^{j2\pi m f_{pn} t} \tag{45}$$

where $f_{pn} = 1/T_{pn}$ is the modulation frequency and G_{nm} is the mth Fourier coefficient of the nth radio-frequency switch, and they are represented as

$$G_{nm} = \frac{1}{T_{pn}} \int_{0}^{T_{pn}} g_n(t) e^{-j2\pi m f_{pn} t} dt \tag{46}$$

$$G_{nm} = \tau_n \text{sinc}(\tau_n m) e^{-jm\pi(2\xi_{\text{on},n} + \tau_n)} \tag{47}$$

where $\tau_n = (t_{\text{off},n} - t_{\text{on},n})/T_{pn}$ and $\xi_{\text{on},n} = t_{\text{on},n}/T_{pn}$ are the normalized switch-on duration and the normalized switch-on instant for each element in the TMLA, respectively.

In the case of MTMF, the array factor can be written as

$$\text{AF}(\theta, t) = \sum_{m=-\infty}^{+\infty} \sum_{n} I_n G_{nm} e^{j2\pi(f_0 + m f_{pn})t} e^{jkz_n \cos(\theta)} \tag{48}$$

$$\text{AF}(\theta, t) = \sum_{m=-\infty}^{+\infty} AF_m(\theta, t) \tag{49}$$

where

$$\text{AF}_m(\theta, t) = \sum_{n} I_n G_{nm} e^{j2\pi(f_0 + m f_{pn})t} e^{jkz_n \cos(\theta)} \tag{50}$$

It is worth noticing that AF_0 is the array factor at the fundamental frequency f_0 and AF_m accumulates different sideband frequencies as it was described in [15, 16]. The idea of using MTMF is to avoid the superposing of sidebands because each array element has its corresponding switching frequency. Therefore, the sidebands of each element could not be superposed with the sidebands of another element; this concept is explained clearly in **Figure 15**.

2.2.1 Power radiations

The power radiation by TMLA with MTMF can be obtained by considering the following assumption:

- The sidebands of each antenna element are not overlapped with the sidebands of the other elements.

In this case, the sidebands power is given as

$$P_{\text{SB}} = \sum_{n=1}^{N} |I_n|^2 \sum_{\substack{m=-\infty \\ m \neq 0}}^{+\infty} |G_{nm}|^2 \tag{51}$$

$$P_{\text{SB}} = \sum_{n=1}^{N} |I_n|^2 \tau_n (1 - \tau_n) \tag{52}$$

where the power radiated at the fundamental frequency is given by

$$P_{f_0} = 4\pi \sum_{n=1}^{N} |I_n|^2 {\tau_n}^2 + 4\pi \sum_{\substack{n,k=1 \\ k \neq n}}^{N} I_n I_k^* \,\text{sinc} \frac{2}{\lambda}(z_n - z_k) \tau_n \tau_k \tag{53}$$

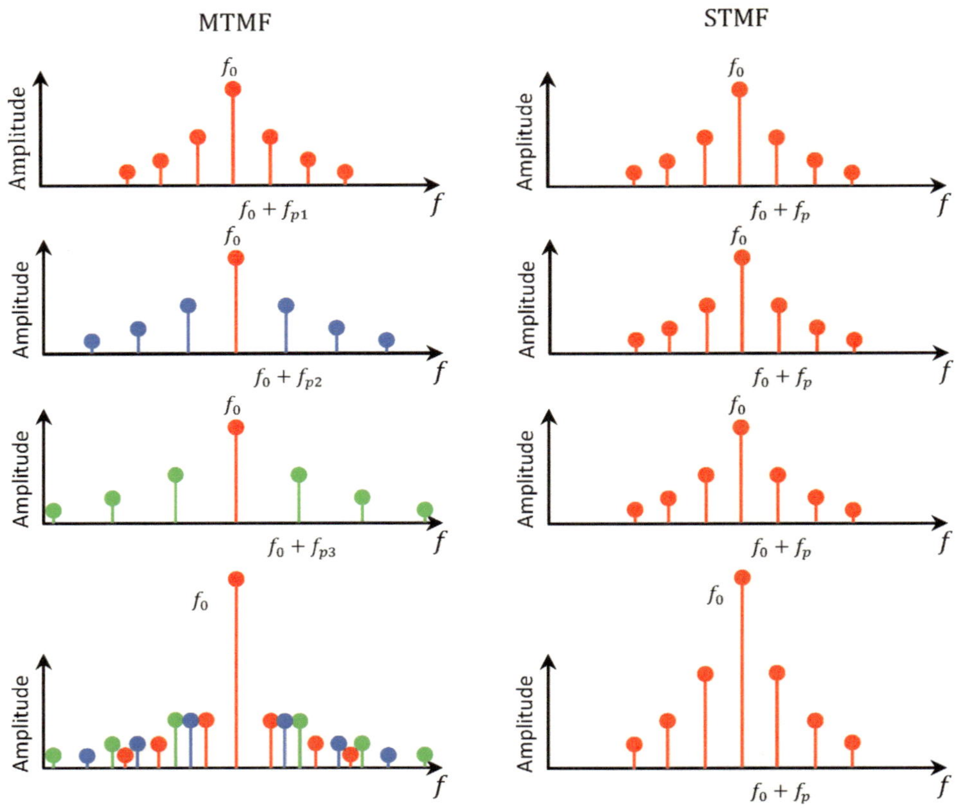

Figure 15.
Illustration of SBL suppression in TMLA with MTMF. The SBLs in STMF are superposed because all the time modulating switchers have the same time modulating frequency, whereas the SBLs in MTMF are not superposed because each element has the corresponding time modulating frequency.

It is worth noticing that relation (51) shows that all the Fourier's coefficients of each element are summed independently because they are located at different frequencies. Also, it should be indicated that for $d = \lambda/2$ the total power relations for the STMF and MTMF are identical. Another formulation for the power relation of TMLA with MTMF is given in [17], where a prime distribution is assumed for the time modulation frequencies. For more details about time modulation frequencies with prime distribution, the reader should be referred to [17]. It should be noted that the directivity of the TMLA with MTMF is identical to the one given in relation (35).

2.2.2 Simulation and computed results

In this section, computed results and examples are considered in order to investigate the benefits of TMLA with MTMF. The same example taken in Section 2.1.4 is considered so as to make a fair comparison between TMLA-STMF and TMLA-MTMF. Let's consider the fundamental frequency f_0 to be 2.6 GHz, the time modulation frequency f_p for the STMF case is 30 MHz, and for the MTMF, f_{pn} are selected as [15]

$$f_{pn} = (30 + 0.5(n-1))\text{MHz}, \quad n = 1, 2, 3, \dots, N \tag{54}$$

The results are plotted in **Figure 16**, where the maximum sideband for the STMF is -12.28 dB, whereas only -35.98 dB is obtained for MTMF case. It is evident that

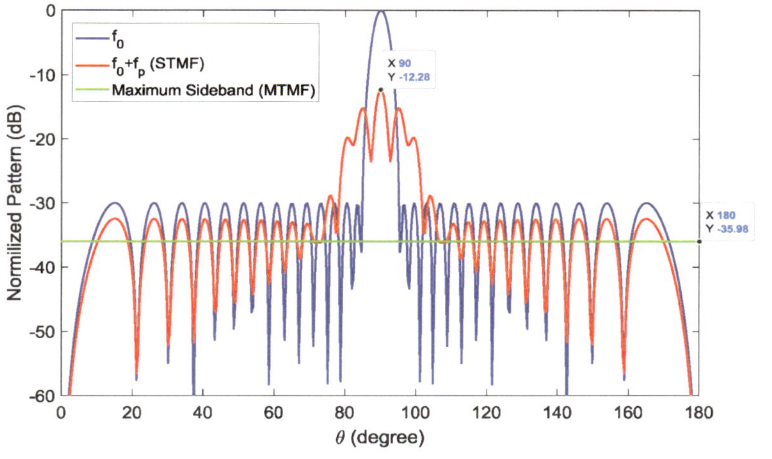

Figure 16.
Normalized array pattern at the fundamental frequency with STMF first sideband and the MTMF maximum sideband level.

the TMLA with MTMF is more efficient than STMF in reducing the sideband levels. However, the power loss in the sideband radiations is the same for the case $d = \lambda/2$; hence, the directivity has remained unchanged. It is worth noticing that the normalized switch-on time is identical to the one given in **Figure 5**, where, in the case of MTMF, each element has its corresponding normalizing period as shown in Eq. (47).

The sideband's power percentages for STMF and MTMF are presented in **Figure 17**. It is evident that the sideband's power of STMF is larger than the sideband's power of MTMF for $d/\lambda < 0.5$. However, the inverse occurs for $d/\lambda > 0.5$.

The optimization techniques were used in order to reduce the SBLs and the SBRs, e.g., the DE method was applied in [16], and very good results were obtained. In [16], the DE method was implemented so as to maintain the SLLs at a given level, whereas the SBLs and SBRs are minimized as much as possible. **Figure 18** shows the results of the DE applied to the TMLA-MTMF in order to maintain the SLLs at −20 dB, where the obtained SBL was −40.70 dB [16]. In order to make a

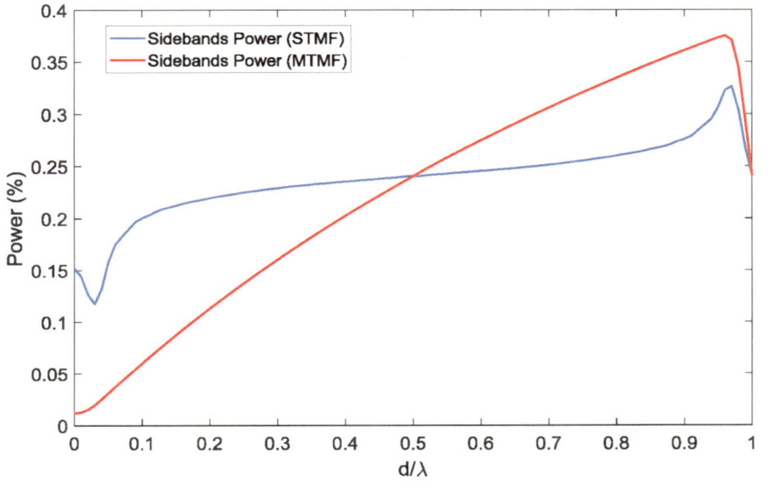

Figure 17.
Sideband's power percentage for TMLA with STMF and MTMF in terms of d/λ (Chebyshev weighting with −30 dB SLLs is considered).

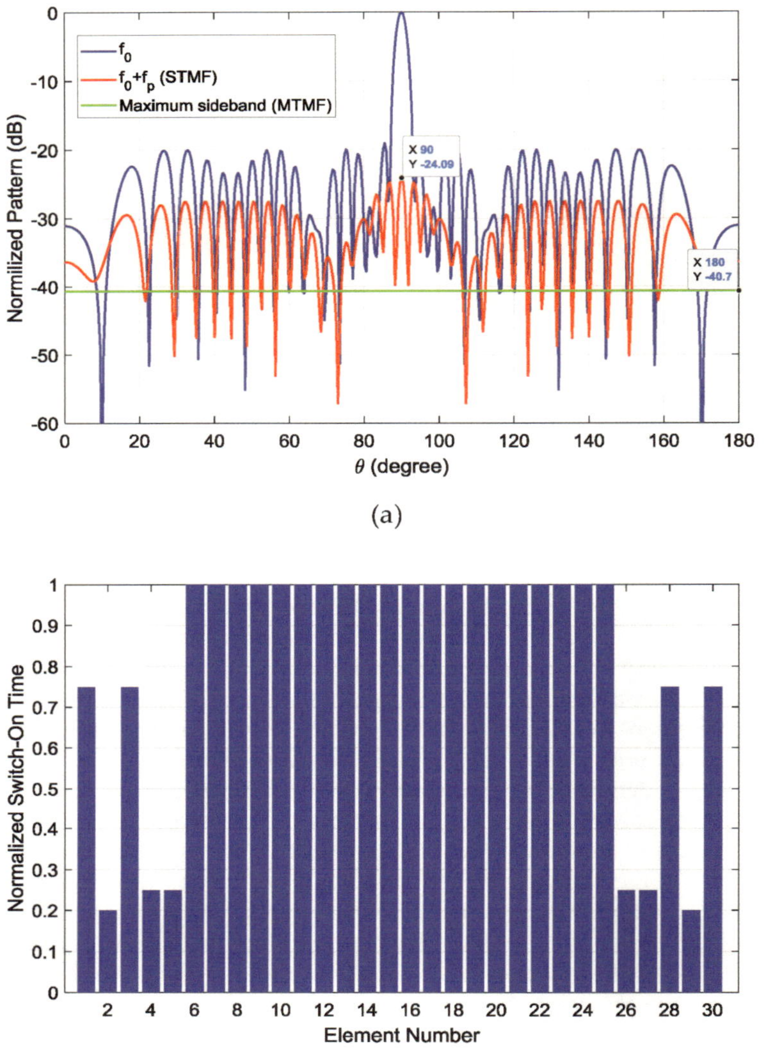

(a)

(b)

Figure 18.
The optimized TMLA by the DE algorithm where the desired SLL is −20 dB (d = 0.7λ). (a) Normalized array patterns. (b) Optimized switch-on time of each element.

comparison between the TMLA-STMF, the first sideband of STMF is also presented. Its maximum level is −24.09 dB.

3. Reducing SLLs and SBLs in TMLA

In this section, an analytical method is used to minimize the SLLs and SBLs in TMLA [18]. The array pattern of the TMLA can be written in the following forms:

For an odd number of elements $N = 2M + 1$ and $z_n = nd$, then

$$AF_0(\psi, t) = e^{j2\pi f_0 t} \sum_{n=-M}^{M} \tau_n e^{jn\psi} = e^{j2\pi f_0 t} \left(\tau_0 + \sum_{n=1}^{M} \left(\tau_{-n} e^{-jn\psi} + \tau_n e^{jn\psi} \right) \right) \tag{55}$$

where $\psi = kd\cos(\theta)$ and by assuming symmetrical excitations $\tau_n = \tau_{-n}$, then

$$AF_0(\psi, t, \boldsymbol{\tau}) = e^{j2\pi f_0 t}\left(\tau_0 + 2\sum_{n=1}^{M}\tau_n\cos(n\psi)\right) \tag{56}$$

Note that $\boldsymbol{\tau} = [\tau_0, \tau_1, \dots, \tau_M]$ is a row vector that wants to be determined to satisfy the desired specifications (SLL, SBL, and directivity), where $\boldsymbol{\tau} \in [0, 1]^M$. We should indicate that $\cos(n\psi)$ can be written as

$$\cos(n\psi) = T_n(\cos(\psi)) \tag{57}$$

where T_n is the first kind Chebyshev polynomial of degree n. The sidelobes' locations of the fundamental array factor AF_0 can be determined by solving the equation $\partial AF_0/\partial\psi = 0$, where

$$\frac{\partial AF_0}{\partial\psi} = -2\sin(\psi)e^{j2\pi f_0 t}\sum_{n=1}^{M}n\tau_n U_{n-1}(\cos(\psi)) \tag{58}$$

where U_{n-1} is the second kind Chebyshev polynomial and it has a relation with first kind Chebyshev polynomial

$$\frac{dT_n(s)}{ds} = nU_{n-1}(s) \tag{59}$$

where

$$\begin{cases} U_0(s) = 1, U_1(s) = 2s \\ U_n(s) = 2sU_{n-1}(s) - U_{n-2}(s), n = 2, 3, \dots \end{cases} \tag{60}$$

The sidelobes are located at $\psi_0 = \pi$ and $\psi_i = \arccos(s_i)$, where s_i are the roots of the following polynomial:

$$P_{\text{odd}}(s) = \sum_{n=1}^{M}n\tau_n U_{n-1}(s) \tag{61}$$

By obtaining the roots s_i, in the region $[-1, 1]$, of the polynomial $P_{\text{odd}}(s)$ in terms of the excitation coefficients τ_n and substituting $\psi_i(\tau_1, \tau_2, \dots \tau_M) = \arccos(s_i)$ into (56), we get the following:

$$AF_0(\psi_i, t, \boldsymbol{\tau}) = e^{j2\pi f_0 t}\left(\tau_0 + 2\sum_{n=1}^{M}\tau_n T_n(s_i(\tau_1, \tau_2, \dots \tau_M))\right) \tag{62}$$

where $\text{SLL}_i = |AF_0(\psi_i, t, \boldsymbol{\tau})|$ is the SLL of the sidelobe located at ψ_i.

For an even number of elements $N = 2M$ and $z_{\pm n} = \pm(2n-1)d/2$, then

$$AF_0\left(\psi, t, \boldsymbol{\tau}^{(1)}\right) = 2e^{j2\pi f_0 t}\sum_{n=1}^{M}\tau_n\cos\left(\left(n - \frac{1}{2}\right)\psi\right) \tag{63}$$

where $\boldsymbol{\tau}^{(\mathbf{1})} = [\tau_1, \tau_2, \dots, \tau_M]$.

The Chebyshev of 3rd and 4th kinds are given as

$$V_n(s) = \cos\left(\left(n+\frac{1}{2}\right)\psi\right) / \cos\left(\frac{\psi}{2}\right) \tag{64}$$

$$W_n(s) = \cos\left(\left(n+\frac{1}{2}\right)\psi\right) / \sin\left(\frac{\psi}{2}\right) \tag{65}$$

respectively, where $s = \cos\psi$.

The expression $AF_0(\psi, t, \boldsymbol{\tau}^{(1)})$ and $\partial AF_0/\partial\psi$ can be written in terms of (64) and (65), respectively, as

$$AF_0\left(\psi, t, \boldsymbol{\tau}^{(1)}\right) = 2\cos\left(\frac{\psi}{2}\right) e^{j2\pi f_0 t} \sum_{n=1}^{M} \tau_n \mathrm{V}_{n-1}(\cos(\psi)) \tag{66}$$

$$\frac{\partial AF_0}{\partial\psi} = -2\sin\left(\frac{\psi}{2}\right) e^{j2\pi f_0 t} \sum_{n=1}^{M} \left(n-\frac{1}{2}\right) \tau_n W_{n-1}(\cos(\psi)) \tag{67}$$

As described before, the sidelobes are located at $\psi_i = \arccos(s_i)$, where s_i are the roots, that are located in the region $[-1, 1]$ of the following polynomial:

$$P_{\text{even}}(s) = \sum_{n=1}^{M} \left(n-\frac{1}{2}\right) \tau_n W_{n-1}(s) \tag{68}$$

and the SLLs are given as

$$\mathrm{SLL}_i = \left|AF_0\left(\psi_i, t, \boldsymbol{\tau}^{(1)}\right)\right| = \left|2\cos\left(\frac{\psi_i}{2}\right) \sum_{n=1}^{M} \tau_n \mathrm{V}_{n-1}(s_i(\tau_1, \tau_2, \ldots \tau_M))\right| \tag{69}$$

It should be indicated that there are no sidelobes contributed by the factor $\sin(\psi/2)$ since its roots are $\pm 2\pi$, which are a worthy advantage in designing even number of elements in TMLA.

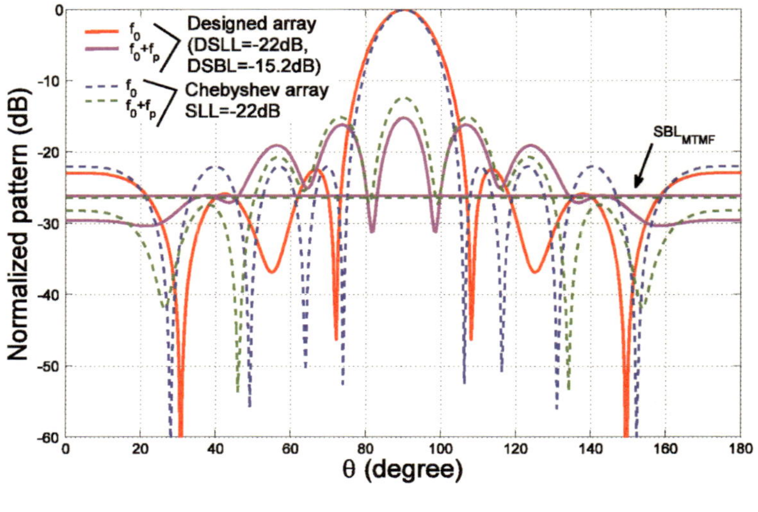

Figure 19.
The designed array ($N = 9$, DSLL $= -22$ dB, DSBL$_{\text{STMF}} = -15.2$) *and Chebyshev array* ($N = 9$, SLL $= -22$dB).

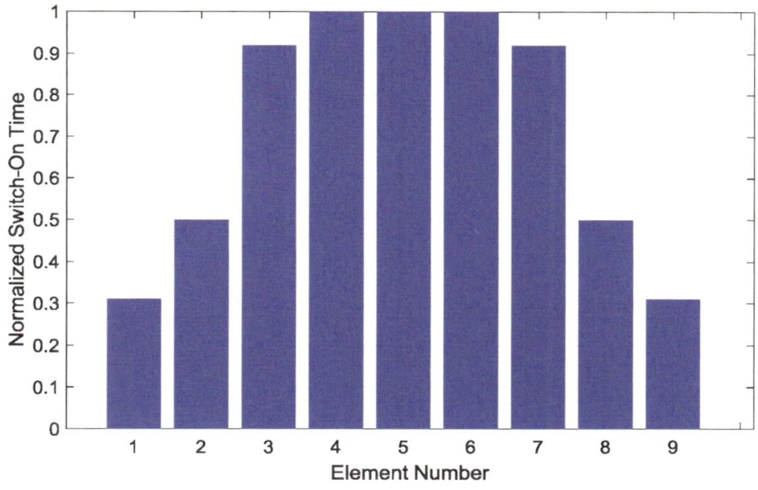

Figure 20.
Designed time sequence distributions of each antenna array element for $N = 9$, DSLL $= -22$ dB, $\text{DSBL}_{\text{STMF}} = -15.2$.

Now let's design TMLA with nine elements to satisfy the specifications; DSLL $= -22$ dB and $\text{DSBL}_{\text{STMF}} = -15.2$. In this case, the SBL for MTMF is -26.15 and the maximum directivity that can be achieved is $D_{\text{max}} = 16.2$ dB. The results are plotted in **Figures 19** and **20**. To investigate the effectiveness of the proposed method, a comparison is performed with nine-element Chebyshev array that has SLL equal to -22 dB. It has SBL equal to -12.4 and its directivity 16.42 dB which is larger than the designed array only with 0.22 dB. Note that $\xi_{\text{on},n} = 0$ for all the cases.

Finally, it should be indicted that the TMLA can be designed by only controlling the time sequence distributions which is a very good advantage as compared to the conventional array under the following reasons:

1. Attain high accuracy in the designed array pattern in the TMLA because the switching distributions can be controlled at very high accuracy.

2. In the conventional array, attenuators and distributors are needed for exciting the array which is not accurate method. Therefore, it causes deviation in the designed array pattern and high SLLs are generated.

4. Conclusion

In this chapter, the main backgrounds and theories of TMLA are derived where different simulation examples are presented and discussed in detail. A comparison between different results given in the previous literature is also discussed. In addition, an analytical method to reduce the SLLs and SBLs in TMLA with maximum achievable directivity has been developed. This analytical method helps us to visualize the relation between switch-on durations, SLL, and SBL, which is an advantage compared to the other designing methods. It was shown that the TMLA has better performance than the conventional array.

Author details

Oussama Gassab[1,2]*, Arab Azrar[1] and Sara Bouguerra[1]

1 The Signals and Systems Research Laboratory, Department of Electronic, Institute of Electrical and Electronics Engineering (IGEE), University M'Hamed Bougara of Boumerdes, Boumerdes, Algeria

2 Key Lab of Ministry of Education for the Design and Electromagnetic Compatibility of High-Speed Electronic Systems, The Center for Microwave and RF Technologies (CMRFT), Shanghai Jiao Tong University, Shanghai, China

*Address all correspondence to: oussamagassab@yahoo.com

References

[1] Ares-Pena FJ, Rodriguez-Gonzalez JA, Villanueva-Lopez E, Rengarajan SR. Genetic algorithms in the design and optimization of antenna array patterns. IEEE Transactions on Antennas and Propagation. March 1999;**47**(3):506-510

[2] Khodier MM, Christodoulou CG. Linear array geometry synthesis with minimum sidelobe level and null control using particle swarm optimization. IEEE Transactions on Antennas and Propagation. August 2005;**53**(8):2674-2679

[3] Gassab O, Azrar A. Novel mathematical formulation of the antenna array factor for side lobe level reduction. ACES Journal. 2016;**31**(12): 1452-1462

[4] Schrank H. Low sidelobe phased array antennas. IEEE Antennas and Propagation Society Newsletters. April 1983;**25**(2):4-9

[5] Shanks HE, Bickmore RW. Four-dimensional electromagnetic radiators. Canadian Journal of Physics. 1959;**37**(3): 263-275

[6] Kummer W, Villeneuve A, Fong T, Terrio F. Ultra-low sidelobes from time-modulated arrays. IEEE Transactions on Antennas and Propagation. 1963;**11**(6): 633-639

[7] Bregains JC, Fondevila-Gomez J, Franceschetti G, Ares F. Signal radiation and power losses of time-modulated arrays. IEEE Transactions on Antennas and Propagation. 2008;**56**(6):1799-1804

[8] Balanis CA. Time-varying and time-harmonic electromagnetic fields. In: Advanced Engineering Electromagnetics. 2nd ed. John Wiley & Sons; 2012, ch. 1. pp. 1-29

[9] Aksoy E, Afacan E. Calculation of sideband power radiation in time-modulated arrays with asymmetrically positioned pulses. IEEE Antennas and Wireless Propagation Letters. 2012;**11**: 133-136

[10] Zhu Q, Yang S, Yao R, Nie Z. Gain improvement in time-modulated linear arrays using SPDT switches. IEEE Antennas and Wireless Propagation Letters. Aug. 2012;**11**:994-997

[11] Yang S, Gan YB, Qing A. Sideband suppression in time-modulated linear arrays by the differential evolution algorithm. IEEE Antennas and Wireless Propagation Letters. 2002;**1**:173-175

[12] Yang S, Gan YB, Qing A, Tan PK. Design of a uniform amplitude time modulated linear array with optimized time sequences. IEEE Transactions on Antennas and Propagation. July 2005; **53**(7):2337-2339

[13] Poli L, Rocca P, Manica L, Massa A. Handling sideband radiations in time-modulated arrays through particle swarm optimization. IEEE Transactions on Antennas and Propagation. 2010; **58**(4):1408-1411

[14] Fondevila, Bregains, Ares, Moreno. Optimizing uniformly excited linear arrays through time modulation. IEEE Antennas and Wireless Propagation Letters. 2004;**3**:298-301

[15] He C, Yu H, Liang X, Geng J, Jin R. Sideband radiation level suppression in time-modulated array by nonuniform period modulation. IEEE Antennas and Wireless Propagation Letters. 2015;**14**: 606-609

[16] Guo J, Yang S, Chen Y, Rocca P, Hu J, Massa A. Efficient sideband suppression in 4-D antenna arrays through multiple time modulation frequencies. IEEE Transactions on Antennas and Propagation. 2017;**65**(12): 7063-7072

[17] Kanbaz I, Yesilyurt U, Aksoy E. A study on harmonic power calculation for nonuniform period linear time modulated arrays. IEEE Antennas and Wireless Propagation Letters. 2018; **17**(12):2369-2373

[18] Gassab O, Azrar A, Dahimene A, Bouguerra S. Efficient mathematical method to suppress sidelobes and sidebands in time-modulated linear arrays. IEEE Antennas and Wireless Propagation Letters. May 2019;**18**(5): 836-840